Also by Benjamin Ehrlich

**The Dreams of
Santiago Ramón y Cajal**

THE BRAIN IN
SEARCH OF ITSELF

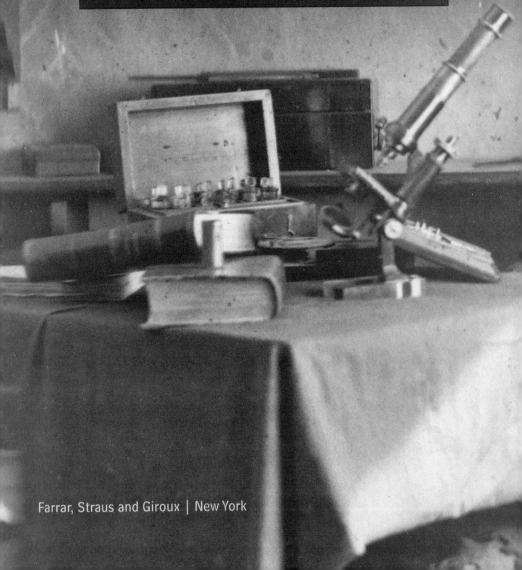

THE BRAIN IN
SEARCH OF ITSELF

Santiago Ramón y Cajal
and the Story of the Neuron

Benjamin Ehrlich

Farrar, Straus and Giroux | New York

Farrar, Straus and Giroux
120 Broadway, New York 10271

Photographs courtesy of the Cajal Institute, Spanish National Research
Council (CSIC), Madrid, Spain, except for the image on page 60, which
appears in Cajal's autobiography and was found on the internet.

Library of Congress Cataloging-in-Publication Data
Names: Ehrlich, Benjamin, 1987– author.
Title: The brain in search of itself : Santiago Ramón y Cajal and the
    story of the neuron / Benjamin Ehrlich.
Description: First edition. | New York : Farrar, Straus and Giroux, 2022. |
    Includes bibliographical references and index. | Summary: "A biography
    of the Spanish neuroscientist Santiago Ramón y Cajal"— Provided by
    publisher.
Identifiers: LCCN 2021049291 | ISBN 9780374110376 (hardcover)
Subjects: LCSH: Ramón y Cajal, Santiago, 1852–1934. | Neuroscientists—
    Spain—Biography. | Neurosciences—History.
Classification: LCC RC339.52.R353 E37 2022 | DDC 612.8/233092 [B]—
    dc23/eng/20211208
LC record available at https://lccn.loc.gov/2021049291

Designed by Janet Evans-Scanlon

Our books may be purchased in bulk for promotional, educational,
or business use. Please contact your local bookseller or the Macmillan
Corporate and Premium Sales Department at 1-800-221-7945,
extension 5442, or by email at MacmillanSpecialMarkets@macmillan.com.

www.fsgbooks.com
www.twitter.com/fsgbooks • www.facebook.com/fsgbooks

10 9 8 7 6 5 4 3 2 1

To Alex and Marilyn Ehrlich, my parents

I know who I am and who I may be, if I choose.

—*DON QUIXOTE*

Every man, if he is so determined,
can become the sculptor of his own brain.

—SANTIAGO RAMÓN Y CAJAL

Contents

CONTENTS

THE BRAIN IN
SEARCH OF ITSELF

"A Vehement Desire of My Soul"

Hour after hour, year after year, Santiago Ramón y Cajal sat alone in his home laboratory, head bowed and back hunched, his black eyes staring down the barrel of a microscope, the sole object tethering him to the outside world. His wide forehead and aquiline nose gave him the look of a distinguished, almost regal, gentleman, though the crown of his head was as bald as a monk's. He had only a crowd of glass bottles for an audience, some short and stout, some tall and thin, stopped with cork, filled with white powders and colored liquids; the other chairs, piled high with journals and textbooks, left no room for anyone else to sit. Stained with dye, ink, and blood, the tablecloth was strewn with drawings of forms at once otherworldly and natural. Colorful transparent slides, mounted with slivers of nervous tissue from sacrificed animals, still gummy to the touch from chemical treatments, lay scattered on the worktable.

With his left thumb and forefinger, Cajal adjusted the corners of the slide as if it were a miniature picture frame under the lens of his microscope. With his right hand he turned the brass knob on the side of the instrument, muttering to himself as he drew the image into focus: brownish-black bodies resembling inkblots, radiating threadlike appendages, set against a transparent yellow background. The wondrous landscape of the brain was finally revealed to him, more real than he could have ever imagined.

In the late nineteenth century, most scientists believed that the brain was composed of a continuous tangle of fibers, as serpentine as a labyrinth. Cajal produced the first clear evidence that the brain is composed of individual cells, later termed *neurons*, fundamentally the same as those that make up the rest of the living world. He believed that neurons served as storage units for mental impressions, such as thoughts and sensations, which combined to form our experience of being alive: "To know the brain is equivalent to ascertaining the material course of thought and will," he wrote. The highest ideal for a biologist, he declared, is to clarify the enigma of the self. In the structure of neurons, Cajal thought he had found the home of consciousness itself.

Santiago Ramón y Cajal is considered the founder of modern neuroscience. Historians have ranked him alongside Darwin and Pasteur as one of the greatest biologists of the nineteenth century and among Copernicus, Galileo, and Newton as one of the greatest scientists of all time. His masterpiece, *The Texture of the Nervous System of Man and the Vertebrates*, is a foundational text for neuroscience, comparable to *On the Origin of Species* for evolutionary biology. Cajal was awarded the Nobel Prize in 1906 for his work on the structure of neurons, whose birth, growth, decline, and death he studied with devotion and even a kind of compassion, almost as though they were human beings. "The mysterious butterflies of the soul," Cajal called them, "whose beating of wings may one day reveal to us the secrets of the mind." He produced thousands of drawings of neurons, as beautiful as they are complex, which are still printed in neuroanatomy textbooks and exhibited in art museums. More than a hundred years after his Nobel Prize, we are indebted to Cajal for our knowledge of what the nervous system looks like. Some scientists even have Cajal's drawings of neurons tattooed on their bodies. "Only true artists are attracted to science," he said.

• • •

HALFWAY THROUGH THE JOURNEY OF HIS LIFE, WHEN HE WAS FORTY years old, it seemed as though Cajal had achieved all that he had ever wanted. He had arrived in Spain's royal capital, where he occupied the chair of anatomy at the Central University of Madrid, the highest appointment in his field. He was married to the perfect wife, in his estimation, and was the father of six children. His finances were finally in order. The university was building him a state-of-the-art laboratory. Three years earlier, he had revealed his "new truth" of the nervous system, revolutionizing the understanding of the mind and brain. His name was renowned in scientific academies across Europe.

From his home in downtown Madrid, Cajal could hear the whistling of trains. The wrought-iron dome of the nearby Atocha Station, renovated to resemble the newly constructed Eiffel Tower, was a suitable landmark for a metropolis that some called "Little Paris." Bright shop windows displayed French jewelry, English biscuits, and Italian opera posters. Even nannies pushing strollers through the elegant gardens of Retiro Park wore the latest fashions in silk and lace. Every morning, men in brass-plated hats washed and swept the streets until the pavement gleamed. On every corner there was a café, where famous artists gossiped, bullfighters recounted their triumphs, and politicians discussed the fate of the nation while lounging on overstuffed couches, sipping liqueurs and chocolate until dawn. The sidewalks were so congested that on Sundays and holidays it seemed impossible to move. Madrid was a city where no one ever asked where you were from. That was one of the reasons Cajal liked living there.

The father of the neuron had no memory of his birthplace. His family had moved away when he was seventeen months old. Where he was raised, people were identified by the names of their native villages. The directory of the school he attended listed "Petilla" next to his name, and any formal petition to his university began with "I, Santiago Ramón y Cajal, native of Petilla . . ." Petilla was a tiny village located in the distant mountains of northeastern Spain, the highlands of Alto Aragon. His draft notice, his doctoral degree, his marriage certificate, the birth certificates of his children all served as reminders that he had almost no knowledge of his own origins.

In spite of the many advantages of his life in Madrid, Cajal could no longer deny "a vehement desire of my soul": he longed to return to his birthplace, to know the first impressions of his own brain, to travel back in time through the landscape of his consciousness until he arrived at its source. "The brain is a world consisting of vast continents and unexplored territories," he said, and he would dedicate his life to charting its geography. Now, in 1892, what he sought was hidden in a village so microscopic that it did not appear on any map.

ACCORDING TO LEGEND, IN THE TWELFTH CENTURY, KING PEDRO II, "the Catholic," of Aragon lost a card game to his royal neighbor King Sancho VII, "the Strong," of Navarre. King Pedro offered four castles, including Petilla, as collateral for the debt of twenty thousand maravedis, a medieval copper coin. Six centuries later, when Napoleon conquered Spain, he divided the country into provinces, preserving its historic boundaries. Petilla became an exclave, surrounded by a foreign kingdom. No roads connected the village to even its closest neighbors. It is hard to imagine a more remote and isolated outpost in all of Spain.

Cajal's journey home could be broken up into three stages, each transporting him back in time and further away from civilization. First, he took a train almost three hundred miles to Jaca, the penultimate station on the northern line, in the shadow of the Pyrenees. Cajal did not mind the long, monotonous journey; he claimed to have once spent twenty hours straight at his microscope, traveling one millionth of a meter at a time. Then he rode in a packed stagecoach westward to Tiermas, a medieval town with ancient Roman bathhouses, famous for its sulfurous, emerald-colored hot springs. The road that he traveled was unusually well maintained, as it was a stage of one of the holiest of Christian pilgrimages, the Way of Santiago, the patron saint of Spain and Cajal's namesake. After disembarking in Tiermas, he hired a guide who knew the way to Petilla, a treacherous, twenty-five-mile trek through Pyrenean hills and gorges. In hemp sandals and knee-breeches, his skin leathery from the relentless sun, the peasant led the distinguished professor on the back of a mule.

The highland climate was volatile. It almost never rained, but when it did, the storms were swift and cataclysmic. A few days before, the guide told Cajal, there had been a deluge, transforming the fields into mud. The peasants, who were desperately poor, had stripped their forests for timber to meet the Spanish navy's demand for material for ships. Without trees to impede them, boulders tumbled down hillsides, damming stream beds and drowning crops. Where he came from, Cajal realized, nothing grew.

A church bell rang in the distance—it was coming from Petilla. The guide halted the mule so that Cajal could listen. Suddenly, he was struck by an "inexplicable melancholy." He was sure that no one would recognize him in Petilla. No one even knew who he was.

But then Cajal and his guide came upon a stream where an old peasant woman was washing her clothes. Turning and seeing Cajal, she cried: "Señor, if you are not Don Justo himself, you must be the son of Don Justo! Do not deny it!" For better or for worse, he would always be his father's son.

The final stretch of the journey was a rough and narrow trail winding precipitously up a steep foothill. The heavily eroded slope had been cut into crude terraces, the villagers' only arable land. Plows could not be used on such precarious terrain; farmers lugged manure and turned the soil using giant two-pronged forks called *layas*. Cajal felt proud to come from such hardworking stock. Farming in Petilla was an act of mythic futility. The deluges came; the retaining walls fell. The Petillans rebuilt them anyway.

As it turned out, the priest and the mayor were waiting for Cajal at the top of the hill. A disfigured rock face loomed over the cluster of cobblestone dwellings like a gigantic tombstone. Windows were lop-

sided holes in the walls, with crudely whitewashed edges. Roof tiles were made of cracked terra-cotta. There were no streets, only the crevices, furrows, and inclines blindly carved by the elements. The Petillans had never even seen a wheeled vehicle before, and they had certainly never heard of the nervous system. The entire population of Petilla was gathered in the square. The oldest among them fondly recalled Cajal's parents. They gave him a tour, which must have been brief. He was touched by their hospitality. Looking out from the railing on the church, the highest point in Petilla, Cajal took in the overpowering scenery of the landscape of his past. From that perspective, looking out over the empty highlands, life seemed infinitely small.

But when the villagers brought him to the former home of the town surgeon, where he had been born, Cajal was shocked. It was in ruins, a heap of stones, a refuge for itinerant beggars. "A voice within," Cajal recalled, "told me that I should never return to these places."

CHAPTER 1

"The Necessary Antecedent"

The Iberian Peninsula is practically an island: 90 percent is surrounded by water, with the only overland route to the rest of Europe, to the northeast, cut off by the "great spiked collar" of the Pyrenees. The mountains protect the inhabitants from the devil, an Aragonese saying goes, but keep out God's love as well. If geography is destiny, then the fate of Spain is isolation.

The Aragonese highlands are a cloistered kingdom existing apart from time: "those hills, those soaring, rocky bluffs / those sunken glades, those harrowing ravines, wastelands and broad plateau" goes the epic *The Song of Roland*. Modern neuroscience is among the most sophisticated, high-technology endeavors in human history, and yet Cajal, its founding hero, was a "peasant genius," to quote his fellow Nobel laureate Charles Sherrington.

The region where Cajal was born, stretching from the Ebro River to the Pyrenees, is known as Alto Aragon, or Upper Aragon, but the identification is more cultural than geographic. Ancient visitors commented on the people's "characteristic of inhospitality" and the "higher degrees of superstition and tribalism in the northern outposts of Hispania than anywhere else." So difficult was it to convert them to Christianity that, according to legend, they nearly drove Saint James to quit his divine mission.

The southern coast of Spain is sculpted to a pinch-point, where only nine miles separate Europe from North Africa. In the year 711, a small force of Berbers—from the Roman word for "barbarian"—crossed the strait in boats and conquered the peninsula within three years. Around the turn of the ninth century, the Frankish king Charlemagne established the territory south of the Pyrenees as a buffer zone, and the highlands became the sacred battleground of a war between Islam and Christianity. Those Moorish rulers who were flourishing in the south considered the frontier region too godless and lawless for anyone but the most devout warriors to survive in. A few highland counties formed a core of resistance, and in the eleventh century became the Kingdom of Aragon, which led the charge to expel the Moors. Finally, in the fifteenth century, Aragon merged with the Kingdom of Castile to form the nucleus of modern Spain. This account of Spanish history, known as the Reconquest, is essentially a romantic myth, a story of national identity conjured and popularized in the second half of the nineteenth century, when Cajal was coming of age.

The northern highlands, guarded by castles, forts, and towers, remained a contested territory, controlled by neither Muslims nor Christians, and it was there, in the year 864, that García de Benavides, nephew of the king of Pamplona, and Ibn Abdalá, son of the ruler of Zaragoza, fought a duel over a piece of territory. According to legend, both men broke their armor and chipped their daggers but continued swinging maces, bare-chested. Abdalá knocked García to the ground and was about to deliver the fatal blow, but García ripped a stone from the ground and smashed his enemy across the face, completely detaching his lower jaw, killing him instantly and scattering his teeth across the battlefield. In Aragonese—a distinct Romance language—the word *caxal*, or *cajal*, means "molar."

As the Greek historian Strabo noted, the highlands were "an exceedingly wretched place to live in," and Cajal's family came from the central Pyrenean region, the most rugged terrain of all. Annual rainfall was so low that grass for livestock dried up, and so few trees grew that peasants were left to gather firewood from bushes and shrubs. In the summer, the sun beat down like a vengeful god; and in the winter, temperatures could plummet to twenty degrees below zero, so cold that some highlanders, afraid of freezing, never took off their clothing. Agricultural plots were no more than a few acres of thin, stone-cluttered soil.

CAJAL'S FATHER, JUSTO RAMÓN Y CASASÚS, WOULD HAVE BEEN THE hero of the family had he never had a son. He was born in 1822 into a family of indigent farmers in Larres, which, with over two hundred residents, counted among the most populous villages in the *comarca*, or district. In those days, school was not compulsory, and so at around seven years old—as was customary for highland children—Justo began to work, both as a farmhand and as a shepherd. The land was still ruled by the medieval law of inheritance, which dictated that, to keep plots intact, property was to be inherited by firstborn sons only.

Justo, the third son, grew up knowing that he would inherit nothing. He could either live under the guardianship of his eldest brother or leave his native village in search of a livelihood, but given the almost complete lack of social mobility in nineteenth-century Spain, his best option would have been to become a farmer, following in the footsteps of his own father, Esteban Ramón, also a younger son lacking an inheritance.

Justo was in love with Antonia Cajal, the only daughter of the town weaver, whose family lived practically next door. Though three years older than Justo, she was confirmed during the same ceremony as he was. In Alto Aragon—a traditionalist, conservative society—where marriage was both a mercantile and sacred contract, whole communities protested the births of bastard children and divorces with extravagant mocking parades. Courtship was a matter of pragmatism, and with neither money nor prospects, Justo could not have been seen as a desirable match.

When he was sixteen or seventeen years old, Justo decided to leave Larres for Javierrelatre, a village about twenty miles away, where he apprenticed himself to a barber-surgeon—a strange choice, given that barber-surgery was one of the lowliest professions in Spain. But Antonia's mother came from the Casa Mancebo, the House of Nurses and Barbers, and Justo may have wanted to impress her family. He promised Antonia that he would marry her when he returned.

Surgical procedures were once the province of the clergy until a twelfth-century papal bull declared the shedding of blood unholy. Already present at monasteries, tonsuring monks' hair, barbers were skillful enough with a razor to open veins. In the Middle Ages, the most common medical treatment was bloodletting, a means of restoring equilibrium among the body's "four humors," an imbalance of which was thought to cause disease. One Renaissance handbook claimed that bleeding "clears the mind, strengthens the memory, cleanses the guts, dries up the brain, warms the marrow, sharpens the hearing, curbs tears, promotes digestion, produces a musical voice, dispels sleeplessness, drives away anxiety, feeds the blood, rids it of poisonous matter and gives long life." Ads posted outside shops depicted barber-surgeons with rolled-up sleeves and blood-soaked hands amputating limbs or bandaging heads.

For the first few years, Justo performed menial tasks, sweeping the shop floor and bringing water from the well to heat curling irons and wash shaving cloths. By watching his master, Albeita, Justo learned to wield the razor, extract teeth, administer enemas, splint fractures, and apply poultices, proving such a quick study that soon Albeita let him treat patients by himself.

Roughly 75 percent of Spaniards over the age of ten were illiterate, and in the highlands that number was closer to 90 percent. Justo either never attended school or left before he could learn to read. Albeita possessed an ample library, where, during his scant off-hours, Justo taught himself to read, probably by matching up the illustrations of barber-surgery practices with his direct experience in the shop. In the process, he discovered that he had been blessed with a miraculous gift: he was able to memorize entire textbooks.

The Spanish word for "bloodletter," *sangrador*, was so demeaning that the *Oxford English Dictionary* notes that, though the term literally means "bleeder," it also means "an ignorant pretender to medical knowledge." To evade the stigma, *sangradores* lobbied the queen for a name change, and in 1836, right before Justo began his apprenticeship, she finally granted their request, reorganizing the medical hierarchy into three classes within the medical profession: first class (physicians), second class (surgeons), and third class (barber-surgeons). With Albeita, Justo had found not only a livelihood but also, perhaps, an inheritance, as he might one day take over his master's shop or open his own. But Justo could not stand the thought of anyone looking down on him, and he believed that, with his extraordinary memory, he could elevate his status by earning an academic degree.

One day, when he was twenty-one years old, Justo shocked Albeita by announcing his departure. With his remaining salary and a small loan from his eldest brother, Justo set out for Zaragoza, walking seventy miles to the provincial capital, toting all his worldly belongings over his shoulder. "If you give an Aragonese man a nail to drive," a saying goes, "he would rather use his head than a hammer."

Justo settled in the working-class neighborhood of the Arrabal, where he apprenticed with another barber-surgeon while attending secondary school, eventually completing his degree. Without telling his master, he applied for a job at the provincial hospital as a *practicante*, a medical assistant, beating out twenty-five other candidates to finish first in the competitive examinations. Though the *practicante* represented the highest achievement for a barber-surgeon, Justo knew that he would remain subservient to the actual surgeons and physicians unless he earned a university degree. He enrolled at the University of Zaragoza for a second-class certificate in surgery, but in 1845 the medical program there was shut down. At that point, he could have returned home and married Antonia, as he had promised. But no desire was stronger in him than professional ambition.

Justo moved to Barcelona, a seven-day walk away, where he continued his training at the university medical school, which boasted the first modern medical program in Spain. The population of Barcelona—the capital of Catalonia and the first Spanish city to undergo industrialization—was two hundred thousand, a thousand times greater than that of his native village. Immigrants from the provinces crammed into foul-smelling slums and shacks made out of garbage. Homeless, Justo wandered the streets for days.

In the village of Sarría, just north of the city, Justo found a barber-surgeon who let him work as his assistant while attending classes, walking to and from the university, an hour each way. He adopted a strict regime of austerity, spending no money, wasting no energy or time, refusing to let anything distract him. On Sundays and holidays, he opened his own portable barber's stall near the port, the most popular destination in Barcelona. As hundreds of ships bobbed in the Mediterranean, thousands of dockhands and sailors roamed the quay in need of a shave.

At Spanish universities, professors recited classical medical texts, which students were then required to memorize and recite back. Justo was the ideal medical student, and he increased his memory capacity with the popular training techniques of Abbé Moigno, a French savant and priest, whose method of associating words with sounds and meanings allowed him to retain up to 41,500 words and up to 12,000 facts. In 1847, Justo earned his licentiate in surgery with highest honors, officially entering a higher class.

Justo kept pushing ahead, enrolling in a doctoral program at the University of Barcelona with the aspiration of becoming a physician. Bad luck struck him almost immediately, however. Justo's boss fired him. Making matters worse, Barcelona was in the midst an economic crisis, resulting in mass unrest, which the government suppressed by firing cannonballs into crowds of protesters. A stray one demolished Justo's stand and wounded his leg. Injured and unemployed, he had no choice but to return to the highlands.

In January 1848, there was an opening for a surgeon in Petilla, where the villagers suffered from high rates of asthma, thought to be the result of exposure to the harsh northern winds. Despite his achievements, the salary was less than half that of a typical rural surgeon. Justo agreed to provide "sanitary services," including shaving the villagers and treating venereal diseases. In exchange, in addition to his meager salary, he would receive thirty loads of wheat per year and would be exempt from taxes.

The *ayuntamiento*—the municipal council—gave him living quarters in a cobblestone building, slightly taller than its neighbors, set on uneven ground, with a main entrance in the back. On the ground floor—usually used for storing animals and tools—Justo established an office: a wooden table on the flagstone floor and two mirrors facing each other, one large and one small. On the second floor was space for a family. Within a year, when Justo had saved enough money to furnish his home, he decided it was finally time to marry Antonia and have children.

On September 11, 1849, Justo Ramón y Casasús Pardo Casasús and Antonia Cajal Puente Marín Satué were married at the same church in Larres where they had been confirmed. Their first child was born on May 1, 1852, at nine in the evening, in a small room in the town surgeon's house, on a simple iron-framed bed beneath a cross nailed to the cracked plaster wall. They named the boy Santiago Felipe—after the patron saint of Spain—and the following day he was baptized. "I cannot complain about my biological inheritance," Cajal wrote. "With his blood [my father] transmitted to me traits of character to which I owe everything that I am." The myth of his father's life, said Cajal, was "the necessary antecedent" of his own.

CHAPTER 2

"Perpetual Miracle"

Most of what we know about Cajal's childhood comes from his auto-biography, *Recollections of My Life*. There are almost no corroborating witnesses to the events that he described. Autobiography is inherently unreliable; the famous nineteenth-century British scientist Thomas Henry Huxley called the genre "a special branch of fiction," and Cajal, who wrote fiction from his teenage years through middle age, knew how to craft a story. When he was young, he imagined himself as the hero of a picaresque novel, a characteristically Spanish genre in which the protagonist—or *pícaro*—is a boy from a lower social class who embarks on a series of loosely connected adventures, surviving on pluck and guile, his behavior ranging from impish to criminal. In his autobiography, Cajal presented himself as he saw himself and had always wanted to be seen.

IN 1853, WHEN SANTIAGO WAS SEVENTEEN MONTHS OLD, THERE WAS an opening for a town surgeon in his parents' native village, Larrés. Justo's contract in Petilla had yet to expire, and the new job was also temporary, but he could not pass up the opportunity to return home. He and Antonia were happy to have family and friends help care for their infant son—nicknamed "Santiagüé"—who, according to his nurse, was exceptionally willful and restless. When he was around three, he almost died when a horse that he hit in the hindquarters kicked him in the head. When Santiagüé was two, a second child was born—his brother, Pedro—who turned out to be far more easygoing and affable than Santiagüé, who described himself as a "wayward, unlikeable creature."

When Lorenzo Cajal, Antonia's father, moved to Larres from his native village of Isín in 1809, he brought the textile trade with him, becoming the town weaver, and as a young child, Cajal spent countless hours in his grandfather's shop. The loom was a flimsy wooden structure, rigged with pulleys and rollers, which creaked and swayed as the weaver pressed his foot to the pedal. Santiagüé's family called him "the devil child"—his earliest memory was of tangling the threads of his grandfather's loom. No one would have guessed that he would one day untangle the impossibly complex threads of the nervous system.

In 1855, Cajal's family left Larres because of tension between Justo and the *ayuntamiento*, most likely over his salary. The same aggression that drove him to succeed earned him enemies wherever he went, and his ambition always trumped the family's interests. They moved to Luna, a larger town with better pay, where Justo worked for less than a year before relocating to Valpalmas, which was smaller and offered less. The likely explanation is yet another conflict.

In 1857, Cajal's sister Pabla was born. All three children looked like their father. With a toddler and an infant, Antonia paid less attention to her eldest, then five years old, and Cajal admits that he longed for more time with her. In her absence, Santiagüé became his father's charge. Regretful of his lack of early schooling, which had hindered his intellectual development, Justo was determined to accelerate his son's education. There was no greater sin in the world than ignorance, he believed. Justo was the kind of man who would stop to lecture other people's children in the street. He thought of boys as young horses, by nature rebellious and wild, in need of discipline. Sometimes it took corralling and whipping to tame them.

In 1857, Spain passed its first comprehensive education reform—the Moyano law—requiring every child to enroll in school at the age of six. Justo started educating Santiagüé a year early. Justo's contract required him to treat patients as soon as he found out that they were sick, and so he would lead Santiagüé away from the town, where no one could find him. In the dry, scraggly fields, they discovered a small, dark cave—so small that Justo, a stout, broad-shouldered man, carrying an abacus and a globe, would have bent down to step through the narrow opening. There were no chairs, only rocks, and Santiagüé and his imposing father must have sat almost knee-to-knee in the cramped space.

Sitting still is the worst torture for a child, Cajal wrote. Santiagüé was desperate to be free. He was disorganized and restless—not a model student. "Only my father could make out in the untamed and chaotic weeds of Santiago's brain the light of an intelligence," his brother, Pedro, later said. Justo believed that it was possible for even the most stubborn mind to learn. Day after day, for the duration of that year, he brought his son to the cave and taught him the basics of geography, arithmetic, grammar, and even physics, nurturing the lofty, seemingly absurd hope that Santiagüé would someday become a great scholar.

It was rare for Spaniards of any age to learn a foreign language; many spoke only a provincial dialect and not Castilian, the national tongue. But Justo taught his son French, the lingua franca of European culture. Geographically speaking, France and Spain could not have been closer, but the distance seemed immense. The French highlands, in the eyes of travelers, stood for refinement and civility, while the Spanish highlands seemed melancholy and savage. "Truth on this side of the Pyrenees," said the French mathematician Blaise Pascal, "error on the other side."

Together, Santiagüé and his father read a 1699 French novel called *The Adventures of Telemachus*, written in simple vernacular prose, telling the story of Telemachus after the return of his father, Odysseus. Santiagüé came to believe that he might follow in his father's footsteps. Even as an old man, Cajal found himself transported back to the cave whenever he saw the cover of that book.

Santiagüé turned out to be a natural reader and writer, able to grasp concepts quickly, and Justo could not help fantasizing that his son would one day surpass him. But Santiagüé showed one trait that threatened to undermine his father's plans: he struggled to recall names and dates, stammering and sometimes failing to retrieve words. He was his father's heir yet had failed to inherit his verbal memory, his family's most valuable asset.

Justo was not a strict Catholic; he believed in the divinity of the human will. So failing to complete his academic mission in Barcelona was more than a practical setback—it had provoked a crisis of faith in himself, and the fact that he never finished his degree had depressed him for more than a decade, according to Cajal. As his family's needs grew, Justo saved money for tuition, imposing the same austerity on his family as he had on himself. It was Cajal's mother who sacrificed the most, spending almost nothing, cooking with fewer ingredients, and mending every garment herself.

In 1858, when he was thirty-five, Justo finally re-enrolled in school in Madrid, where he took the requisite courses in public hygiene, legal medicine, and toxicology. He found another surgeon to replace him, asked the man to look after his family, and left six-year-old Santiagüé in charge of all correspondence. He gave half his savings to Antonia and kept the rest. The family moved to Larres to live with Cajal's grandfather. Justo would visit only during breaks and for the birth of his fourth child, a daughter named Jorja, in the spring of 1859, the only child who did not resemble him. "Our father was the best example of conduct in life," Pedro said. But for two years, while Justo was gone, Antonia was the children's sole parental influence.

• • •

ANTONIA WAS DEEPLY RELIGIOUS. WHEREAS HER HUSBAND HAD grown up illiterate, she had learned to read in girls' school as a child, and she loved novels. When her sweetheart first left Larres in the hope of making his fortune, then nineteen-year-old Antonia occupied herself with cheap romances and tales of knights errant until his return. Antonia, according to her daughter Pabla, had astonishing mental coordination and perceptiveness. But Justo, a strict utilitarian, denounced works of fiction as harmful distractions and allowed only medical texts in the house, and so Antonia hid her novels in a trunk and secreted the books to her children when their father was away. Santiagüé and his siblings savored these stories. Little is known about Antonia Cajal, and nothing survives from her own hand. But her inner life is hinted at by her choice of novels, two of which feature oppressed women as protagonists: Catherine Howard, the fifth wife of Henry VIII, who was accused of adultery and then beheaded; and Genevieve of Brabant, also vilified for infidelity, who escaped execution and lived alone with her son in a cave.

With his father gone, Santiagüé was free to indulge "one of the unbridled tendencies of [his] spirit": exploring and admiring the "perpetual miracle" of nature. "I never tired of contemplating the splendours of the sun," Cajal recalled of his hours wandering around the outskirts of town, "the magic of the twilight, the alternations of the vegetation, with its gaudy spring festivals, the mystery of the resurrection of the insects, and the varied and picturesque scenery of the mountains."

Throughout his childhood, Santiagüé took a special interest in birds, searching for nests and studying them for hours, hoping for a glimpse of a wagtail, chaffinch, linnet, or cuckoo. One time, he got stuck halfway up a tree, and a search party from town did not find him until after dark. His long absences always concerned his mother, who was constantly worried about his well-being. Santiagüé began collecting bird eggs in a thin box that he divided into labeled compartments, a hobby that his father encouraged. At first, Santiagüé was attracted to the eggs only because of their colors—aqua, speckled gray, cream brown, white with red stains—but after leaving his collection outside on a summer day and returning to find a liquefied, stinking mess, he began incubating them. Nothing delighted him more than witnessing the metamorphosis of the newborn birds. From Humboldt to Darwin to Cajal, all great naturalists share one essential quality: they would rather be observing nature than doing anything else.

IN JANUARY 1860, WHEN CAJAL WAS SEVEN YEARS OLD, THE SPANISH army conquered the city of Tetuán, in the North African territory of Morocco. For over a year, Spain had been fighting the "African War" against Berber tribesmen, descendants of their perennial enemies, the Moors. The mission inspired a massive enlistment of volunteers. Ten thousand Spaniards died in the conflict, and Alto Aragon had been a hotbed of recruits. After Spain's victory at Tetuán, Morocco sued for peace, resulting in the Treaty of Wad-Ras and a resounding Spanish victory.

In the town square of Valpalmas, people danced the *jota*, an intricate folk dance set to twanging guitars with brisk, clapping refrains, as women whirled and ruffled their skirts and men roasted chunks of mutton over a bonfire, which they passed to Santiagüé, along with leather pouches full of sweet black wine. This first encounter with patriotism—which Cajal recalled as a sense of collective uplift—would imprint itself upon him for the rest of his life.

Alto Aragon was a devoutly Catholic place, and Santiagüé grew up with his teachers and parents—and even his more freethinking father—telling him that God was a loving Father who would always maintain order in the universe and protect him from evil. He was required to read the Bible, presented to him as the ultimate source of moral authority. In 1851, the year before Cajal was born, the Spanish government signed a concordat with the Vatican, in reaction to the anticlerical Liberal government of the decade before, reestablishing Catholicism as the official religion of Spain. Article II of the document reads, "Teaching in universities, colleges, seminaries, private and public schools of all types will conform in every respect to Catholic doctrine."

Santiagüé attended the one-room schoolhouse in Valpalmas, which had bare wooden benches and a picture of Jesus Christ on the wall facing the students. One afternoon, the class was reciting the Lord's Prayer when the sky suddenly darkened and the students heard a deafening crash. The school had been struck by lightning. In his autobiography, Cajal unbelievably claimed that the bolt entered through a window in the school's attic, destroyed the ceiling of the first, and then headed straight for the portrait of Jesus, smiting the Lord and Savior before exiting through a mouse hole. He described children racing outside, covered in plaster, where they saw the scorched body of the town priest dangling from the bell tower. While his own godlike father was absent, Santiagüé, for the first time, questioned the existence of God. According to Cajal, the lightning struck at the *exact moment* when he and his classmates uttered the words "Lord deliver us from all evil."

That summer, scientists predicted a total solar eclipse, and people from all over the world traveled to Spain for a glimpse of the rare phenomenon. The path of the eclipse's shadow was illustrated in Alto Aragonese newspapers—Valpalmas happened to be right on the fringe. On the morning of July 18, 1860, Santiagüé stood on the top of a tall hill alongside his father, who had just returned from Madrid, having finally completed his degree. Justo explained to his son that scientists could calculate exactly when and for how long the sun would disappear. Santiagüé stared through smoked glasses at his father's watch to see if the prediction would come true. Severe thunderstorms darkened the sky, and a heavy mist lingered well into the afternoon, until suddenly the sky turned from pure azure to indigo as the sun began to slowly vanish. As darkness enveloped the earth, Cajal said that, surprisingly, his mind was absolutely calm. Reason served him as a shield against superstition. He had lost one faith and found another.

CHAPTER 3

"Plunging into Social Life"

The fortunes of Cajal's family changed dramatically with the achievement of his father's new degree. People now addressed Justo as *Don*, an honorific with echoes of blood aristocracy, and his credentials now qualified him for higher paying, more stable positions. Later in 1860, he moved the family to Ayerbe, a town of three thousand residents located on a main road between three larger population centers, with weekly agricultural and livestock markets attended by people from all across the district.

Ayerbe lies south of the Pyrenees, closer to the fertile plains of the Ebro River where there are only the inklings of foothills. Land deeds—not academic degrees—served as proof of real status, and the Ramón y Cajal family still rented their home. They lived in a tiny, narrow apartment on the second floor of a multifamily house, where all four children shared a single bedroom. Justo had to compete with other surgeons, which meant that he was often away from home, too busy to tutor Santiagüé.

Justo moved the family to Ayerbe in part because its school was better than the one in Valpalmas. Because of his early lessons, Santiagüé was ahead of his peers; his father sent him to school mainly for discipline, but with only one teacher and one assistant to control a hundred students, real supervision was impossible. At that time, primary-school teaching was a profession for those who lacked either the resources or the intelligence to pursue another career. Students were drilled on multiplication tables and forced to do penmanship exercises, and Santiagüé was routinely asked to recite from textbooks, exposing the weakness of his memory. For a shy, sensitive child, public failure was a nightmare, and furthermore, Santiagüé knew that poor academic performance would disappoint his father. He began to act out in class, sometimes skipping school altogether.

Santiagüé had no friends other than Pedro, who followed his brother everywhere. "Things always interested me more than people," Cajal wrote. It would have been difficult for Santiagüé to make friends, even if he had wanted to, since Ayerbe was his fourth home in eight years. He noticed that whenever someone spoke to him, their words instantly supplanted his own thoughts, almost as though they were trying to dominate his mind. He was determined to remain "the owner of his own solitude." Whenever he could, Cajal said, he wandered into the countryside by himself, drawn inexorably toward the mountains, following the courses of streams and stealing into orchards. The Somontano district, where Ayerbe is located, was the scene of many battles during the Reconquest, and the ruins surrounding the town thrilled Santiagüé more than any school history lesson ever could. "From the heights of the mountains," Cajal recalled, the ruins "seemed to tell us heroic stories and legends of distant grandeurs." In his imagination, the structures of the past came back to life. In the future, he would perform a similar act of imagination, deducing the workings of cellular life by examining dead tissue samples.

The first time he set foot in the town square of Ayerbe, Cajal recalled, the other boys mocked him. He wore the wrong clothes. He spoke the wrong dialect. Since his father was a medical practitioner, the town boys assumed that Santiagüé was wealthier than they were. They punched, kicked, and threw stones at the new boy, whom they nicknamed *forano*, the Ayerbean word for "foreigner." Suddenly, Santiagüé felt the "necessity of plunging into social life." He wanted to prove himself in front of the boys who derided him. The hyperbole in his autobiography is almost comical: he claimed the flutes that he carved reached notes no other flute could, and children followed him around when he played as though he were the Pied Piper; the slingshots that he made, using leather from his own shoes, were so fine that all the local shepherds clamored to buy them; the bows that he shaped were perfect, and the arrows that he carved flew straighter and farther than anyone else's. Even at a young age, Pedro said, his brother was driven by a "blind desire to overcome, to be first in everything without making amends for anything in order to achieve it."

Every fight, theft, and act of vandalism seemed to lead back to the home of the respected surgeon, where neighbor after neighbor—including the mayor and the priest—appeared at the door, accusing Santiagüé of a litany of transgressions. Santiagüé's dismal academic performance and unruly behavior, in his father's assessment, were symptomatic of a grave disease that required harsh treatment.

When Justo returned home after a long day to find that his sons had misbehaved, he would become enraged, screaming at Antonia for her "excessive softness," Cajal recalled. The tools of a barber-surgeon—such as artificial leeches, tonsil guillotines, hemorrhoid forceps, skull saws, and mouth gags—easily doubled as torture implements. Patients, bound to the operating chair, were known to thrash and shriek and die *como chinches*—like bugs. There is good reason why, in the eighteenth- and nineteenth-century popular imagination, barber-surgeons were the archetypal villains. Cajal's father would whip him until the skin on his back became raw and bled, then beat him with a club, and sometimes pulled at his flesh with tongs, which could be heated to even more painful effect.

Cajal recalled feeling "absolute terror." Yet he devoted only a few lines to this trauma in his autobiography, and he never wrote about these experiences again. His response to the violence, as a child, was to affect numbness. "I am used to it," Pedro recalled his brother telling him. "The blows anesthetize me, and I no longer feel pain." During an attempted escape—one of many—Santiagüé and Pedro skipped school and headed into the hills, where for a few days they lived in the wilderness. One night, they were sleeping in an old lime kiln when their father appeared, shook them awake, tied their arms together, and paraded them back to town, leading them through the square so that people could gawk and jeer.

It was around this time that Santiagüé suddenly began to experience what he later called an "irresistible mania": the impulse to draw. He scribbled constantly on every surface he could find—on scraps of paper and school textbooks, on gates, walls, and doors—scrounging money to spend on paper and pencils, pausing on his jaunts through the countryside to sit on a hillside and sketch the scenery. Unable to focus in the classroom, Santiagüé could draw for hours, his chaotic energy channeling itself onto the page. He entertained his classmates with caricatures, and they began to recognize him as an artist, earning him the validation that he desperately sought. While drawing, he would concentrate so intensely on an object or a fantasy that he later described his mind as a well-fortified castle.

One day, when he was ten years old, Santiagüé marched into the living room and declared that he was going to become a professional artist. His father considered artistic expression a developmental defect, an illness of the will, and artists themselves itinerant deadbeats. Justo assured Santiagüé that once he became a doctor, he could waste his time on any dilettantish hobby he wanted. In the meantime, Justo confiscated Santiagüé's art supplies and tossed his drawings into the fire, so Santiagüé began using wadded-up bits of paper as a brush, extracting blue and red pigments from the bindings of cigarette papers by soaking them in water, and hiding his drawings between rocks in the fields. To Santiagüé, art was a portal to a fantasy world of heroes and legends of bullfights and battles; catapults, armor, and charging steeds; and sailors, cannons, and the open seas. Cajal "entered the Castle of Science," Pedro later said, "through the Door of Art."

Nonetheless, seeking a second opinion on his son's talent, Justo showed Santiagüé's drawings to the most prestigious art critic that

Alto Aragon had to offer: a plasterer hired by the *ayuntamiento* to whitewash the walls of a local church. A transient worker, who depended on community leaders like Don Justo for business, the plasterer knew exactly what to say when Justo asked if his son had any talent. The plasterer replied, "None," Cajal recalled, then proceeded to mock the child. Standing beside his father, hearing the verdict about his artwork, Santiagüé, who never broke down during beatings, could barely stop himself from crying.

Some of Santiagüé's childhood drawings have survived. A simple pencil sketch, made when he was eight years old, depicts an Aragonese peasant in full costume, leaning on a walking stick, the folds of his blouse and sash expertly contoured and shaded. Particularly striking is a cobblestone wall in the foreground of a watercolor of the Ayerbe church. The wall is a conglomerate, like cellular tissue, a unified structure composed of distinct, individual rocks, each of which Cajal renders true to its unique form. Within his youthful artwork are the outlines of his scientific drawings.

CHAPTER 4

"A Castle of Dreams"

Cajal finished primary school in 1861, when he was nine years old, and Spanish law did not require children to continue their education. Secondary schools, or *institutos*, existed to bestow titles, confer social status, and groom elites, and only around 5 percent of boys between the ages of ten and sixteen attended, most of whom were children of aristocrats or the bourgeoisie. Recent liberal reforms, however, offered rural children newfound possibilities for advancement. Justo was determined to send Santiagüé to secondary school regardless of how uninterested he was in his studies. A baccalaureate degree—awarded after a six-year course—was a prerequisite for university enrollment, and so Santiagüé could not become a doctor without one. More than that, the choice of school would define his role in society. "You are where you did your baccalaureate," wrote one twentieth-century Spanish novelist.

Parents often looked to religious boarding schools to straighten out their wayward children. Justo decided to send Santiagüé to the Institute of Jaca, a private school run by Aesculapians, a sect of Jesuit priests devoted to teaching medicine, especially renowned for its program in Latin; physicians and surgeons distinguished themselves with their knowledge of the language, while barber-surgeons—also known as *romancistas*—spoke only in the vernacular. Santiagüé declared that sending him to Jaca would be a waste of money and begged to be sent to an *instituto* that offered drawing classes, but his father refused.

When Santiagüé left home at the end of the summer, Antonia cried and pleaded with him to obey his teachers. She was afraid of the treatment that her son might receive. The motto of the Spanish education system was *la letra con sangre entra*—"knowledge comes with blood."

IN THE HIGHLANDS, THERE IS NO REAL AUTUMN; NEAR THE END OF September, winter was already looming, and Santiagüé had never been away from home. The journey took about ten hours, the road heading north into the heart of the Pyrenees. Santiagüé sat at the front of the carriage on a mattress spread out over his luggage to cushion him from the bumps in the road. His father accompanied him, to ensure that he made a good impression. Justo knew the landscape intimately, since the towns of his youth were located not far from the road. Santiagüé pelted his father with questions. They shared a curiosity about nature, which became the best—if not the only—way for them to bond. As long as Justo held forth while his son dutifully listened, their relationship was relatively peaceful. Justo began to tell his son the kinds of stories that he loved to hear, tales of heroism and war, of highland battles that took place on the very terrain that they rode along during the 1808 War of Independence, when Napoleon invaded Spain. The French were better disciplined and better armed, but the Aragonese possessed one unassailable advantage, according to the revolutionary general Francisco Espoz y Mina: the guerrillas were "commanders of themselves."

Sunk in a furrow between two heavily eroded glaciers, Jaca, in geological terms, is founded on a depression. At twenty-seven hundred feet, the medieval city, hemmed in by turreted walls, has an air of gloomy detachment. The peak of Mount Oroel—where it is said that bonfires were lit to signal the start of the Reconquest—was visible from any point in the city, its stern gaze seemingly following the residents of Jaca wherever they went. The institute was an old building with a peeling facade and no windows; the priests, in their black hats and black robes, resembled executioners; and the dormitory, with its rows of plain metal beds, looked less like a home than a barracks. Santiagüé would live with his uncle Juan, his mother's brother, whose house was a few minutes' walk from school.

Santiagüé was given an entrance examination in three parts, including Castilian grammar and religion. "Who was incarnated?" the test asked. "Who was Jesus Christ? What does Jesus wish to say?" Santiagüé made no errors, raising the expectations of his new teachers. Justo tried to explain that his son might not perform as well in class, that he learned differently, but that, though he might seem shy and lacking in confidence, Santiagüé would produce the right answers if given a chance. At the same time, Justo encouraged the priests to punish Santiagüé for even the slightest infraction. Through gritted teeth, Cajal recalled, he vowed that his oppressors would never break him.

Forty boys, looking gaunt and homesick, sat crammed together on short benches in a classroom. Secondary-school teaching was like "cerebral injection," a contemporary of Cajal's explained, with rules and formulae jammed into the heads of students. The Spanish state dictated a national curriculum, discouraging individuality and creativity and punishing critical thinking and innovation. Pupils absorbed lectures in the morning and were expected to regurgitate them in the afternoon. Brandishing rulers, whips, and straps, Santiagüé's teachers stalked the room, primed to lash out at the slightest error. Cracking sounds echoed through the hallways and thumped in his head "like those of a door knocker in an empty house." In his famous painting *La letra con sangre entra*, Goya depicts a typical classroom scene: some crying students are being whipped while the rest of the class, seeming not to notice, go about their studies.

With its numerous conjugations and declensions, Latin was an especially tortuous subject for a boy like Santiagüé, who struggled to recall words. According to Cajal, his Latin teacher once threw a student against the blackboard, shattering it and injuring two more students with the flying shards. The teacher imposed a daylong fast for every error, and when the list grew longer than the number of days in the term, he tacked on beatings and public shaming. Santiagüé had to dress as the "King of Cocks," donning an oversize feathered robe and running back and forth between two lines of students, each of whom— along with the teacher himself—punched him as he passed. The fathers began locking him in the freezing classroom after school for hours, without food. Santiagüé's response was to hide any evidence of pain by staring directly at his abusers and holding back his tears.

Cajal recalled that he approached the classroom each day "trembling with fear." He struggled to concentrate. The more hostile the environment, the deeper Santiagüé retreated into fantasy, drawing more feverishly than ever, conjuring worlds where mythical heroes lived forever and evil was always conquered by good. "For if the world does reject or bore us," Cajal said, "we may build a castle of dreams inside ourselves."

Outside school hours, Santiagüé took long walks alone in the wilderness. "Before the grandeur of the tremendous mountains which surround the historic city," Cajal said, "I forgot my humiliations, discouragements, and sorrows." Sitting in an abandoned fort, he pretended to be a medieval watchman, surveying the vast, empty plain for hours on end.

In keeping with family tradition, Cajal's uncle Juan was a weaver and owned a respectable shop, but since his eldest son, Victoriano, had unexpectedly left home, Juan's business had begun to struggle without his help. He had accrued a considerable debt, most of which was owed to his brother-in-law, Don Justo, who agreed to wipe the slate clean in exchange for his son's room and board. Worried about the future, Juan spent long nights at the loom, oblivious to the affairs of the house. During Santiagüé's stay in Jaca, Juan's wife, Orosia, died. Cajal claims that she passed away before his arrival, and though he misremembered other names and dates in his autobiography, just as he did when he was a student, this discrepancy seems especially significant. The death of his aunt brought about the return of his cousin Victoriano, who remained a dear, lifelong friend. Curiously, all that Cajal mentions about the homecoming is that the quality of his meals improved.

Victoriano was the older brother Santiagüé never had, a handsome, strapping young man in his early twenties, headstrong and rebellious, with independent opinions that he never cared to hide. The stories that Victoriano told were as entertaining as any adventure novel: he had set out alone as an adolescent, resourcefully taking on odd jobs and bravely meeting every challenge. Santiagüé wished that he could escape his father, and while he had tried and failed to run away from home, Victoriano had actually succeeded. He never became a weaver. He was the rare example of a firstborn son who had defied the law of inheritance.

By the end of the year, Santiagüé had stopped going to school. The fathers threatened to expel him, leaving him with no choice but to write to his own father, who was furious at first, but who then asked the teachers to be more lenient, seeing that their regime of terror had failed. His son could never become a doctor, Justo reasoned, if he was too afraid to attend class. Santiagüé passed his examinations at the end of the year in Latin I, Castilian I, Principles and Exercises in Arithmetic, and Christian History and Doctrine, earning the lowest possible grades—no doubt aided by the fact that Justo had performed a lifesaving surgery on the wife of one of the examination judges.

Santiagüé returned home to Ayerbe that summer looking so sickly and battered that his mother hardly recognized him. She nursed him back to health, cooking hearty stews and serving them to him by his bedside.

Before long, Santiagüé was back on the streets causing trouble. He taught himself to make gunpowder, and, with an assortment of junk, he was able to construct a cannon, which he fired at the neighbor's house. The police appeared at the door and arrested Santiagüé, and he spent a few days in the town jail, sleeping on a moldy straw pallet infested with bugs while townspeople gathered at the window grate to point and laugh at him. Justo ordered the constable to starve his son, but his mother recruited a friend to sneak him fruit, meat, pies, cakes, and biscuits. In Cajal's personal mythology, his mother existed as an unconditional source of love and care.

Justo wanted to send his son back to Jaca, despite the obvious damage the teachers there had caused. Records show that Santiagüé's tuition was paid, but he failed to appear for registration. The following day, he wrote a letter to the director of the Institute of Huesca formally requesting admission. Some friends in Ayerbe, who studied in Huesca, had sold him on the school. Teachers punished students but did not torture them, and since the institute was not geared toward medical study, Latin was not emphasized. And while religious classes were held every day in Jaca, in Huesca they were held only three times per week. Santiagüé never accepted Catholic dogma. "Jesus did not prophesy mathematics or thermodynamics nor use the microscope . . . ," Cajal later wrote. "The whole Bible seems to ignore science and is full of singular contradictions, starting with Genesis." He drew wicked caricatures of the church ushers who tried to discipline him. "Through every fanatical Catholic," he later wrote, "one can always make out a financier."

Justo, who almost never reversed his judgments, especially not as a concession to his family, finally agreed to Santiagüé's transfer. It must have been Antonia who convinced him. She had a way of wisely choosing moments to undermine his tyranny, and her intercession saved Santiagüé from another year of torture. He never discussed the conflict between his parents, which must have been ferocious. Some things, Cajal admitted, he would "prefer to bury in the shadows of the unconscious."

CHAPTER 5

"The War of Duty and Desire"

The story of his youth, Cajal said, was one of "a reaction against the overly utilitarian and positivist tastes and culture imposed on the author by fathers and teachers." In "the war of duty and desire," Santiagüé's transfer to Huesca constituted his first victory. Located south of Ayerbe, Huesca was even farther from the highlands, even closer to the plains, both geographically and symbolically in the opposite direction from rugged Jaca. Framed by reddish peaks and propped up by a gentle hill, its buildings chalk-white and crumbling, the ancient city of Huesca was arranged in the shape of a crude oval, as though drawn by an unsteady hand, its Gothic spires visible from a distance like the masts of a great ship. Santiagüé was always hungry for new, impressive sights, and Huesca promised him a panoply. On this second journey away from home, there was no sadness, only joy.

Fearful that his son's "artistic instincts" would return, Justo installed Santiagüé in a quiet boardinghouse in the shadow of the cathedral, owned by a religious widow and frequented by seminarians and priests. Justo paid an older boy—a family friend—to drill Santiagüé in Latin until he mastered every nuance of Horace and Virgil. As soon as his father was gone, Santigüé became the commander of himself once again, and his first mission was to find the local art supplies store.

To European travelers in the nineteenth century, Huesca seemed like nothing more than a large town, but in the enchanted eyes of a country boy, it qualified as a metropolis. Towns in the highlands had only rugged mule tracks, while Huesca's streets were paved, and its buildings, unlike the austere houses of his childhood, rose above two stories and had elaborate facades. The population of a highland town—at most—was in the low thousands, and Huesca's was ten thousand and growing. In the decade before Santiagüé's arrival, new cultural institutions had been founded, including centers for dance and music and an intellectual organization for study and debate. For the first time ever, Santiagüé saw a bookstore, which appeared to him "like an open window to the universe."

While awaiting the completion of his school transfer papers, Santiagüé was left with a few weeks to occupy himself. The library at the institute was endowed with thousands of books, and he chose to borrow *Memories and Beauties of Spain*, a well-known, richly written travel and history book, featuring beautiful illustrations, to guide him while he explored. The Isuela River flowed through the center of the city, watering the *alamedas*, groves that Aragonese cities often planted to compensate for their lack of trees. Santiagüé liked to sit among the butterflies and birds, sketching every rock, flower, and tree, imagining the secret lives of insects and plants, contemplating the world from their perspectives. He visited San Pedro de Viejo, the eleventh-century monastery in the central plaza, where it seemed as though, on holidays, all of Huesca knelt as one in prayer. In the bowels of the cathedral, illuminated by lamplight, he came face-to-face with the tombs of Aragon's medieval kings, heroes from the legends of his childhood.

The Institute of Huesca was housed in an old palace whose byzantine lobby was adorned with portraits of its famous alumni. The classrooms were dreary, and the textbooks the same as in Jaca, but the atmosphere in Huesca could not have differed more, as students sat in the back of the classroom smoking cigarettes, playing cards, and reading novels. The institute did have a "jail," out by the stables, where repeat offenders like Santiagüé might serve sentences of up to twenty-four hours. He relished this confinement. With smuggled chalk and char-coal, he could draw all over the walls.

At the Institute of Huesca, Santiagüé finally found a subject that he enjoyed: geography. He had learned the basics from his father years before, but it was more than the names of countries and seas that now engaged him. Geography class involved drawing. Atlases, globes, and maps, which the department had in abundant stock, served as course materials. Students were required to copy maps in detail, and though he struggled to remember the spelling of words or their order within a sentence, Santiagüé never forgot an image. He relied not on abstract thinking but on direct experience. His visual memory was every bit as impressive as his father's verbal one, and his talent allowed him to re-produce even the most intricate maps to perfection.

Santiagüé barely passed his examinations at the end of his first year in Huesca, though not failing felt like a major success. After a year in Huesca, where he was the "absolute master of his own actions," Santi-agüé returned for the summer to Ayerbe, reuniting with his old gang to show off his drawings, brag about his exploits, and wreak havoc in the streets. The idyll lasted only a few days. Having missed a whole semester, Santiagüé had fallen behind in the curriculum, and his father ordered him inside to study. "Such a decision was like a jug of cold water poured over my head," Cajal wrote with typical flair, "which was aflame with eagerness to give joyful rein to my natural inclinations."

Around that time, a classmate in Huesca began lending Santiagüé novels and volumes of poetry, the kinds of books that his father banned. Santiagüé's favorite poet was José de Espronceda—sometimes referred to as the "Spanish Lord Byron"—who wrote Romantic odes to perse-cuted revolutionaries. "Every soldier is an absolute king," reads one of his refrains. "The world is mine: I am as free as the air!" reads another. Santiagüé also read *Mournful Nights*, the 1774 poem by José Cadalso, infamous for its treatment of necrophilia and suicide and a cult sensa-tion among the youths throughout the country, who organized midnight readings by candlelight. Approaching adolescence, Santiagüé began painting darker scenes, including illustrations from an Espronceda poem in which the narrator recites his morbid fantasies. Santiagüé already had the soul of a Romantic; now he had officially discovered Romanticism.

Santiagüé asked his father if he could study in an abandoned pigeon shed on top of the barn, a request that his father granted. From the doorway, Santiagüé could see whether anyone was watching him, and behind the nearby chimney, he hid his paper, pencils, watercolors, and novels in a niche that he built out of sticks and boards. Whenever he heard someone coming, he would race across the roof tiles and back to the empty coop, where he'd pretend to read his textbooks. His algebra book presented polynomials as reflections of divine truths from a metaphysical realm; his Latin book assaulted him with declensions; and his history book was an endless march of incidental names and dates, utterly barren and bloodless.

One day, reconnoitering his private realm, Santiagüé happened to look through a window into the neighbor's attic, where, along with pieces of old furniture, he spied trays of sweetbread and candied fruit. The neighbor, who owned the town confectionery, was a man of refined taste. Examining the room more closely, Santiagüé found a trove of books, including histories, novels, stories, poetry collections, and travel narratives. Though the sweets tempted him, literature proved even more seductive. Santiagüé, the veteran orchard thief, devised a plan: before dawn, while the neighbors were still asleep, he tiptoed across the roof of the barn, edged past the chimney, crawled onto the neighbor's roof, and lowered himself down through the window. He did this all summer, taking one book at a time, figuring that if he returned it promptly, the neighbor would never notice. No book would be gone for long, since Santiagüé devoured each one quickly.

The protagonists of Romantic novels were like kin to the heroes of Aragonese legend: self-centered, larger than life, and impossible for society to contain. Santiagüé recognized his own qualities in them, albeit exaggerated for dramatic effect. His favorite books offer a window into his psyche: *The Hunchback of Notre-Dame*, whose hero is a disfigured outcast; *The Three Musketeers*, wherein a poor, clever, and fearless young man leaves his provincial home and joins an illustrious band of swordsmen in Paris; *The Count of Monte Cristo*, about a falsely imprisoned man exacting his revenge; and *Robinson Crusoe*, in which a young man defies his father, sails across the ocean, and, through his own ingenuity, survives a shipwreck on a deserted island. Cajal sought out Romanticism, he later said, as a reaction against his utilitarian upbringing. Reality was too harsh; he preferred imagining himself as a character in a novel.

During his summer of literary awakening, Santiagüé also discovered Cervantes's *Don Quixote*, a book that would inspire him for the rest of his life. Santiagüé sympathized deeply with Quixote, who, like himself, hailed from a rural village and became obsessed with romances and books of knight-errantry. The nature of Quixote's madness is that he cannot distinguish fiction from fact. Santiagüé preferred the world of fiction, and he was indignant when, at the end of the book, the "Mournful Knight" is forced to renounce his life of chivalry.

In his neighbor's attic, Santiagüé also encountered *The Swindler* by Francisco de Quevedo, the classic picaresque novel from the seventeenth century, the golden age of Spanish culture. The protagonist's father sends his son to religious school, where priests starve him nearly to death, and Santiagüé could surely recognize Quevedo's descriptions of the boy's empty stomach, dizzy head, dusty mouth, and rattling bones. Santiagüé's cousin Victoriano had made himself into a *pícaro*, and the stories he told him upon returning home from his travels exemplified the genre. Santiagüé wanted to transform himself into a *pícaro* too. He drew no distinction between that fantasy and his own life.

At the end of the summer, Justo decided that Santiagüé was not mature enough to return to school. For most of 1865 he remained at home, and then returned to Huesca in 1866 to begin what he called "the most disturbed and unfortunate period of my student life." Pedro, now eleven years old, joined Santiagüé at the institute. More doubtful of his eldest son's prospects than ever, Justo pinned his hopes on Pedro, who was obedient and steady, the model student. Still, the brothers were both spitting images of their father, with the same intense expression and prominent nose, so much so that they were often confused for each other. Spanish boys were called by their fathers' surnames; Santiagüé and Pedro both went by "Ramón" and were known as "the sons of the surgeon of Ayerbe." They were the only two people in the world who shared the burden of their father's expectations.

Worried that Santiagüé would corrupt his younger brother, Justo housed the two boys separately, sending Pedro to live in a quiet boardinghouse and Santiagüé to apprentice with a barber, where the work would be so demanding that he would have no time to cause trouble or make art. If he did not manage to finish secondary school, which seemed more and more likely, at least his son would have acquired a vocation, his father thought. "And at what a time!" Cajal recalled, with customary floridity. "Exactly when my soul was still vibrating with the tremendous jolt which it had received from its sudden impact with the romantic!" He believed that he was destined to become a great painter, the next Titian, Raphael, or Velázquez, and tasks like sweeping up hair and lathering beards seemed well beneath the dignity of an artist. But the barber, Acisclo, whom Cajal described as a kind of ogre, with coarse features, jaundiced skin, and a violent temper, offered him an indispensable piece of advice: forget how you are feeling and focus on the task at hand.

Barbershops were social clubs for the lower class. Neighborhood characters like the militiaman, the cobbler, and the milliner, their faces masked with soap, showed off their scars and told tales of swashbuckling exploits. Justo's plan had backfired; he wanted to sever his son's contact with fiction. Instead, Santiagüé soon found himself in a Romantic novel come to life.

Talk at the barbershop revolved around the conduct of the rural police, a militarized force formed by local elites to protect their interests and property, which constantly harassed and abused the towns-people. Officers in brown uniforms, who earned twice as much as the average day laborer, delighted in issuing heavy fines for the slightest infraction of the most obscure laws. The barbershop gang, all political rebels, discussed the prospect of revolution in hushed tones. Liberal generals were rumored to be returning from exile in France, crossing the Pyrenees and heading south through the highlands, enlisting vol-unteers. Santiagüé understood nothing about politics at that time, but he longed to take part in a rebellion, and, with his "inborn dislike" of authority and hunger for individual freedom, democracy instinctively appealed to him.

In the nineteenth century, Spain—"the most reactionary nation in Europe"—experienced eighteen different governments, three consti-tutions, and more than two hundred uprisings, "a tangle of revolu-tions," to quote a historian of the day. Students at the institute divided themselves into Liberals and Reactionaries for their war games. The Spanish Liberal party had emerged soon after Napoleon invaded Spain, in 1810, when a group of parliamentary delegates began calling themselves Liberales. Their 1812 Spanish Constitution advocated for equality before the law and a constitutional, representative govern-ment. Liberal ideas spread from the intelligentsia to politicians, the military, some clergy, certain professions such as lawyers, and rentier landowners, who argued that the Old Regime was economically inef-ficient and unjust and that the Church and nobles enjoyed unfair priv-ileges. Ayerbe and Huesca were traditional Liberal strongholds. Santiagüé's schoolyard allegiance was never in question. He always fought on the Liberals' side.

When he was not at the barbershop, Santiagüé could be seen wandering around lost in thought, brow furrowed, face burdened by a per-petual frown. A former classmate described Santiagüé as "capricious"—referring not to flightiness but to a type of goat that roams the hills of the Italian island of Capri, alone. Santiagüé appeared suspicious and untrustworthy in the eyes of his classmates, who interpreted his rare smile as a sign of sarcasm, not happiness. He had a tendency, after not speaking for long periods of time, to burst into uncontrollable laughter, irritating his teachers, who assumed that he was mocking them. He never followed their instructions and attended class only when it suited him. Well-behaved students knew to steer clear of Santiagüé, but some classmates—the more wayward boys—liked to trail him on his rambles, not knowing where he was going or why, responding to the mysterious, commanding power in his silence.

Every day, before, after, and in between classes, Santiagüé and his friends staged battles in the alleyway next to the school, leaving him cut up and bruised, with lumps on his head so large that he was sometimes unable to put his hat on. When he was around fourteen years old, the future author of some of the greatest works in the history of scientific literature produced his first book: a slingshot manual outlining projectile selection, firing strategies, and battle plans, complete with his own illustrations. He called it *Estrategia lapideria* (Lapidary Strategy); Latin class had finally come to good use.

During his teenage years, Santiagüé also wrote and illustrated a novel, a facsimile of *The Adventures of Robinson Crusoe*, and he convinced some classmates to act out the plot. The Isuela River became the Atlantic Ocean, its *alamedas* a great jungle, and wearing makeshift loincloths, with mud smeared on their faces, the boys shot bows that Santiagüé had shaped out of greenwood branches and arrows he had tipped with shoemakers' awls, blunted for safety. He led a group of boys—fellow academic underachievers—in a failed attempt to run away and, like Crusoe, become sailors. "Cajal was a novelist of action," a fellow classmate later recalled, after Cajal had achieved fame as a scientist. He "believed and made us believe in the possibility of his novel being realized in actual life."

THAT SUMMER, SANTIAGÜÉ RETURNED TO HIS FAMILY, WHO HAD moved to a town called Gurrea de Gállego after a fight between Justo and the *ayuntamiento* of Ayerbe. Santiagüé's grades at the end of the year showed no improvement, and now more furious than ever, his father decided to teach him the ultimate anti-romantic lesson: he apprenticed him to a cobbler. The word that Cajal uses to describe his experience is *antiesthetic*; the shoemaker confiscated his paper and pencils and did not even allow him charcoal to draw on the walls of the barn. He slept on the vermin-infested floor of a dark garret, but at least it was a room of his own, and after dinner he would rush back there and spend all night staring at the stains and cobwebs on the walls and ceiling. He discovered that the longer he concentrated, the more the forms seemed to become animated, "transformed, by the power of thought, into the wings of a magic stage, across which raced the cavalcade of [his] fantasies." For Santiagüé, imagination was nothing frivolous—it was necessary for his survival (along with the meat and pies his mother smuggled to him so that he did not have to eat the cobbler's disgusting stews). "Never did I live more prosaically," he recalled, "or dream more beautiful, noble, and consoling dreams."

A quick learner and skillful with his hands, Santiagüé excelled as a shoemaker. In the fall, when Justo resumed his position in Ayerbe after reconciling with the *ayuntamiento*, he apprenticed Santiagüé to another shoemaker, a friend of his nicknamed Pedrín. Justo told Pedrín to starve Santiagüé if he did not eat the food that he was served. Santiagüé worked hard and did not complain, and Pedrín was thrilled with his new apprentice. One day, Justo visited the shop to check on his son's progress. "So? Are you chastened?" Justo said, according to Pedro. "Do you want to come back home?" "No," Santiagüé replied, "I am more than fine here. I like the job and don't want any other." He said that he would rather become a factory foreman than an academic of any kind. Santiagüé learned to make boots, trim heels, and fashion ornamental toecaps, which the local aristocrats fancied, and by the end of the summer he was able to buy more paper and another pencil with the tips that he'd received. He considered himself a shoe artist.

That summer, the battalion of Republicans descending from the highlands to overthrow the queen met the royal Spanish army in a battle that happened to take place just outside Ayerbe, in Linás de Marcuello. Since Linás belonged to Ayerbe's medical district, Justo was charged with treating the wounded royal soldiers, and while fulfilling his official responsibilities he secretly tended to the injured rebels, hidden in surrounding villages, whose cause he supported. He asked Santiagüé to accompany him. Battlefield surgery would serve as the perfect introduction to medicine, Justo thought, understanding his son well enough to know precisely how to engage his attention. Though he had read about death in Romantic poems, Santiagüé had never seen a real corpse.

CHAPTER 6

"The Nasty and Prosaic Bag"

After Santiagüé's year as a shoemaker's apprentice, Justo proclaimed his son cured of his "artistic madness" and sent him back to school in Huesca for the 1867–1868 academic year. Santiagüé promised to focus on his studies if his father enrolled him in a drawing class, arguing that, during the Renaissance, drawing was regularly taught to engineers and even doctors, and that over the course of the previous decade, linear and figure drawing had come to be accepted as an industrial profession in Spain. Housed in a separate building, drawing class was not part of the institute's standard curriculum and would cost an additional fee, which Justo agreed to pay, so long as Santiagüé also got a job at a barbershop.

Though only a few years old, the arts program in Huesca had already acquired an outstanding reputation. Its founder, León Abadías, was a respected Aragonese painter, having trained in Madrid at the Real Academia de Bellas Artes de San Fernando, the school once directed by Francisco de Goya. The purpose of art, according to Abadías, is not to produce work but to cultivate self-awareness, "to learn to realize what you do and why you do it." "What I really want to make clear to you," he said, "is the necessity of discovering how to acquire an *artistic soul*." Abadías was the mentor Santiagüé had been waiting for. He was living proof against Justo's claim that all artists are failures—he was a respected professional whom the provincial and city governments commissioned to paint murals and restore frescoes in local cathedrals. Drawing class provided Santiagüé with an "intoxication of the aesthetic," and Abadías called Santiagué his most brilliant pupil ("more than once," Cajal emphasized). Drawing was the only class in which he received a grade of excellent, and his work was even awarded a prize.

IN THE SUMMER OF 1868, JUSTO DECIDED THAT IT WAS TIME TO START Santiagüé's medical education, beginning with osteology, the study of bones, the first course in the standard university curriculum. For centuries, due to taboos against handling blood, only barber-surgeons were allowed to perform dissections—perhaps the only advantage of their lower status—and while at university, Justo studied with professors who were known to accompany their surgery lessons with autopsy reports. The word *autopsy* means "seeing for oneself." Justo told his son, who hated reading textbooks, that the only way to learn about surgery was from direct experience.

Throughout the Middle Ages, human dissection remained forbidden, except in special cases, when a judge might allow for a criminal to be "anatomized" after hanging. The shame of dissection was so intense that, immediately after public hangings, friends and relatives would rush to the corpse of an executed man to protect it from so-called anatomizers. A decrease in the number of executions in the seventeenth and eighteenth centuries led to a dearth of cadavers, while medical schools proliferated, and the demand for anatomical material increased. The first known "body snatchers" were caught stealing corpses from a cemetery near a university in Italy, where public dissections for the sake of teaching anatomy had recently begun. Men called "resurrectionists" in Great Britain were paid to exhume and deliver bodies, often working in teams, sending women as scouts to funerals, and casing cemeteries for fresh dirt.

In the middle of the night, Santiagüe's father led him from their house to the cemetery on the outskirts of town. Santiagüe was no stranger to trespassing; having his father beside him as an accomplice, however, was exceedingly strange. On the other side of the high brick wall, they landed in a hollow, which served as the public ossuary, where old remains were deposited to make room for the newly dead. Heaps of bones, coated with gravel and nettles, lay half buried in the scraggly grass. Justo instructed Santiagüe to pick out the most intact pieces, and they filled their sacks with crania, ribs, pelvises, and femurs. Cajal recalled the haunting sound of bones clattering behind him.

Back at the family barn, Santiagüe and Justo emptied their sacks and started sorting through the contents. Justo opened *The Complete Course in the Anatomy of the Human Body*, the Spanish textbook that he had used in medical school, a massive, five-volume tome describing every known structure of the human body, and proceeded to hold each bone up to the lamplight, examining every inch from every possible angle, noting the smallest details, no matter how insignificant they might seem. The most important skill for anatomists is observation, and it was Cajal's father who taught him how to observe. Justo, busy with his surgical practice, devoted all his free time that summer to teaching his son osteology. According to Cajal, his father "experienced an incomprehensible pleasure in awakening childish curiosity and hastening intellectual development." When Santiagüe gave a correct answer, his father would look up at the ceiling or raise his hand to his lips, barely able to contain his excitement. Santiagüe finally absorbed his father's lessons, and his father no longer dismissed him as foolish and lazy. Justo even asked his son to rattle off the names of bones to impress his friends.

The cemetery raid—clandestine, macabre, and daring—felt like a scene from a Romantic poem. Osteology offered Santiagüe clear *visual* perceptions—"fragments of solid reality"—allowing him to establish natural, logical connections between words and images, which he then had no problem recalling. Skeletons were not unlike the other intricate mechanisms that captivated him, like cannons and guns, and when he took the bones apart and put them back together again he experienced a similar thrill. Once more, his father tried to recruit him to the ranks of surgery, comparing the human body to a battlefield and legendary physicians to surgeons. Those arguments had no effect on Santiagüe. He had no interest at all in being a surgeon. "If things are looked at in their true light," Cajal later admitted, "osteology constituted one more subject for [my] pictures."

One afternoon, in his final year at Huesca, a classmate of Santiagüe's asked if he wanted to see a secret. They walked along the edge of the city to an abandoned church—one of many properties confiscated decades earlier by the Liberals during the *desamortización*, a systematic redistribution of ecclesiastical property—and descended the broad stone steps underground to the vault. Santiagüe's friend knocked, and the door opened. Inside were trays of strange liquid and an eerie red light. Students from the institute had set up a darkroom.

Even in Alto Aragon, Santiagüe had encountered traveling photographers, unmistakable in their pinstripe suits and button-down vests, watch chains dangling as they knelt to set up box cameras under black tents. Exposures were long and awkward; subjects had to sit still for several minutes, stared at by a giant accordion camera, until the bitter puff of magnesium smoke punctuated the interaction.

In the secret darkroom in the vault of the church, Santiagüe witnessed the intricate process of wet collodion, the dominant method of photography at the time, an improvement over previous techniques such as the daguerreotype and calotype. First, the photographer smoothed a glass plate with a polishing stone and then treated the surface with a solvent, before brushing the plate absolutely clean, since even the smallest particle of dust would appear as a dark spot on the final image. A solution of syrupy liquid—collodion—was then poured onto the center of the surface and guided to the corners, photosensitizing the whole plate. Cajal recalled the aroma of collodion, slightly

sweetened by the smell of ether, which is part of the solution, as "delicious." The process itself was thrilling, even perilous, since gun cotton,
added to the collodion, was highly flammable. After the plate was
exposed to light and removed from the camera, pyrogallic acid—the
developing agent—was poured in a sweeping motion over the plate.
As the image slowly appeared, seemingly out of nothing, Santiagüé
was "positively stupefied"; the photograph was so accurate that even the
finest details were reproduced. His peers cared only about the commercial possibilities of photography, according to Cajal, but he was
fascinated by the underlying principle, which he called "the theory
of the latent image"—that hidden on the plate itself, there is a germ of
some invisible structure, waiting to be exposed.

From that moment on, Santiagüé became a lifelong devotee of
photography, recording his life in images, constantly experimenting
with techniques. He was his own favorite subject. His self-portraits
were composed and staged in order to project an image of himself as
he desired to be seen by others. They capture his fantasies and aspirations, and, as with every story that Cajal told, there is more than a hint
of mythology. He used the most accurate method of portraying reality
in order to communicate a carefully curated fiction. In one image from
his adolescent years, the future Nobel laureate is shown wrestling,
tautly muscle-bound and wearing a loincloth, facing off against another boy clutching a spear.

EARLY ON THE MORNING OF SEPTEMBER 18, 1868, THE RESIDENTS OF
Cádiz, a city on the southern coast of Spain, were awakened by a
twenty-one-gun salute. They gathered at the port, where the royal
Spanish navy had arrived to declare an insurrection against the queen.
Celebrations swept through the streets, and soon news of the rebellion
spread to nearby towns, then on to cities like Málaga, Granada, and
Seville, before reaching the northern parts of the country a week or so
later. Cajal recalled waking up early one morning in Ayerbe with an

eerie sense of restlessness. He walked down to the square, where he found the townspeople wildly cheering, shouting the names of the great Liberal saviors whose portraits he had drawn while apprenticing at the barbershop, his first artwork ever to be hung on a wall. Revolutionary proclamations were read aloud: demands for universal suffrage; freedom of the press, trade, and religion; and the abolition of the death penalty. They invoked morality and enlightenment, heroism and honor, echoing the rhetoric of Romantic novels. The rural police laid down their uniforms and peasants took up arms, wielding sickles and daggers.

The September Revolution—which Queen Isabel called "the mortal enemy of tradition" and Liberals referred to as "the Glorious Revolution"—divided Spanish society between property owners and laborers. Rebels raised the red flag of socialism and burned a portrait of the queen in the town square. A mob paraded through the streets, shouting and jeering, chanting, "Down with the Bourbons! Death to the Conservatives!" and Santiagüé heartily joined in.

IN SANTIAGÜÉ'S FINAL YEAR OF SECONDARY SCHOOL, LATIN AND Greek—his least favorite courses—were replaced on the curriculum by natural history and a combined course of physics and chemistry. The science department at the institute was spectacularly equipped; the natural history reading room in the library featured thousands of volumes as well as a cabinet with sixteen hundred specimens, and there were a host of gadgets in the physics and chemistry room—levers, pulleys, and gears; pumps, cranes, and propellers; magnets, magic lanterns, and turbines—the kinds of strange contraptions that would captivate a young, curious mind. Classroom lectures had always seemed dry and pedantic; now they sparked and crackled.

Nonetheless, his academic reputation was far from stellar. Cajal "was the typical student who was inattentive, lazy, disobedient, and annoying, a nightmare for his parents, teachers, and patrons," one teacher at Huesca recalled. He "will only stop in jail," predicted another, "if they do not hang him first."

In September 1869, at seventeen years of age, Santiagüé finally graduated with his baccalaureate degree from the Institute of Huesca. Three days later—not wasting any time—Justo accompanied Santiagüé to Zaragoza to enroll him in the medical school there, his own alma mater. Fearing a return of his son's artistic demons, Justo arranged for him to stay with a local surgeon, a friend and former classmate named Mariano Bailo, who lived in the Arrabal, the same working-class neighborhood where Justo had spent his student days. Santiagüé also served as Bailo's assistant, to gain experience in the field of surgery, which his father expected him to pursue as a career. "Farewell to ambitious dreams of glory, illusions of future greatness!" Cajal recalled lamenting. "I must exchange the magic palette of the painter for the nasty and prosaic bag of surgical instruments."

ZARAGOZA, THE CAPITAL OF ARAGON, WAS A LEGENDARY CITY OF RE-sistance. During the French siege of 1808, a small force of Aragonese volunteers managed to hold off Napoleon's much greater army for days, armed with only a hodgepodge of old cannons and defending the thresholds of their homes with knives. So many Zaragozans were wounded that virtually every building was transformed into a hospital, yet as the city crumbled around them, they fired away with abandon, even glee, declaring the Virgin Mary their captain general and dancing and singing in the streets.

Santiagüé was eager to reunite with his old friends from Huesca, now studying in Zaragoza, but they treated him coldly. After his transfer and apprenticeship, he had fallen a year or two behind his peers, who had since formed new friendships and abandoned him. Since he had been bullied in Ayerbe, Santiagüé had done everything possible to make himself indispensable to his gang. Despite not feeling brave, he noticed that when he acted bravely—during fights, for example—more and more people seemed to *believe* that he was brave. He now realized that his bravery had been a performance. Was he like Don Quixote? Had he been delusional the whole time?

The Arrabal was located on the left bank of the Ebro, directly across a fifteenth-century stone bridge, which offered the best view of the city. "The sight of it was a great pleasure," Cervantes wrote of Don Quixote's arrival in Zaragoza. Santiagüé sat listening to the hissing and lapping of the shallow waves. Suddenly, he was overcome by the impulse to follow the course of a river. He hiked fifteen miles upstream and ten miles down, but his expeditions failed to satisfy him. What he wanted was to discover new territory, a virgin plot of land with "sylvan glades and idyllic wild-flower beds unprofaned by the footsteps of man." He wanted to be like Robinson Crusoe, exploring his own island, far away.

CHAPTER 7

"A Myth Concealed in Ignorance"

In September 1869, when Santiagüé began his first term of medical school, the classrooms were half empty, since many students were still involved with the revolution. The new Republican government had declared that university teaching should be free of state supervision, and so, just as Santiagüé arrived, the medical school in Zaragoza had become the Free School of Medicine, unaffiliated with the public university system. Supported by only the provincial and city governments, the Free School was left with limited means. There was no physiology laboratory or obstetrics clinic, standard in European medical schools. Professorships were awarded by a committee of local politicians who lacked current medical knowledge and favored clinicians with archaic views. The Free School was housed in the provincial hospital, where students could expect to gain clinical experience. Training in scientific research did not exist.

Cajal's natural history professor, Florencio Ballarín, was a memorable exception. He held his class at the Zaragoza Botanical Garden, which he had single-handedly revived after its destruction during the French siege. Long stone benches faced a tranquil pond, inviting passersby to sit and gaze at the scenery. Paths wide enough for visitors to stroll along wove through a garden with hundreds of species of exotic flowers, with more sitting on the asphalt-covered roof of the building.

The basement had a state-of-the-art greenhouse, incubating a thousand plants, equipped with lighting and ventilation systems, and even toilets. On the first floor was the lecture hall, with room for a hundred students and ample cabinet space for all the seedbeds and dried plants necessary for a comprehensive modern curriculum. Ballarín's collection of specimens, including vertebrates, invertebrates, plants, minerals, and fossils—especially those excavated in Aragon—had won him a national award, and he enhanced his teaching with new samples every year.

Cajal's natural history teacher in secondary school had forced him to memorize Latin nomenclature, but Ballarín introduced classifications *by sight*, holding up specimens as he named them, pointing out their distinguishing features, as Justo had done with bones. "[Ballarín] was the first person whom I heard defend with true conviction the necessity of teaching objectively and with experiment," Cajal wrote. An old man by the time Cajal encountered him, Ballarín had suffered through the absolutist repression of King Ferdinand VII, who exiled, imprisoned, or forcibly removed liberal academics and physicians during his reign from 1813 to 1833. Ballarín was as passionately devoted to science as any artist was to the Romantic ideal.

Another crucial influence on Cajal's scientific development was Bruno Solano, his chemistry professor, a freethinking polymath who had studied literature, history, philosophy, fine arts, sciences, and law; he had once taught mathematics, and was also fluent in English and German. A large, charismatic man with wild black hair and a beard, Solano was a fixture at local cafés, where he debated so theatrically that strangers gravitated to his table. He had a reputation for challenging his students' assumptions, demanding them to question their habitual ways of thinking. His greatest gift, according to Cajal, was the "exquisite sensibility to perceive excellences even in the commonest things," a trait that Solano shared with Ballarín, Justo, and Abadías, his drawing teacher. Solano's class was a "temple for the worship of science," where he preached observation and experimentation as gospel, urging his students not only to *study* science but to let science guide their lives.

Solano's way of teaching chemistry was to tell stories, the kinds of adventure tales that Cajal loved, with atomic elements as protagonists—oxygen as "a kind of Don Juan," conquering virgin elements, and hydrogen "a jealous lover" stealing partners from other compounds. By anthropomorphizing the chemical world, Cajal said, Solano managed to convey "the most difficult points or the most uninteresting and abstruse ideas." In his class, Cajal realized that science could engage his imagination as raptly as literature could, and that his penchant for fantasy, denigrated by his father and teachers, could finally be applied toward more useful ends.

CAJAL COMPLETED HIS PREPARATORY COURSE, EARNING HIS STRONGest grades to date with no disciplinary incidents on his record. Justo, wary as ever, insisted on supervising his son in person. He applied for a position with the provincial medical service in Zaragoza, a public clinic for the poor, and soon after his arrival he was appointed interim professor of dissection at the Free School. Cajal left Bailo's house and moved in with his parents—he had no choice. Now he saw his father both at school and at home, where he constantly talked about medicine, even at the dinner table. In addition to teaching at the Free School, Justo also audited courses—he and his son might even have attended the same class.

To reach the provincial capital was a highland family's dream, and after decades of scrimping and saving, the Ramón y Cajals had finally arrived. With two salaries—from the medical school and the clinic—Justo could afford to buy a comfortable home and even a *finca* (country house) outside Ayerbe. Money and success served to further buttress Justo's autocratic reign, emboldening his cruelty, especially toward Antonia. One time, in the middle of winter, she and their daughters forgot to inform him about an emergency call, so Justo made them accompany him on his rounds in the freezing cold to ensure that they would never be so negligent again.

THE DISSECTION ROOM OF THE FREE SCHOOL WAS HOUSED IN A DIlapidated round building—more like a hut—in the garden between the hospital and the morgue. Patients too poor to afford burials or with no relatives to arrange their affairs often became corpses for anatomy classes. Justo wanted to teach his son anatomy, continuing the course they began in the family barn. Once again, he devoted all his free time to their lessons together, and since he no longer played war games in the afternoons, Cajal had nowhere else to go but the dissection room. Surrounded by cadavers—mutilated, eviscerated, and flayed—he and his father, draped in long smocks, stood side by side at the marble slab. On a steel tray nearby lay an array of sharp instruments: hooks, blades, and scissors; pincers and probes; hammers, saws, calipers, and shears; special rib cutters, and two wide, flat scalpels, wielded like spatulas, to remove the brain.

Leonardo da Vinci, who studied anatomy as training for drawing human forms, warned that dissection was not for the faint of heart. "You might be deterred," he wrote in his diaries, "by the fear of living in the night hours in the company of those corpses, quartered, flayed, and horrible to see." The sight of entrails sickened Cajal at first, but, over time, the gruesome grew ordinary. Page by page, they forged through the same mammoth textbook, meticulously dissecting muscles, blood vessels, nerves, and viscera, checking the authors' findings against their own. His father's explanations were pithy and serious, stripped of all rhetorical fat, leaving nothing but the bare bones. Sometimes they would come upon a new detail that no author had noted before. No pleasure could compare to the joy of discovery, Cajal said. It was a thrill that he first shared with his father.

With each anatomical structure their scalpel uncovered, Cajal did what he always did when he encountered new things: he drew them.

Only now, instead of rocks, flowers, and trees, Cajal turned his eye toward ganglia, sheaths, and plexuses.

Cajal's notebooks contained hundreds of drawings at varying stages of completion, from simple pencil sketches to finished illustrations embellished with watercolor and colored pencil. Justo finally realized the value of his son's artistic skill, and as Cajal's portfolio of anatomical images swelled, his father grew prouder and prouder. At some point, Justo also adopted the practice of drawing. One of his private students recalled that Don Justo would grab colored chalk, one in each hand, and, with incredible speed and dexterity, draw intricate pictures on the blackboard, almost like he was a machine. His former student concluded that Cajal must have inherited his artistic skill from his father, like so many other traits, but it is far more likely that, for once in Cajal's life, the son had taught the father.

Years later, Justo tried to publish an anatomical atlas, illustrated by his son. The project failed, according to Cajal, because they did not have access to advanced color-printing technology, though the more likely reason was probably conflict between father and son. Some of those anatomical images survive today. In their intensity and exaggeration, Cajal's close-up depictions of corpses, epically posed, echoed the portraits of heroes he once drew as a child—only now with their skin peeled back and their insides exposed.

Cajal, who always loved disassembling things and piecing them together, was a naturally gifted anatomist. With his outstanding visual memory he could see the body even when he was not observing it, creating a vivid map of its anatomy and storing it in his mind. He was motivated by his father's approval; for the first time ever he excelled at an academic subject, so much so that his professor appointed him assistant dissector, tasked with preparing material for classes. This new credential allowed Cajal to tutor students privately. Father and son both taught in the family home on San Jorge Street—Justo in the basement, Cajal on the second floor. Tutoring earned him a little bit of money, bringing him closer to independence from his father, who, though no longer able to physically punish him, still held the purse strings.

Cajal's responsibilities in the dissection room often caused him to miss other classes, which hardly bothered him, since he had no interest in the courses required for practicing clinical medicine. In his Topographic Anatomy and Operations course, however, he won an award for his description of the inguinal ring, the fascia of the abdominal muscles, among the most complex structures in the human body, which he had dissected with his father and drawn enough times to add down-to-the-millimeter diagrams to his paper. According to Cajal, the work was so advanced that some judges assumed he must have plagiarized.

AS A UNIVERSITY STUDENT, CAJAL WENT THROUGH THREE INTENSE phases, or "manias," as he called them. The first was weight lifting. Moving to larger towns and cities, amid more cultured and educated crowds, Cajal felt increasingly self-conscious about his intellectual upbringing. To compensate, he boasted about his physical strength, a habit that persisted until the day he died. He believed that, as a highlander, his constitution was more robust than that of the coddled city boys. (The dig appears in his autobiography, written decades later, a testament to his enduring insecurity.) One day, in university, Cajal walked into class and declared himself the strongest boy in school, and another boy immediately challenged him to an arm wrestling match and defeated him. Obsessed with revenge, Cajal spent two hours each day at a local gymnasium, receiving training in exchange for tutorials in muscular physiology, which the owner felt would enhance his scientific credentials. In his autobiography thirty years later, Cajal could still recall the exact weight of the wheat sack he threw to win a strongman contest, as well as the 112-centimeter circumference of his chest.

Cajal's autobiography did not include a photograph from his Nobel Prize ceremony, but it did include one of himself at the pinnacle of
his bodybuilding phase, when he was around twenty years old. He has
one arm behind his back while a thick vein on his biceps bulges, and

his broad chest juts out past his chis-
eled ribs, straining his neck muscles.
In the caption, Cajal went further,
insisting that his strength was even
greater than the picture suggests, but
that certain muscles could not be conveyed because of inherent defects in
the medium. He acknowledged that
his stories of vigor and prowess might
be "wearisome" to the reader, yet he
included them anyway. He could not
help it.

The second mania of Cajal's university years was a desire to write novels, folktales, and poetry, imitating the
Romantic literature he loved when he
was younger. Some of the pieces are
conserved in his notebooks, including
two romances called "Romance of Mari-Juana" and "The Pastor and
the Serrana," which were written in neat script, as though Cajal intended to share them. A poem, "Ode to the Student Commune," is a
satirical tribute to a group of classmates who, inspired by the Paris
Commune of 1871, went on strike against an unjust professor. "The
Commune," Cajal began, "Who does not feel his chest burning with
joy, upon the trembling of this beautiful phrase?" A few of these poems
were printed in the student newspaper because a friend of Cajal's knew
the editor, but Cajal dismissed them as no more than "servile imitations." Two love verses also survive: one a limerick addressed to a blond
girl with an angelic smile, the other an ode to a brunette girl with

beautiful eyes, composed in hexasyllables, a meter typical of French epics.

When Cajal was a teenager, he had fallen in love with a friend of his sisters', who would come to the house during wintertime, joining the knitting circle by the fire. He never summoned the courage to declare his feelings for her, instead penning a secret acrostic:

MARIA

M y heart was free
A nd then I saw your eyes
R iotously laughing at the sun
I watched in eternal spring
A nd I dreamed, happy and content.

Girls liked boys with handsome faces, noble bearings, and fine manners, according to Cajal, and he regarded himself, with his "massive and ungainly carcass," as "no Adonis." He claimed that his hands crushed his friends' hands when he shook them. He strutted through the streets "like a sideshow Hercules," toting an iron bar instead of a walking stick, which he dragged along the pavement, its clanging announcing his presence.

Cajal's third mania was for reading philosophy. The University of Zaragoza library, like most of the city, had been demolished at the beginning of the century by the French, but after the war, some volumes were recovered, and the estates of dead professors provided the rest. By the time Cajal was enrolled at the Free School, the shelves had been fully replenished with thirty thousand books, most on theology, philosophy, and law. His readings brought him into contact with great thinkers such as Kant, Berkeley, Fichte, and Hume, whose writings Cajal struggled to understand. Philosophy was another sport at which he tried to dominate; he treated philosophical thought with the same intensity and aggression as he did lifting weights. His favorite gag was to claim that no world existed outside of his own consciousness, insisting on his friends' unreality while staring them in the face. But he never really believed in solipsism—how could he, when he spent hours each day in the dissection room picking apart cadavers? "The direct observation of phenomena has an indescribably disturbing and leavening effect on our mental inertia," Cajal wrote. No abstraction could ever outweigh a "fragment of solid reality."

WHEN CAJAL WAS A STUDENT, SPANISH MEDICAL SCIENCE, BY AND large, was still dominated by the venerable theory of vitalism, which held that life is a function of some immaterial energy or force. In the Judeo-Christian tradition, the idea originates in the Book of Genesis, when Adam appears as a clay figurine until God breathes *neshamah*—wind or spirit—into him. Ancient Greeks believed that the body was animated by *pneuma*—meaning breath—which Galen identified as the mysterious intelligence behind all vital function. According to vitalists, this force distinguished living from nonliving matter but could not be analyzed by physicists or chemists; no reaction that took place in a flask could ever stand in for one that occurs naturally in the human body. Vitalistic theories dominated Western medicine for at least a millennium, and they remained the bedrock of scientific thought well into the nineteenth century. The chair of clinical medicine at the University of Madrid, who determined the national curriculum, was an avowed vitalist, as was Cajal's physiology professor.

After Napoleon's invasion, at the beginning of the century, Spanish science had collapsed, with scientific institutions either vegetating or disappearing within a half century and almost no scientific teaching taking place in universities. It was difficult for Spaniards to receive information about what was happening in the rest of Europe. Then came the 1868 Revolution, which ushered in an unprecedented era of free thought. Spanish scientists returned from exile, bringing foreign texts, which were translated into Castilian. Among these was *Cellular Pathology*, an 1858 book by the German researcher Rudolf Virchow, presenting an alternative view of disease based on the cell theory, established two decades before. When the body is ill, vitalists believed, a healing response is generated by the soul, a position Virchow rejected. "We must not transfer the seat of real action to any point beyond the cell," he wrote. When Cajal encountered Virchow's book in university, it struck him as nothing less than "revolutionary."

Virchow was not just a physician; he was a political radical who fought during the revolution of 1848 in Berlin, alongside students and workers on the barricades. His persistent advocacy for sanitation rights bothered Otto von Bismarck so much that the German chancellor challenged the scientific researcher to a duel, which never took place. Biology and politics were inextricable for Virchow—both were subject to the same laws. He refers to vitalism as the "monarchical principle," assuming the existence of a central authority, meting out healing force as though by divine right. "As an investigator of nature," Virchow declared, "I can only be a Republican"—and it was immediately after the Glorious Revolution that Cajal encountered his views. Biologists began to conceive of life in terms of cells at around the same time that the term *individualism* came into use. "Each cell forms an independent, isolated whole," one biologist wrote, which "nourishes itself" and "builds itself up." The cell was portrayed as a "distinct individuality," with "an independent existence," living "a life of its own," existing "individually of its own account," exercising "functional autonomy." "The body is a state in which every cell is a citizen," Virchow famously said. His anthropomorphizing captivated Cajal in much the same way as his chemistry professor Solano's had. Each cell, in Cajal's mind, was an "autonomous living being, the exclusive actor in pathological events, the protagonist of the ultimate epic: the history of life."

With his disdain for authority and "repugnance for all kinds of dogmatism," Cajal was inclined to rebel against vitalism. His reaction was in keeping with an intellectual trend; in the nineteenth century, by and large, conservatives and traditionalists upheld vitalism, while liberals and republicans, who trusted empirical science, dismissed the theory as metaphysical. One day in pathology class, heeding the cell theorists' call to arms, Cajal could not help challenging his professor, who happened to be the dean of the medical school—the man who had helped his father secure his job. Cajal attacked the professor, a symbol of both Justo's tyrannical authority and the illegitimate, absolute reign of scholasticism, which had oppressed him since the beginning of his schooling. Vitalism was "a myth concealed in ignorance," Cajal said, explaining its failure later with a parable: an Italian prince, traveling through China in an automobile, passed a village whose inhabitants had never seen a car before. "You can't fool us that easily," the villagers said. "There's a horse inside that car!"

Reading about ideas was never enough for Cajal—he needed to see cells for himself. Virchow cited recent findings aided by the microscope, an instrument that his father's anatomy textbook mentioned only in passing. Justo may have encountered microscopy during his own education (there is evidence that, in addition to attending the universities of Zaragoza, Barcelona, and Madrid, Justo also earned a degree at the University of Valencia, where he studied with two professors known for working with microscopes to research disease). Though microscopes were fixtures in anatomy departments throughout Europe, most Spanish universities had yet to embrace them. A prominent professor during Cajal's era disparaged microscopy as "celestial anatomy," a fanciful pursuit. Cajal described the Zaragoza faculty as "academic reactionists" who dismissed the new truths of microscopy as "pure fantasy."

The only microscope in the city of Zaragoza belonged to the physiology department, and an assistant invited Cajal to view some preparations. On top of a glass slide, a frog lay on its back, completely paralyzed, respiring only through its skin—an effect of curare, a plant-based poison with which some indigenous peoples in South America coated their arrowheads. Through a small incision in the abdomen, the coil of its intestines had been removed and extended carefully over a wooden cylinder, revealing the frog's mesentery, the folds of tissue affixing its intestines to its abdominal wall. Two days before, a solution flecked with carmine—a pigment obtained from dried female cochineals (parasites that feed on cacti), discovered by the Aztecs and Mayans—had been injected into the frog's lymph sac, circulating through the bloodstream and staining its cells bright red, allowing observers to track them more easily. Air irritated the tissue exposed by the incision, triggering an inflammatory response, which was traceable in real time under a microscope.

Red and white cells, only a few microns (millionths of a meter) in diameter—thinner than a strand of hair—rushed through the frog's bloodstream. The walls of the capillary were narrower than the cells themselves, forcing their elastic bodies to shift and contort. Cajal was "enraptured and tremendously moved" by "the sublime spectacle," he later recalled, its liveliness and clarity leaving him speechless. "I felt as though I were witnessing a revelation. It was as though a veil were suddenly lifted from my soul."

AROUND THAT TIME, CAJAL WROTE HIS SECOND NOVEL, A SCIENCE-fiction story inspired by Jules Verne, the favorite author of Cajal and his friends at that time. In Cajal's own "biological novel," a scientist arrives on the planet Jupiter, inhabited by beings ten thousand times as large as he is. He is the size of one of their cells. Equipped with scientific instruments, he decides to explore a giant creature's anatomy, slipping through its skin and entering its bloodstream, where he sails on a red blood cell and witnesses battles against parasites. He continues traveling along the sense pathways until finally he arrives at the brain, where he discovers the secret of thought and will, though he does not disclose what that secret is. "Think of it!" the great brain scientist later wrote, reflecting on his fantasy. "There was nothing trivial about it!"

CHAPTER 8

"Humbled by My Failure"

While Cajal was coming into his own in Zaragoza, Pedro, in his final year of secondary school at Huesca, failed a course and decided to run away rather than face his father's wrath. He walked 250 miles to Bordeaux and stowed away on a ship to South America. Though he could seem meek, Cajal explained, Pedro was just as rebellious as his older brother was. The two were similar in many respects—they read the same books and planned adventures together, confiding only in each other—but when Pedro left home, their paths diverged. Cajal had also failed a course in his final year at Huesca and had attempted to escape as well, with a group of classmates, but after one night his comrades succumbed to homesickness, forcing him to turn back. Now Pedro succeeded where his brother had failed.

On his journey to the New World, Pedro was discovered by the crew of the ship and punished by keelhauling—a line was tied to his arms, he was thrown overboard, and then dragged underneath the ship. If he drowned, the sailors said, then he belonged to the sea; if he lived, the crew would accept him. Pedro survived. The ship landed in Uruguay, a country in the throes of revolution—the Revolución de las Lanzas, or Revolution of the Lances. Pedro fell into a series of picaresque odd jobs—barber, cowboy, boatman, bread deliveryman, and bricklayer—before enlisting as a soldier and serving as the personal secretary of the illiterate revolutionary general Timoteo Aparicio.

When Pedro and a friend decided to leave the army, they stole the general's horse and revolver but were captured and sentenced to death. Pedro's friend sent a telegraph home to his native Italy, whose embassy intervened. The boys' lives were saved, a miracle that Pedro believed was due to direct intercession by the Virgin Mary. After Pedro returned to Spain, riding the train to Zaragoza, the car stopped for a few minutes at the last station before the city, where he could barely glimpse the outline of the Basilica of Our Lady of the Pillar, built on the site of Saint James's vision of the Virgin Mary. Inspired, Pedro asked his fellow passengers to join him on the platform to say a prayer. No one listened, and so he went out alone. While he was down on his knees, there was a mysterious explosion, killing everyone in his car.

No one had heard from Pedro since he disappeared, and Justo and Antonia were terrified. That year, Cajal earned his most outstanding academic grades, as though trying to make up for his brother's absence. Like all stories in the Cajal family, Pedro's epic tale may have been embellished. But for the first time, it was Cajal who looked up to his younger brother, not the other way around. "I was left as it were," Cajal wrote, "humbled by my failure to do anything so great."

ONE AFTERNOON IN SEPTEMBER 1873, NOT LONG AFTER HIS COLLEGE graduation, Cajal walked with a friend along the Promenade of Nightingales, one of his favorite paths in Zaragoza, shaded by a thick canopy of poplar leaves. The path skirted the Imperial Canal, built to connect landlocked Aragon to the Atlantic Ocean and the Mediterranean Sea, and the air was stale with the smell of stagnant water. "Zaragoza is a desert," Cajal lamented to his friend. His father planned to keep him home all summer so that he could study for teaching examinations. But, though he was qualified to practice medicine, Cajal had no interest in a career for which he would have been unprepared anyway, having skipped or ignored his clinical courses. What he dreamed of was an exciting, new world.

In June 1873, when Cajal had graduated from the Free School with a license in surgery, he was expected to begin a stable career—and yet Spain teetered on the brink of anarchy. In February of that year, King Amadeo I had abdicated the throne, and the Cortes, the Spanish legislature, voted overwhelmingly to form a republic, a federation of provinces modeled after the United States. On the floor of the Cortes, Liberal politicians, long persecuted, hugged one another and wept. *Al fin los hemos logrado!* they cried out. *We have done it at last!* The representative from Zaragoza, Emilio Castelar, rose to the podium. "Gentlemen," he declared, "let us salute [the republic] as the sun rising by its own gravitation in the sky of our nation." Castelar was the most famous orator in Spain, renowned throughout Europe and hailed by liberal republican students like Cajal as a conquering hero.

The republic's first president-elect compared the new era to "a rainbow of peace and concord for all good Spaniards." Storm clouds gathered almost immediately. Spanish provinces, once independent kingdoms with their own distinct history, language, and culture, resisted integration. Some wanted the country divided into sovereign districts, or cantons, as in Switzerland, and three major cities rebelled to form their own. The Carlists, a conservative movement initially resisting the succession of Isabel II to the throne, shifted toward a generalized reaction against the liberalizing, centralizing, and secularizing Republican policies, while Carlos VII, the royal pretender, stood waiting in the wings. The first Republican government failed, and the president was so fed up that he went out for a walk and, without telling anyone, boarded a train to France.

Desperate for a leader, the Republicans turned to Zaragoza's own Castelar, who took control of the government in September. To combat the Carlist and cantonal threats, he reinstated the military draft. Castelar's *quinta*, or lottery, was the biggest mass conscription in the history of Spain, summoning 175,000 Spaniards to military duty. When Cajal received his draft notice, he was relieved and thrilled. He happily traded his academic gown for a military outfit. His boyhood dream had come true; he would fight in a war.

Cajal joined his comrades in basic training, sleeping in the barracks, eating in the mess hall, and performing drills, but when the opportunity arose to join the army medical service, he applied to become an assistant physician, earning the higher rank of lieutenant. The examination was held in Madrid, and for two months Cajal lived in the capital. He spent time studying in the Café de la Iberia, a famous meeting place for the nation's leading progressives, figures whom he had only heard and read about. As he entered the room, thick, sour tobacco smoke stung his eyes. The bright lights transformed the café into a theater, illuminating the grand wall mirrors that faced each other, conjuring the illusion of eternity. History was being made by well-groomed men in fancy coats, shouting and gesticulating wildly all through the night.

Cajal passed the examination to be a military physician. After lingering in Madrid for a few more days, he returned to Zaragoza, flaunting his crisp uniform with polished brass buttons, soaking up the admiration of his peers. Once a juvenile delinquent, Cajal was now a proud army medical corps lieutenant.

EVERY SEPTEMBER, FOR AS LONG AS HE COULD RECALL, CAJAL HAD found himself either trapped in school or slaving away at an apprenticeship. Now there were no books to study, no classes to attend. He was dispatched to Catalonia, a Carlist stronghold, where rebels slashed telegraph wires, burned bridges and railway stations, and tore up rails. Traveling east from Zaragoza, the train passed warm, brown earth accented by scarlet poppies, purple-white buckwheat, and gold irises. In the city of Lérida, Cajal joined a regiment of fifteen hundred men, including battalions of light infantry and artillerymen, and a squadron of cavalry known as *cuirassiers*, their gleaming lances torn straight from the pages of *The Three Musketeers*. Waking to the sound of the reveille, day after day, he marched back and forth across the countryside with his battalion, only to find, in each town they arrived at, that the rebels had just left.

Secretly, Cajal hoped to be attacked, but he joined the conflict a few months too late, when the Carlists were already on the retreat. In eight months, he never saw the enemy, or heard gunfire, or even treated an injured soldier—not counting the odd case of venereal disease or an officer who had fallen off his horse drunk. The closest he came to a battle wound was the cough he developed one night after sleeping outside in the dew.

As the country was mired in domestic conflicts, Spain faced insurgency abroad. After the 1868 revolution, a coalition of wealthy Creole plantation owners and Black laborers in Cuba, resentful of high taxes and the unequal distribution of wealth, rose up against the Spanish colonists. "When a people arrives at the extreme of degradation and misery in which we see ourselves," said Manuel Cespedes, the leader of the Cuban revolution, "no one can condemn another who takes up arms to escape from a state so full of ignominy." With the Spanish army occupied on the home front, the Cubans intensified their revolt, which jeopardized the sugar trade and threatened to destabilize the Spanish economy. The government ordered a surge of reinforcements to the island, where men were dying faster than they could be replaced. Soldiers were chosen by lottery, and in April 1874, while languishing in Catalonia, Cajal found out that his name had been drawn.

His family reacted as though he had received a death sentence. Justo begged his son to request a discharge, but Cajal ignored his father. The army promoted Cajal to the rank of captain. "To be perfectly sincere," he wrote in his autobiography, "I acknowledge now that, beside the austere sense of duty, I was drawn overseas by the bright visions of novels that I had read, the irresistible craving for wandering in search of adventure, the longing to see for myself the people and the customs of foreign lands."

On June 15, 1874, the steamship *España* set out for the Caribbean from Cádiz, its deck bustling with soldiers smoking cigarettes, playing cards, and shooting the breeze. The Transatlantic Company—which ran twice-monthly voyages carrying passengers and mail and was subsidized by the government to transport soldiers—offered all sorts of onboard pastimes and diversions, but Cajal stood apart from the crowd, near the edge of the boat, transfixed by the vastness of the sea and the strangeness of its creatures: sharks, flying fish, and jellyfish. When the seas calmed, he could see phosphorescent glimmers, which seemed to appear in swarms, a phenomenon that Jules Verne described as the "milk sea." These marine animals—sea sparkles—remain invisible, gliding beneath the swell, until the moment that they are disturbed by the churn of the ship's propeller. Only then did they reveal their hidden brilliance. "What a great alarm for the soul, and an instigator of energy, is pain!" Cajal later said, using the sea sparkles as an analogy for his creative psyche, formed by the abuses of his childhood.

After sixteen days at sea, the *España* docked in San Juan, Puerto Rico, and Cajal wandered the city, admiring the all-white houses and their terraces. Two days later, the ship entered Havana, its harbor open and inviting, backgrounded by gentle green hills sprinkled with huts, palaces, and villas, some steepled and domed, with lush gardens and lonesome palms between them. The plants were gigantic, he said, the women mesmerizing, the fruit exotic, the flowers strange. "I lived as if in a dream," Cajal recalled, "and as if under a sort of spell."

One month passed before Cajal and his fellow military physicians were summoned to the quarantine station to receive their assignments, either as a hospital instructor, regimental physician, or director of a field infirmary. Hospital instructors had the cushiest job, farthest from combat; regimental physicians might come under fire, but they could return to the capital every once in a while to rest and collect their paychecks; while directors of field infirmaries were exposed to the most inhumane conditions, at the highest risk of both violence and disease. Cajal could have shown his superiors the letters of recommendation that his father had given him, but he never did. He wanted to see the most action. And with his father on the other side of the world, Cajal felt more emboldened than ever to defy him. Looking back, Cajal called his mindset "foolishly quixotic."

The armored train, headed for the heart of the island, was packed with troops, fresh bodies meant to replace casualties. Cajal's new base was located deep in the jungle, severed from the rest of the world, and rations arrived only once per month. Its name, Vista Hermosa, was darkly ironic, for there was no "beautiful view"; the tangled growth was too dense to see through, and if not cut back frequently, the jungle would overwhelm the fort. Official military strategy was to blindly hack away to form a path, then charge deeper to drive the insurgents back. Two guard towers faced the green vastness, where guerrillas hid less than a kilometer away, the two sides exchanging daily volleys of fire. The infirmary at Vista Hermosa was a wooden barracks covered by a thatched palm roof, infested with cockroaches, mosquitoes, fleas, and ants. Two hundred patients lay on cots, stricken with malaria and dysentery. Cajal read, drew, and took photographs in his free time, trying to establish a normal routine while sleeping in a room stocked with jars of quinine, tins of biscuits and sugar, cartridge boxes, and the muskets of dead soldiers. Nothing separated him from his infected patients but a makeshift partition.

It took less than a month for Cajal to start losing his appetite and strength. Then his skin turned yellow, and his face grew spectral and emaciated. Dizzy and anemic, he could barely drag himself around the barracks. At the time, doctors knew that swamps were hotbeds of malaria, but they had not yet identified mosquitoes as the carriers. Heavy doses of quinine failed to improve his condition, and before long he was bedridden. In the middle of the night, the guerrillas staged a surprise attack; a few weeks earlier, they had ambushed a nearby outpost and massacred the entire garrison. This was the drama Cajal had been hoping for. Suddenly energized, he leaped out of bed and ran through the infirmary in a delirium, rousing the patients, before aiming his Remington through the window and firing into the darkness.

A few months later, the health inspector finally granted Cajal a leave of absence, and he withdrew to Puerto Príncipe, the district capital, to recover—"the pleasantest period of my stay in Cuba," he recalled. He was reassigned as a hospital instructor, joining a staff with friends and former classmates who went to casinos, cafés, and private gatherings together in their free time. Cajal took pride in not drinking, smoking tobacco, gambling, or visiting prostitutes. The others mocked him for having the "scruples of a nun."

After six weeks, Cajal was tapped to replace the dead director at the infirmary in San Isidro, the most dangerous part of the war zone, in a swamp that sucked up every fort that the Spaniards tried to build there. The camp was where officers were exiled as punishment for insubordination. Years later, Cajal explained why he had been targeted: he had mocked an anatomical illustration that the health inspector had drawn. This story did not appear in Cajal's autobiography, and he seems to have told it to only one person, a fellow anatomist and good friend. The anecdote may not be true, and there may have been other reasons for the health inspector's animus, but Cajal's tale serves as a statement of principle: that he was the kind of person who would rather die than silence his opinions about a bad illustration.

The hospital in San Isidro was a giant open shed where three hundred patients lay dying. Two-thirds of the garrison was ill at all times, which kept Cajal extremely busy, though the work itself was easy, since every case was more or less the same.

His malaria worsened every day, and he was forced to take to bed. Neither quinine, nor tannin, nor even opium offered him any relief.

With his mother far away, and no friends or family to take care of him, Cajal felt abandoned, which made him feel anguished and enraged. Eventually, he had no choice but to apply for a medical discharge, which would strip him of his rank and send him back to Spain a failure. Leaving home was meant to be an act of defiance, and now he suffered the indignity of asking his father to pay for his ticket home.

With his father's money, Cajal traveled in a stateroom while, wretched and miserable, the other sick and wounded soldiers were crammed into third class. Once his condition began to improve, aided by the ocean air, he brought medicine to his comrades and chatted with them to lift their spirits. Some recovered, though others were not so fortunate. Through his stateroom window at dawn, Cajal could see the corpses of young men like him being thrown into the sea.

CHAPTER 9

"Cells and More Cells"

Near the end of June 1875, Cajal finally returned to Zaragoza, relieved to be home and embraced by his family, whom he once feared he would never see again. He had left Spain at the peak of his physical powers, and he had returned sick and enfeebled. His officer's stripes had been stripped from his sleeves. His wide-eyed idealism had warped into bitter disillusionment; in scandal-ridden Spain, there were no knights or heroes, only "blundering generals and egotistical ministers." One hundred and eighty thousand men had been sent to war; half of them never came back.

Before leaving for Cuba and near the end of his stint at university, Cajal had begun a romantic relationship with a young woman in Zaragoza, whose letters, he said, were all that had buoyed his spirits while he languished in the jungle. As soon as he regained enough strength, he paid her a visit, anticipating a classic romantic scene: the weary soldier returns from battle in a faraway land to marry his faithful sweetheart. She received him coldly; each time he called on her, her demeanor turned colder. To test her affection for him, he decided to conduct an experiment by kissing her and observing how she reacted. At twenty-four years old, he had yet to have his first kiss, and it took all his courage just to peck her on the cheek. She recoiled, he recalled, with a look of disgust, even loathing, condemning the kiss as "sinful." Later, he learned from a friend that, because of his sickly state, she no longer loved him, only pitied him. She was an orphan who lived with her uncle, and she feared being left prematurely widowed and poor.

Cajal could not deny that his father had been right, that the overseas adventure had been a disaster, robbing him of his health, his career, and his love. When he returned to the cafés of Zaragoza, nothing had changed; young people like him lounged on the couches without a care in the world. Stripped of his uniform, he felt he lacked an identity. "I had to reshape my life," Cajal recalled, "directing it once more toward its old course." Cajal now turned his attention to studying anatomy and forging a career path. He vowed to adopt a new, more stoic attitude so that he could better confront life's vicissitudes without succumbing to despair.

In the war of duty and desire, Cajal now surrendered to his father, reopening his old textbooks and attending dissections at the Free School once again. Justo brought his son along with him on house calls as an assistant, to show him how to treat patients and to keep him busy. Over the years, Justo had amassed a substantial clientele, and he intended for his eldest son to inherit it. Cajal was the apprentice; his father, the master. Justo soon let his son treat patients by himself, but Cajal's bedside manner could seem aloof and sometimes tactless. Cajal claimed that he rejected clinical medicine as a statement of his independence from his father. However, those who knew him wondered if he might be too sensitive to cope with patients' deaths after his traumatic experience caring for doomed soldiers. Moreover, the neediest and poorest patients often lived on the top floors of buildings and in garrets, and Cajal was now too weak to make it up the stairs.

Justo hounded his son constantly about finding a job, and in November 1875—four months after returning from Cuba—he was appointed interim assistant professor of anatomy at the Free School, a position that his father had arranged. The annual salary was one thousand pesetas, a modest sum. Colleagues and former classmates at the Free School, who had watched him flaunt his achievement and rank, delighted in finding him back in the desert he had so desperately tried to escape.

"Nothing worth relating happened in 1876 and 1877," Cajal claimed—this was the same man who narrated seemingly every trivial episode from his childhood. The banality of professional life, as far as he was concerned, was unworthy of remembering. All that he relates is that in May 1876 he became a first-class medical assistant at the provincial hospital, where the medical school was located, and the next year he was named interim professor of anatomy. In 1877, his father, then fifty-five years old, finally achieved his ultimate goal, earning his medical doctorate. His dissertation, "Considerations About the Organicist Doctrine," promoted the theory known as organicism, which holds that everything in nature can be understood only as part of an organic whole. The dissertation was written in Cajal's hand, as though he were still working as the family scribe. The thirty-one-page work ends with a plea to the judges, citing Justo's advanced age, the demands of private practice, the obligations of family life, and his own intellectual shortcomings as reasons for leniency. "You are always great with the small, indulgent with the weak, and charitable with the naive," Justo dictated, and Cajal faithfully copied his every word.

Justo pressured Cajal to pursue his own doctorate, a requirement for a professorship—which, given his ill-suitedness for clinical practice, seemed like his only hope for a stable career. The examination was held in Madrid, along with the three preparatory courses, but rather than register his son at the University of Madrid, where the majority of doctoral students studied, Justo kept Cajal by his side in Zaragoza, preparing his son himself, as wary of his son's waywardness as ever.

In advance of the examination, each candidate chose a topic from a predetermined list on which to write his thesis. Still enchanted by the spectacle of blood flow, Cajal wrote "The Pathogeny of Inflammation," about the organism's response to injury or illness, a hot topic on which vitalists and cell theorists disagreed. Cajal described inflammation using military terms: vitalists imagined that the soul, acting as the body's "central authorities," deployed forces to the site of a foreign invasion, whereas cell theorists saw disease as a "trivial frontier skirmish" that local militias could handle without intervention from the national government. Cajal passed his doctoral examinations with an average grade.

While in Madrid, Cajal visited Aureliano Maestre de San Juan, the chair of anatomy and histology, who judged the examination in histology, the microscopic study of tissue treated with chemicals to expose its underlying structure. Considered the founder of Spanish histology, Maestre was the man for whom the university chair had been created, the first Spaniard to publish a microscopic anatomy study and the founder of the Spanish Histological Society. Inside the histology laboratory were tables covered with glass slides stained with colors and cabinets full of bottles whose contents appeared to glow. The histology textbooks that Cajal studied began with chapters on microscope technique, but he had little experience actually using one. Maestre's assistant, Leopoldo López García, showed Cajal the basics: bringing the lens into focal position by adjusting the rack and pinion and turning the screw on the side to increase the magnification.

A former student recalled that López was "in love with his profession, and devoted to it heart and soul, willing to sacrifice everything." He had begun his research career in a makeshift home laboratory before bringing a microscope to the attic of the anatomy department, where he spent entire days alone, staring at tissues. "His colleagues labeled him a madman," the former student said, "more predisposed to censure him than follow his example."

CELLS—WHICH WERE COMING TO BE KNOWN AS THE MOST BASIC structure of life—are composed mostly of water, so that when light hits tissue no elements or structures can be discerned within them. But, like fabric, tissue absorbs certain dyes, and different elements of cells have various chemical affinities. In 1858, a German anatomist named Joseph von Gerlach walked into his local drug store and asked the apothecary to recommend a chemical agent with which he might be able to color blood vessels, and the apothecary suggested carmine. When Gerlach injected a solution with the powder into an animal's bloodstream, he was amazed to see that nearby cells absorbed the dye as well. Different staining techniques, such as dahlia violet and methylene green, yielded different colors; hematoxylin stained cells' nuclei blue, and eosin stained them pink. Tissue could also be counterstained, meaning that some elements were stained with one color and the rest with another for contrast. Histology is optical dissection—the dye serves as the scalpel. "To be an histologist," one historian of science later said, "became practically synonymous with being a dyer."

Looking through the microscope in Maestre's laboratory, Cajal saw the intimate structures of cells, previously invisible, now unmistakable in vibrant color—the ultimate latent image, the hidden picture of life itself. Cajal questioned López about every detail and could hardly wait to draw the strange forms. That night, he could not stop thinking about the images, and he returned the next morning to see more. Though he had failed in his adventure to the Americas, Cajal still dreamed of exploring new worlds, and he thought of microscopic anatomists as "Columbuses."

Histology was a relatively low-cost endeavor, requiring only some chemical reagents and a microscope in order to begin. Wandering the streets of Madrid before his examination, Cajal had stopped into medical supply shops to look at microscopes, but he could not bring himself to buy one at the time. He had no disposable income, and it would be difficult to justify the expense of a microscope, especially to his father. Cajal's desire was the same as ever: to see new sights.

In his autobiography, Cajal sometimes misremembered the addresses of the places he lived and the birthplaces of his children, but he could never forget 25 León Street. There, on the ground floor, a man named Francisco Chenel owned a medical supply shop, which had recently released a new catalog advertising all the world's leading microscope brands. Cajal chose a first-rate model from Verick, a cutting-edge French manufacturer, with a lens that magnified objects up to eight hundred times—the maximum possible at the time—and a water-immersion lens for increasing resolution. Then he wrote to Chenel to work out a deal. The price was seven hundred pesetas, more than half his yearly income from the university and private tutoring combined. When he was discharged, the military had paid him six hundred pesetas (after a significant bribe), which would almost cover the cost of the microscope, and Chenel, who happened to also be a veteran, allowed Cajal to pay in four installments. "Pain is a necessary stimulant to creativity," Cajal later wrote. If he had never suffered through his ordeal in Cuba, he could not have begun his scientific career when he did.

NO SINGLE PERSON INVENTED THE MICROSCOPE; THE TECHNOLOGY evolved over time, beginning with ancient magnifying lenses—the word *lens*, in Latin, means "lentil," a reference to its convex shape. According to legend, in a lensmaker's shop in Middelburg, in the Dutch Republic, two children held two lenses against each other and saw that the nearby church suddenly appeared larger and closer. Town officials sent a letter to The Hague heralding the discovery of a "magical device." At first the instrument was treated as nothing more than a novelty item, a toy for wealthy children hawked by spectacle makers and traveling salesmen. The first microscope ever constructed—according to another tale—was a six-foot-long gilt copper tube held up by pillars shaped like dolphins. Galileo called his microscope the *occhialino*, or "little eye," "for observing at close quarters the smallest objects." Fleas became as large as lambs, he said, and so a term for early microscopes was "flea glasses." In the evenings, well-to-do families would gather in the parlor and look through the microscope for entertainment.

Most owners of microscopes were content to observe things superficially and had little interest in their scientific uses. In 1665, the British naturalist Robert Hooke published the first scientific treatise on microscopy—*Micrographia: or some Physiological Descriptions of Min-*

ute Bodies made by Magnifying Glasses with Observations and Inquiries thereupon—and the book became a bestseller. The illustrations, when unfolded, were the size of multiple pages. "By the help of Microscopes," Hooke said, "there is nothing so small, as to escape our inquiry." He suggested that, more than just satisfying curiosity and providing amusement, the microscope could be a means for transforming individuals and humanity. In the preface to *Micrographia*, he wrote:

> *It is the great prerogative of Mankind above other Creatures, that we are not only able to behold the works of Nature, or barely to sustain our lives by them, but we have also the power of considering, comparing, altering, assisting, and improving them to various uses . . . By the addition of such artificial Instruments and methods, there may be, in some manner, a reparation made for the mischiefs, and imperfection, mankind has drawn upon itself.*

Some contemporaries mocked Hooke's obsession as "trifling"; one critic dismissed him as "a Sot, that has spent 2000£ in Microscopes, to find out the nature of Eels in Vinegar, Mites in Cheese, and the Blue of Plums." But his devotion to the seemingly insignificant revealed a hidden truth about the nature of life. When he examined a thin sliver of cork under the microscope, he noticed what looked like a sequence of empty boxes, which reminded him of the living quarters of monks—called *cellula*—from the Latin *cellula*, meaning "little rooms," and so the microscopic structures became *cells*.

Around the same time, in the Dutch city of Delft, well outside the scientific mainstream, a draper named Antony van Leeuwenhoek taught himself to grind lenses, eventually making glass beads whose diameter was one quarter that of a pea that rendered objects 450 times larger. Spurred by pure curiosity, Van Leeuwenhoek trained his microscope on a sample of lake water, and he saw floating green particles shaped like spirals, which reminded him of serpents. "All the water seemed to be alive," he wrote. In his own bodily substances, such as saliva, blood, sperm, and dental plaque, he found "many very little things, very prettily a'moving" that were "so exceedingly small millions might be contained in one drop of water." Van Leeuwenhoek called them animalcules—"little animals."

About two hundred years after Leeuwenhoek, Cajal waited eagerly for the package to arrive, a magical box with velvet-lined compartments. According to Pedro, he stayed up all night playing with the microscope "like a new toy." Still living in his father's house in Zaragoza, he chose the attic for a laboratory, like López had—an upgrade over the pigeon shed in Ayerbe, if only slightly. All that mattered was that it was Cajal's own cell.

"Come with us to this laboratory of microscopy," he wrote shortly after beginning his research:

> *There, upon the stage of the microscope, tear up the petal of a flower without consideration for its beauty or its perfume; then take a fragment of animal tissue; tear it apart without respect, although its contractile fibres pulsate and tremble at the touch of the needles. Afterwards, look diligently through the window of the ocular and—a remarkable fact, a stupendous discovery— the leaf of the plant and the tissue of the animal will reveal to you in every part the same structure: a sort of honeycomb built up of little cells and more little cells . . . Now examine a drop of saliva, a little of the epithelium which covers your tongue, a drop of your blood, the mould upon decomposing organic substances, etc.—and always the same architecture appears: cells and more cells.*

Cajal came to think that all doctors should use microscopes, and he wrote a series of articles that he called "The Marvels of Histology," aimed at popularizing microscopy, which ran in a weekly Zaragoza medical journal called *La Clínica*. In hindsight, he dismissed the articles as "philosophic-scientific temerities," "cloying flights of the imagination" that are "overflowing with fantasy and ingenuous lyricism." Despite his self-criticism, he reprinted excerpts from "The Marvels" in his autobiography, because they retained "something like a comforting aroma of youthful confidence and robust faith in social and scientific progress." He was in the infatuation stage, falling in love with biological research. (Cajal may have also wanted to forget his early essays because, in them, he perpetuated the vitalist idea of the "protoplasm" inside cells—not the cells themselves—as the only essential component of all living beings, even arguing that the cellular theory should be renamed the "protoplasmic theory." He also advocated for "spontaneous generation," which violated the rule of cell theory *omnis cellula ex cellula*, or "all cells come from cells." Nowhere in his later writings do such ideas appear.)

Cajal's "Marvels" are anthropomorphic, like Virchow's *Cellular Pathology*, only more so: the body is "our organic edifice" inhabited by "thousands of microscopic workers to whose work all organic activities are in debt." The muscle fibers are soldiers "in compact battalions"; epithelial cells are merchants; lung, kidney, and liver cells are police. "It is certain that millions of autonomous organisms populate our bodies," he wrote, "the eternal and faithful companions of glories and of toils, of which joys and sorrows are our own." The tissue is a neighborhood; the bones are buildings; blood is the streets; skin is the walls; the intestines are sewer systems; and the skull is a palace, where the brain sits as though on a throne.

Cajal told the kinds of stories that he had loved since he was a child, only this time the setting was the body, not the highlands of Alto Aragon, and the protagonists were cells, not medieval warrior kings. The boy with a "blind desire to be first" cast sperm cells as mythic heroes, "only one of [whom], the strongest or the most fortunate, will survive." He glorified the "homeric struggles" of cells threatened by cancerous invasion, and portrayed the fight against germs as a battle in "the incessant war carried on between the small and the great." Cajal imagined the politics of organic "societies" as similar to human ones, relatively harmonious as long as labor is divided, laws are followed, and common resources remain unthreatened. "Cells are what live," he said; "cells are what get sick and die." For Cajal, cells were more than just inanimate objects of study—they seemed like microscopic versions of himself.

CHAPTER 10

"The Irremediable Uselessness of My Existence"

Cajal's "honeymoon with the microscope" was disturbed in the spring of 1877 when a notice appeared in a local medical gazette announcing *oposiciones*—competitive examinations for public appointments—for vacant anatomy chairs in Zaragoza and Granada. Candidates in anatomy were required to take oral and written tests and to perform demonstrations before a tribunal. The format was artificial; none of the exercises resembled actual medical practice. *Oposiciones* were displays of verbal memory and rhetorical skill meant to showcase academic knowledge alone. The ritual was inherited from medieval times, when doctors of canonical law held public debates to settle theological conflicts. *Oposiciones* were quintessentially Spanish, one poet said, since they included aspects of bullfighting, the Inquisition, and the national lottery. Though the word is untranslatable, the English cognate *opposition* conveys the combative spirit. The grueling examinations lasted for days on end, and the pressure was so great—with lifetime appointments and social reputations on the line—that even the strongest candidates were prone to fainting. *Oposiciones*, Cajal says, were "cruel and always bitter"; a younger friend and colleague of his called them "cancers."

Though he was never one to shy away from a contest, Cajal knew that he was far from ready to compete. Two years removed from Cuba, he still showed grave effects of his illness: his hair, once thick and slicked straight back, was now thin and tufted. His face had grown longer and more angular, his skin paler, his eyes sunken and faded. The dark suit that he wore to teach classes hung loose on his frame. He used to pose for photographs proudly, with his chin up and his chest puffed out; now his body was perpetually slouched. In his role as auxiliary professor, Cajal was overworked; he taught as many as three classes per day, and his schedule was liable to change at the last minute, forcing him to improvise lessons. Preparing for an *oposicion* demanded sacrifice, and Cajal did not want to leave his laboratory. At the same time, he felt the need to fight for tenure and financial security. After years of doubt, Justo had renewed faith that Cajal could become a professor. Certainly, Cajal wanted to become independent. He also did not want to disappoint his father. But Cajal failed the *oposicion*, receiving a single vote from one of seven judges.

A few weeks later, back in Zaragoza, Cajal sat at the Café de la Iberia on Paseo de la Independencia, where he liked to spend his time when he was not teaching. The newly built boulevard was lit with electric streetlamps and lined with young trees, its arcade spanning as far as the eye could see. It was warm enough in March for Cajal to sit outside with a friend from the army and play chess—his only vice. Café waiters delivered boards and pieces to the table ceremoniously. Cajal played the game with the same boldness and intensity that had propelled him throughout his life, staging daring attacks into enemy territory, often at the expense of his own pieces. As the match progressed, the atmosphere at the table became increasingly tense. By the end, Cajal had become completely absorbed in contemplating his next move. Suddenly, he felt a sharp burst in his chest and tasted blood. He immediately recognized the symptoms of a pulmonary hemorrhage.

Cajal did his best to hide the incident from his friend and hurried home in a panic. At the dinner table with his family, he ate almost nothing, avoided conversation, and went immediately to his room. Soon after that, he felt foaming blood rush up his throat, threatening to choke him. He called for his father, who saw that his son's face was flushed, heard bellowing sounds in his chest, and felt a rapid pulse. Because of Cajal's history of malaria, and his persistent paleness and emaciation, Justo suspected the worst. Cajal's high fever, profuse sweating, and shortness of breath matched the classic symptoms of tuberculosis. "A physician rarely deludes himself about his own condition," Cajal said.

In the eighteenth and nineteenth centuries, tuberculosis killed millions—it was by far the largest cause of death in the United States and Europe. Cajal had read about the disease in medical school, and he had known veterans who had returned from overseas weakened by malaria, as he had, then died of tuberculosis contracted at home. Soldiers from the countryside who fought on the front lines were thought to be especially vulnerable to infection, having lived together in the barracks, coughing and sneezing and breathing the "bad air." The disease, which was responsible for one half of all deaths of those between the ages of fifteen and thirty-five, was called "the robber of youth." Cajal was twenty-five.

Justo prescribed the usual treatment for pulmonary hemorrhage, which in those days included a few spoonfuls of salt to speed up the clotting process and bed rest to lower blood pressure. Cajal was left without a pillow, not allowed to even raise a hand or speak unless absolutely necessary, in which case he would have to whisper. For two months, he was trapped alone with his dark thoughts. At times such as these, Cajal admitted, he wished that he had religious faith. His stoic facade was collapsing. He had just found his purpose, setting up a laboratory and devoting himself to scientific research. He could not stop thinking about his humiliating public failure in the *oposicion*. "This idea of the irremediable uselessness of my existence plunged me into the utmost anguish," he said. He was convinced that his heroic journey was over.

By summertime, Cajal's condition had stabilized enough for his father to send him to Panticosa, the Pyrenean baths once frequented by Roman emperors, which were famous throughout Europe for their supposed ability to treat consumption. Drinking, inhaling, and bathing in the nitrogenated waters—at natural temperatures between eighty-five and ninety-five degrees—was seen as a way to calm the nervous system. At a mile in elevation, the pure air was thought to promote an easy and active mind and to open the soul. In Cajal's time, Panticosa had become a popular destination for affluent patients, some of whom went there for vacation. Cajal was admitted in July, the peak season, when all the resort's one thousand beds were constantly occupied. His sister Pabla was sent to act as his companion and nurse. The Panticosa treatment required between six and eight weeks of isolation from friends and family at home, and doctors maintained that it was vital not to hurry the cure.

"The plague of all plagues," as tuberculosis was called, was considered a badge of honor by the Romantics, who treated the disease as though it were the latest fashion. "I should like to die of consumption," said Lord Byron, who limited himself to only three meals per week consisting of just rice, water, and vinegar in order to appear more sickly, "because the ladies would all say 'Look at that poor Byron, how interesting he looks in dying.'" Some associated the illness with genius; to console his friend John Keats, fellow poet Percy Bysshe Shelley told him the disease was "particularly fond of people who write such good verses." George Sand remarked that Frederic Chopin, her lover, looked like a "poor melancholy angel," coughing "with infinite grace," while Edgar Allan Poe similarly portrayed his young dying wife as "delicately, morbidly angelic." In *Les Misérables*, one of Cajal's favorite books, Victor Hugo described the consumptive Fantine as having a "strangely luminous face," like an angel "more likely to soar away than die."

Even in summertime, the Pyrenean peaks still wore tattered robes of snow. The ride was easy and picturesque, as the carriage passed through hillocked valleys, thick with velvety meadows and dotted with grazing goats and sheep. The spa, located two hours from a tiny village of the same name, was so remote that it could be reached only by mule. Two stone buttresses marked the start of El Escalar—"The Staircase"—a dangerous, zigzagging trail, hemmed in by tall boxwoods, that wound tighter and tighter along the scarred face of a steep, gray cliff. Every week, women from the village braved El Escalar to deliver strawberries, milk, chicken, and eggs to the spa-goers. The path was so narrow that the only way to avoid a mule train from the opposite direction was to press oneself against the rocks and wait.

After one final, sharp turn, a dark blue lake appeared, fed by white cascades and surrounded like an amphitheater by high granite walls. The surrounding buildings, whitewashed with green shutters, looked less like hotels than like barracks. Their layouts were uniform: all the corridors were equally long; each small, whitewashed room contained a simple iron bed; and the floors were kept finely polished at all times. Long, oval tables in the dining room were meant to foster social interaction. The staff was pleasant and accommodating, with porters and concierges at the desk happy to answer any questions. Guards were stationed at the entrances to ensure the guests' privacy, and at the exits to ensure that nobody left.

The international team of expert balneologists—bath doctors—at Panticosa recommended that patients drink between twenty and twenty-five glasses of the waters every day, but to take care before bathing, since some people experienced a mildly sedating effect. The sulfuric water, smelling slightly like rotten eggs, was not easy to digest.

Fussier patients diluted it with whey or milk and sipped the mixture daintily through glass straws. Spas like Panticosa enforced strict schedules: wake up before dawn, drink waters before breakfast, take morning baths for fifteen minutes to an hour, walk, drink more waters until lunch, nap for an hour, walk more, drink more waters, snack at five-thirty, engage in leisure activities such as rowing and pistol shooting, eat dinner at eight, and then play games or sports. Some guests woke up to bathe at three in the morning to beat the crowds, which had become so large that the spa constructed turnstiles where patients would queue up, as though the springs were theaters on opening night.

Cajal felt confined by his illness, and he detested the regimen and resented his doctors. The victim of his angry outbursts was Pabla, who found her brother so insufferable that she considered abandoning him. He set up a makeshift photography studio and tried to enlist her as an assistant, yelling at her and ordering her around. In his despair, he wrote poems imitating his idol Espronceda, and he contemplated suicide. "Existence is a blind and unshakeable force indifferent to feeling," wrote the poet Giacomo Leopardi, whom Cajal was reading obsessively at the time. The words rattled around in his head like bones in a grave robber's sack.

On nights in Panticosa the glowworms flickered and guests haunted the grounds wrapped in black beaver hats and heavy cloaks. Cajal noticed that many patients harbored delusions that they were, in fact, well. Their minds were, in their own way, resilient, and he felt ashamed at the weakness of his own will. They wanted to live, yet they were dying; he wanted to die, yet nonetheless he continued to live. "When least expected," he recalled, "the horse which we considered contemptible and weak turns out more spirited than the rider." From then on, he simply decided not to be ill. He stopped dieting, bathing, and drinking the waters and resumed his normal, active life. He drew, took photographs, and walked, activities that "still the vibrations of sorrow" and "free us from our own ideas." After a few weeks, he stopped coughing up blood, his fever went down, and his lungs and muscles became stronger. When he felt his heart pounding or heard his lungs wheeze, he ignored the symptoms rather than interpreting them as signs of impending death. By thought alone, Cajal believed, one can transform one's own mind.

CHAPTER 11

"Not for the Living but for the Dead"

In October 1878, Cajal returned to Zaragoza "in almost flourishing health," six months after his diagnosis. Recovering from tuberculosis, especially after such a short period of time, was extremely rare, and Cajal attributed his resilience to his highland constitution. There is another factor underlying Cajal's illness that seems impossible to deny: he came down with the ailment just weeks after failing his *oposicion*, which he felt had ruined his life, and—"most painfully"—disappointed his father. Still, it would be unlike Cajal to invent a bleeding lung. He may have had an ailment that was less serious than tuberculosis, and his father, out of an abundance of caution, might have misdiagnosed him.

Justo had given up hope that his son—who had earned only a passing grade on his doctoral examinations, failed the *oposicion* spectacularly, and nearly died of two illnesses—could ever have an academic career. He started pestering Cajal about practicing medicine, arguing that the open-air life of a rural surgeon would be better for his health. One evening at the family dinner table, Justo announced to his son, "They have named you doctor of Castejón de Valdejosa," referring to the small village about thirty miles from Zaragoza. Justo had arranged for the position. "I doubt it," Cajal responded, "since I never applied."

Cajal took the job to spare his father public embarrassment, before leaving after a few months and returning to his parents' home, where he locked himself in his attic laboratory. "The histologist is a physician," one history of science textbook says, "not for the living but for the dead."

For histologists, death represents both an opportunity and an impediment. As soon as an organism dies, its body starts to decompose. To preserve the organic structure of tissue in as lifelike a condition as possible, the histologist must halt the natural process of decay, just as photographers must stop the development process, treating their plates with chemicals to stabilize an image and avoid overexposure. Histologists and photographers use the same word for their respective chemical processes: fixation.

Dissected tissue must be transferred into a fixative bath immediately, since, with every passing moment, its structure further degrades until every cell is broken down completely. Fixation is a temperamental process; in cold weather, it might take up to twenty-four hours longer, and the duration depends on the thickness of the tissue, because the chemicals need time to penetrate every layer. If a sample is too thick or immersed for too short a time, then the tissue might not be completely fixed, and the inner layers will continue to decay, compromising the microscope image. Thinner slices might take minutes, while bulkier tissue may require hours, and some fixatives work over the course of months. Every fixation method has its drawbacks. The first chemical fixative used by histologists was alcohol, which was not ideal, since alcohol absorbs water, causing the tissue to shrink, warping its image under the microscope. Osmic acid, a popular fixative, is fast-acting, taking between fifteen minutes and three hours, depending on the sample size, but it colors all proteins in the tissue black, which blocks the absorption of the stain later.

After fixation, the tissue sample must be further hardened so that it can be cut into slices thin enough for light from beneath the microscopic stage to pass through. But if the tissue is hardened for too long, it will not absorb the stain. Then the sample has to be washed to eliminate any chemical residue, but if it's washed for too long or at too low a temperature, the tissue becomes too pale to be easily studied.

Next, water must be removed from the tissue, displacing excess fixative and further solidifying the sample. If the tissue is dehydrated for too long, however, then the tissue will shrink and become brittle. The dehydration process is then repeated, this time with a solvent, "clearing" the tissue of the dehydrating fluid rather than of water. Each chemical used in the process has to be miscible with the last so that the proportion of the chemical added can be steadily increased until it replaces the preceding one. Finally, to stabilize the tissue, the sample is infiltrated with paraffin wax, filling all the space that had been occupied by water, and then embedded in a paraffin block with the help of a mold. The tissue—encased in and infused with paraffin—is then cut serially into sections, which are stained and mounted on glass slides so that the observer can adjust the slide and see images of consecutive sections, as though moving through the tissue sample. In Cajal's day, the slides were rinsed with turpentine and xylene to clear away the paraffin and then covered with Canada balsam to make the sections stick. The tissue has to be cut so that the thickness of each section is uniform—otherwise the depth of the image will not be consistent, which might threaten the accuracy of any observations.

"The laboratory is the ideal sanatorium," Cajal wrote. He spent more and more hours locked in the attic, taking up his histological work with newfound vigor and self-belief. Working alone is the best way to be oneself, Cajal said. No assistants or collaborators bustled around, so there was no noise to disturb him, and, with no one waiting to use the microscope, he felt no need to rush. While universities were closed at night, his attic always remained open. The best place to train would have been in the laboratory of a famous scientist, but while some students traveled abroad for the experience, Cajal did not come from a wealthy family, nor did Spain offer research grants. The only encounters he could have with great histologists took place in the pages of journals and books—the ideal mentors, Cajal said, because they are forever wise and calm—and unlike people, he joked, books know how to keep quiet after saying their piece.

Now Cajal was master and apprentice both. He set no rules for himself; practice alone was his teacher, he said. He began his research by reading textbooks chapter by chapter, replicating the authors' experiments to verify their findings, exactly as his father had taught him. "For the union of two minds to occur and generate fruitful results through a book," he said, "the reader must become fully absorbed in what the master has written, must penetrate its meaning fully, and finally must develop an affection for the author." He read technical books as passionately as he had always read fiction.

Cajal's work in the laboratory revealed his own strengths and weaknesses as much as it illuminated the world around him, and without a teacher to guide him, he had no choice but to address his own shortcomings if he wanted to make progress. He channeled two of his most powerful character traits—his obsessiveness and his competitiveness—into his scientific practice, calculating the amount of time and energy that a task would require, then devoting ten times more to it, ensuring that he would outwork all his peers. For the time being, there was no one to surpass but himself, and he considered that triumph to be the only one worth celebrating. He felt a distinct sense of pride when learning a new skill or fact for himself without anyone teaching it to him. Before discovering anything, Cajal wrote, we must first discover ourselves.

CAJAL'S MISSION WAS TO EXPLORE ALL THE TISSUE OF THE HUMAN body, stopping for a while when he produced a clear slide in order to photograph or draw it. Bone tissue was the most difficult to cut, because it contains mineral crystals, and so he started with epithelial tissue, which is easier to prepare. (The epithelium covers all the surfaces of the body, lines cavities and hollow organs, and makes up the majority of tissue in the glands.) He preferred notebooks with thick sheets of smooth paper and two brands of Faber pencil—one hard for subtle outlines, and one soft for shadows—and kept red, yellow, and blue pencils on hand to capture the various stains. "Needless to say," he wrote in his histology textbook, "the pencils should be perfectly sharpened." He kept an eraser close by.

Before the invention of photography, microscopists had to either reproduce images by hand or share actual tissue samples with other researchers in order to disseminate their findings. "In making [illustrations]," Hooke writes, "I endeavoured (as far as I was able) first to discover the true appearance, and next to make a plain representation of it." Van Leeuwenhoek, on the other hand, blamed some inaccuracies of observation on the fact that he "carn't draw" and paid others to execute his illustrations. Without drawing, said Georges Cuvier, the founder of paleontology, there would be no anatomy. And if the sight of dissected bodies did not scare people away from anatomy, da Vinci wrote in his diary, then lack of drawing talent might. "No matter how exact and minute the verbal description may be," said Cajal, "it will always be less clear than a good illustration." Even with the option to photograph his slides, which offered a more exact reproduction of nature, he preferred to draw. "Drawing develops understanding" had been Abadías's motto; Cajal believed the same.

Laboratory work consumed Cajal. He went to sleep late and woke early, unable to divert his mind from the microscope in the other room. He stopped going to the café, almost completely withdrawing from the world and interacting only with Pedro. Cajal was desperate to do something momentous, something that would outlast "the whims and fads of the day." In the end, he says, he decided that the rifle was no better than the microscope or the pen. Colleagues, classmates, and professors, who acknowledged his hard work and talent, nonetheless mocked him for his quixotic, solitary pursuit. "Fortunately," Cajal said, "I have endured the absence of social life quite well."

EARLY ONE MORNING, BEFORE DAWN, CAJAL'S FATHER RECEIVED AN emergency call from a family whose sons had gotten into an argument and shot each other with blunderbusses. Justo asked Cajal for assistance, and he obliged. One of the brothers had a deep hole in his buttocks that was bleeding profusely. After much time and effort, applying sustained, tight pressure, Cajal and his father managed to stop the bleeding. By law, Justo was required to inform the police, but the parents convinced him not to do so. For eight months, Cajal and his father continued treating the young man, who developed fevers and abscesses from the shrapnel, until he finally healed. "Don't worry about the payment right now," Justo told the parents. "I'll bill you one of these days." "Don't worry," the patient's father replied, "we don't intend to pay you." Pedro recalled that their father related the story at the dinner table, saying that the patient's father had warned Justo to "forget it—or else." Cajal growled, "I never will," under his breath.

From then on, Cajal did not help his father with any more patients. The conflict between them escalated, even approaching violence. One night, Justo asked Cajal to assist him with a cesarean section, and Cajal refused. During the ensuing fight, he stormed out of the house and did not come back until a few days later, out of concern for his mother. He and Justo stopped talking to each other and communicated only through Pabla, who kept relaying medical calls from her father to her brother. According to her, Cajal responded only once, for the sole purpose of insulting his father. Justo later secured another clinical position for his son—in the town of Corella, in Navarre—but this time Cajal refused. None of these incidents is mentioned in Cajal's autobiography, though his dislike of surgery persisted for the rest of his life. Decades later, in the prologue to a former student's book, he dismissed a surgeon's career as "nothing more than lancing boils," the kind of insult he might have leveled at his father.

In March 1879, Cajal became director of the Anatomical Museum in Zaragoza, a full-time faculty position at the university, albeit a junior one. Despite their feuding, his father was responsible for getting him the job. Cajal's annual salary, now 7,500 pesetas, combined with his income from private tutoring—about 750 pesetas—allowed him to finally move out of his parents' home. It was best to distance himself from his father, Cajal thought, to avoid Justo's fits of rage. Cajal took up residence in a nearby boardinghouse, but it was noisy there, and he could not work.

One afternoon, Cajal was walking through the Zaragoza neighborhood called the Torrero, where the open air reminded him of the countryside, when he saw a young woman whose beauty stopped him dead in his tracks. She reminded him of a painting of the Madonna, or a color illustration of Margarete, the love interest in Goethe's *Faust*. Her hair was also blond, and Cajal made no secret about his preference for blondes.

Cajal followed her home and started talking to her. Her name was Silveria Petra Rafaela Josefa Florentina Fañanás y García, and she had been born in Huesca. In fact, they had already met: Cajal recognized Silveria as the "fair and slender little thing with great sea-green eyes, lips and cheeks like geraniums, and huge braids the color of honey," one of the girls he had seen outside school, hurrying past him to avoid getting hit by a stone. Her father and uncle had cursed Cajal, "the son of the physician of Ayerbe," for disturbing their afternoon siestas with his slingshot battles.

Cajal's sole romantic relationship to date had resulted in such extreme feelings of shame that he had vowed never to expose himself to rejection by a woman again. In one of his notebooks from his twenties, a line from Lord Byron appears: "Marriage is born from love like vinegar from wine." But now, having suffered from two serious illnesses, and living apart from his mother, Cajal suddenly wanted to get married. In fact, he was utterly incapable of taking care of himself. Still, he knew from experience that intense passions could unsteady and overwhelm him, and he feared that falling in love again might deplete his energy, making it impossible to work. "I have a brain that is enslaved to my heart," he later admitted.

He needed to choose his partner wisely. He earned a modest income from teaching, and, after he paid for rent and food, any extra money would be spent on resources for his laboratory. A hypothetical wife would have to forgo buying clothes and jewelry, eating at restaurants, and going to the theater. The most important feature of a successful marriage, according to Cajal, was that a wife offer "a sensitive compliance with [her husband's] wishes, and a warm and full-hearted acceptance of [his] view of life."

Silveria struck him as honest, intelligent, and wholesome. Her father, a municipal functionary, had died prematurely, leaving mother and daughter in dire straits. She and her mother had moved to Zaragoza to be near relatives and had lived off Silveria's father's humble pension. Marrying well was her only chance of improving her life, and Cajal was far from the ideal candidate. He was known as a rebel and a dreamer, a delinquent and an invalid, at risk of leaving her a young widow. But at twenty-six years old, Cajal still had an air of roguish charm, with a wispy beard and a deep, longing gaze. Silveria became enamored of him—even though, as a pious Catholic, to throw in her lot with a scientist was an act of pure faith of another kind.

The wedding on July 19, 1879, took place early in the morning, almost in secret, in a church that was empty save for Silveria's mother and a few relatives. Cajal's family—even his devoted mother—boycotted the ceremony at the command of Justo, who strongly disapproved of his son's decision. He feared that his son, finally gaining momentum in his career, was once again throwing away his future for

a Romantic adventure. The only person in the church to support Cajal was Pedro, who had just returned from South America.

Honeymoons in Paris and Switzerland were for "ordinary bourgeois newlyweds," Cajal believed. In truth, he did not have enough money for such a trip. The couple instead moved to a shabby building a few minutes away from the university. Cajal could only afford to furnish the apartment piece by piece with the cheapest items available. The only well-appointed room in Cajal's home was his laboratory. After paying for food and rent, he spent the rest of his money on expensive foreign books and journals. The laboratory also served as his "Academy of Anatomy," which he advertised in the newspaper, charging students to help prepare them for their doctoral examinations. He started teaching more private classes, supplementing his income while practicing his own histological techniques. Sometimes he would come across a new detail or an unknown fact of anatomy, which might cause him to mumble excitedly to himself, though his students never knew why. On long winter evenings he locked himself in the laboratory, where he would sit in a tall leather chair, losing himself in the life of the infinitely small.

One of the only distractions that Cajal allowed himself was photography, not only a source of "undeniable satisfactions and consolations" but also a potential business venture. Sometimes he would take his camera out into the city, accompanied by Silveria. One day, they went to the local bullfighting ring in a wealthy neighborhood, where Cajal climbed the balcony and managed to snap a picture of the provincial governor, seated in the presidential box, enjoying the event surrounded by beautiful women in their fancy *mantillas* and *peinetas*. Few Spanish photographers at that time were aware of gelatin-bromide plates, a popular new technique that allowed for exposures of only fractions of a second, but Cajal was. After a local newspaper printed Cajal's picture from the bullfight, photographers in Zaragoza and throughout Spain contacted him to inquire about his new method. Cajal considered starting a manufacturing company with Silveria as his assistant, whom he showed how to weigh and mix the necessary chemicals. In the candlelight, filtered through red-stained glass, their faces glowed like "goblins or necromancers." The neighbors wondered what the newlyweds were doing out in the barn so late at night.

Oftentimes, in the early days, Cajal found himself deeply unsure of his choices. His father, still furious at his decision to marry, had cut him off financially, and Antonia and his sisters were forbidden to help him in any way. He could not rely on any support from Silveria's family, who were poor. "The sensible person pursues a career and establishes himself," Cajal wrote in a notebook from around this period, ". . . guiding his work by the compass of what is convenient for his life, by tranquility and calm." He continued:

> *What foolishness to seek poverty for the sake of glory! And what if it does not come? And even if it does come, it comes later, and*

it generally happens for old men and the dead, and in our case it would be the same. And what do praises matter to me? When they applaud me I will not exist, I will not know any of them, I will take no pleasure. Is it not better to employ my talent and energy for all the good and happiness possible, escaping the contradictions and spending my life the most happily that one can?

Not long after their wedding, Silveria became pregnant with their first child, a daughter, whom they would name Fe, the Spanish word for "faith." In the year after his marriage, with his domestic life stabilized, Cajal gathered the courage to publish his first scientific paper, a version of his dissertation, titled "Experimental Investigations on Inflammatory Genesis and Especially on the Emigration of Leukocytes."

IN THE 1870S, AS CAJAL EMBARKED ON HIS RESEARCH, INTELLECTUALS were engaged in "the polemic of Spanish science," a debate about Spain's cultural values. Some argued that the outmoded state of Spanish science resulted from defects in the Spanish character; the French encyclopedist Nicolas Masson de Morvilliers called Spain "probably the most ignorant country in Europe," whose people are "indolent, lazy, and apathetic." This view of Spain was later termed "the Black Legend"; the narrative can be traced back to the sixteenth century, Spain's Golden Age, when foreign propaganda attacked the empire as "inquisitorial, ignorant, fanatical, incapable of taking a place among the cultured nations, always inclined toward violent repression, an enemy of progress and innovation." Cajal experienced these stereotypes of Spanish character as personal insults. "I heard in derogatory foreign judgments the pain of the Black Legend," Cajal said. "So deep in the core of our race was the conviction of our sad and radical incapacity for the cultivation of science." He doubted that he had anything worth contributing to the world of science, but he felt an obligation to try. "One should publish the results of one's personal experience, for whatever it is worth," Cajal said, "if for no other reason than to show that Spanish researchers care."

No topic was hotter than inflammation, and so Cajal decided to expand on his dissertation. When the body becomes infected, white blood cells, also called leukocytes, form pus that accumulates at the infection site, acting as an emergency barrier. Researchers in the 1870s disagreed about the origin of those leukocytes and how they travel to their destination in the body. Studying the formation of pus was seen as crucial to understanding disease. The relevant experiment, on the mesentery of a frog, was the same one that Cajal had seen demonstrated as a medical student in the physiology laboratory at Zaragoza.

Examining the frog under the microscope, Cajal observed every detail that he could, from the speed of blood flow, to the velocity of red and white cells relative to each other, to the dilation of the capillaries, arteries, and veins. At first the cells were almost invisible, their contours blurred by the rushing of blood, but once the frog's circulation slowed, Cajal could make out the bodies of the leukocytes well enough to follow them. After one or two hours of observation, Cajal said, the phenomenon became clearer, though some cells could not be perceived until six or seven hours later. He observed their different shapes and marveled at their elasticity, how their bodies flattened in order to fit through a breach and then reassumed their previous form, almost magically, on the other side. He noted the places where leukocytes most frequently got stuck, tried to figure out why, and looked for patterns in how they changed shape.

He drew the strange forms in his sketchbook, portraying the deformity of cell bodies with different intensities of pencil shading. In other illustrations, Cajal outlined veins in pencil, yellowed the leukocytes to make them stand out against the page, and colored their centers pinkish red, representing their carmine-stained nuclei. Where leukocytes overlapped in three-dimensional space, he established depth by darker shading. The leukocytes in the bloodstream that Cajal drew look, he wrote, "like pebbles caught up in the force of a torrent," an accurate scientific observation, and one evocative of a scene from his childhood. He repeated his observations multiple times, changing the experimental conditions, the lens magnification, the room temperature, and even the part of the frog, studying the lungs, tongue, corneas, and rib cartilage.

Why was he so captivated? He saw the leukocytes not as passive elements of the inflammatory response but as active participants in "excursions." Sometimes they moved sluggishly, dragging themselves along the vascular wall and stopping altogether until some red blood cells crashed into them, forcing them to rejoin the flow. Sometimes they sped up, avoiding obstacles, taking shortcuts, and slipping into open spaces "like an army charged and diffused on a battlefield." Or they would escape through a wall "like a prisoner who files through the grating of his cell." In his mind, he was watching scenes from an adventure novel.

At the end of his study, Cajal concluded that his hero, Rudolf Virchow, had suffered from "a defect of interpretation" in his conclusions about the subject. How could a great scientist, devoted to the truth, arrive at a false understanding? Cajal blamed "the innate tendency in us to go beyond, swept away by the imagination, from the reduced circle in which the facts force us to enclose ourselves." This is a surprising insight from someone who had spent his youth dreaming of other worlds—but starting in the mid-nineteenth century, there was a movement toward objectivity in science, with the goal of removing the observer from the process of observation. At the very start of his career, Cajal felt the need to abide by these principles; Spaniards were stigmatized as emotional and irrational, and he wanted to counter the stereotype. "It is a general ailment of the human spirit that always tends to project onto the works of nature an image of one's own intelligence," he wrote. "Let us aspire to comprehend organization such as it is, not as we would wish it were."

That would prove easier said than done. Every facet of histological observation seemed to conspire against clear seeing. Human beings perceive the world in three dimensions, which the focal plane of a microscope necessarily collapses into two. Under higher magnification, some tissue samples only became less clear, and some images became so blurred that little could be discerned from them. If the object under the microscope stage were moist or oily, then light would be reflected too brightly, causing a simple membrane to appear decorated, like a fancy silver cloth, as one seventeenth-century investigator warned. If the microscope light itself were too strong, or the sample too thick, or elements spread too unevenly or too far apart, then striations, globules, or waves might appear, which could be easily mistaken for anatomical elements. Cajal taught himself to recognize when grease, hair, air bubbles, or traces of cotton, hemp, silk, or wool appeared on the slide. It took time and caution to learn not to be deceived. "Because we love them," Cajal said about microscopes, "we appreciate their fine points, we are aware of their defects, and we avoid the traps they occasionally set for us." Every time they looked through the lens, microscopists had to ask themselves whether they were seeing fact or fiction.

CHAPTER 12

"The Role of Don Quixote"

The Zaragozan medical press *El Católico Diario*, in which Cajal published his first scientific paper, had almost no circulation. Self-publishing was the only other option, but it was a luxury that Cajal could barely afford, and the expense was difficult to justify, given that so few people would read his work. He decided to print one hundred copies of his article as a monograph, which he did not sell but rather gifted to friends and colleagues, most of whom were uninterested in histology. "I was excessively surprised by the almost total absence of objective curiosity on the part of our professors," Cajal recalled, "who spent their time talking to us at great length about healthy and diseased cells without making the slightest effort to become acquainted by sight with those transcendental and mysterious protagonists of life and suffering."

For his second scientific paper, published in 1881, Cajal turned his attention to muscle fibers. Anatomists since 1840 had known that certain muscle fibers had an alternating, striated pattern, but they did not know how they contracted. Histological techniques were employed to determine the structures behind the function. "If we would know with exactitude the relations that exist between muscle and nerve," Cajal wrote, "the investigation of the nature of muscular acts would be easier, and perhaps the key to the particular way the nervous current determines the phenomenon of contraction." A few years later, in a letter to a former student, Cajal announced what he believed was his first discovery: "a striated thread that stitches the cells together, impeding their retraction." The original cover of his first textbook bore an image of this phenomenon in an epithelial cell, shaped roughly like a square, with a striated thread tying it to its neighbor, as if it formed part of a patchwork quilt. But upon using a more powerful microscope, Cajal soon realized that he had been wrong. He removed the image from subsequent editions of his textbook and expunged all descriptions of the cellular stitching. He had been fooled by the idea of connections between fibers, known as the reticular theory, or network theory, of muscle fibers. "We saw networks everywhere," Cajal later said of that time.

AT THE END OF 1883, CAJAL RETURNED TO MADRID TO COMPETE IN HIS third *oposicion* in five years, for an opening in Valencia. He had taken second place in an 1880 *oposicion*, a moral victory when compared with his previous debacle, but his financial difficulties remained. This time, when Cajal read a summary of his merits, he was able to add two scientific publications. It did not matter that the scientific community had ignored his work; none of the other contestants had produced original scientific research. Cajal won the Valencia *oposicion* unanimously. What distinguished Cajal, a judge from that 1883 tribunal recalled, was "the *simplicity* in the exposition and the *force* of the reasoning." The anatomical drawing he prepared as part of the exam was so exquisite that a judge took the blackboard home with him to cherish Cajal's chalk sketch as a work of art.

When Cajal arrived in Valencia, in January 1884, he felt as though he had traveled to another country. Valencia, on the eastern coast of the peninsula, is often called "the garden of Spain." The Moors, who ruled over the principality in medieval times, considered it their paradise, a piece of heaven fallen down to earth. The air was soft and delicate, so pure and dry that cooks could leave salt and sugar out in the open for months without them forming clumps. Snow existed only in fantasies; hoarfrost had been reported only twice in the previous five centuries. You could sit for hours reading and writing without a fire on the coldest day of the year; one nineteenth-century traveler called life in Valencia "one continual delicious spring." At sunset, the golden light was tinted with the purple of figs. The Mediterranean breeze was scented with the smell of fresh oranges, and strong winds blew fruit from the trees onto blankets that peasants held waiting below.

Cajal was accompanied by his children Fe and Santiago, born in 1882, now about three and two years old, and Silveria, who was nine months pregnant with a third child. Of all the boardinghouses in the city center, Cajal chose to lodge his family at the one nicknamed "the Aragonese," a new building with large windows facing the fifteenth-century Silk Exchange and Market Square. Little stands flaunted heaps of shiny oranges the size of cannonballs. Peasants with tanned faces, heads wrapped in colorful silk, bustled down dark, narrow lanes, their shawls laden with dates and grapes. Every Thursday at noon, a tribunal of twelve peasants convened on a long pink couch in the doorway of the cathedral to decide how to irrigate the land. Water coursed toward the fields through irrigation canals like veins of silver. In Aragon, half the land was barren, while every inch of Valencia was fertile.

A few days after they arrived, Silveria gave birth to their third child, a daughter named Enriqueta. Records show that on that same day, Cajal requested admission to the Agricultural Society of Valencia, a social club for scientists, doctors, and other intellectuals and academics. Cajal began participating in *tertulias*, small intellectual gatherings like Parisian salons, which he had encountered on his first trip to Madrid. (In the sixteenth century, King Philip II of Spain used to organize discussions of ancient texts, especially those of his favorite author, Tertullian; *tertulia* also became the Spanish word for the highest balcony in a theater, an area restricted for literary criticism and intellectual debate.) Cajal refused to allow his family obligations to interfere with his intellectual pursuits. "Children of the flesh should not drown out the children of the mind," he declared.

After a few weeks at the boardinghouse, Cajal and his family moved to a third-floor apartment on nearby Las Avellanas Street, which had a room large enough for a laboratory. As a full professor, Cajal now earned 3,500 pesetas per month, more than five times his previous salary. He was determined to remain as frugal and disciplined as ever; every peseta saved meant more for his laboratory. Silveria kept the books; because of her economy, there was enough money left over to hire a nanny.

The streets of Valencia were like a stage where family disputes, romantic encounters, and friendly rivalries played out in public. After finishing their chores, Valencian women liked to go outside and mingle on the promenade, chatting with one another and enjoying the sun and breeze. Silveria preferred to stay inside. She did not know a soul in Valencia and had never lived away from her mother. She was hesitant to leave her own patio.

THE FACULTY IN VALENCIA DID NOT LIKE CAJAL AT FIRST. THE atmosphere in the department was friendly and informal, but the new professor seemed to avoid social interactions. He did not speak much, and his face always wore an odd, serious look. Having come across him, as judges or fellow candidates, in previous *oposiciones*, they knew his reputation for being smart, hardworking, and ambitious. The Valencia faculty viewed him as a competitor in both the academic and professional arenas, but they quickly realized that he had no interest in a clinical career; he would rather lose money and become "a millionaire in time" instead. The only work that interested him was research in the laboratory, sacrificing frogs and guinea pigs, experimenting with staining techniques, and scrutinizing tissue through the microscope.

Despite his single-mindedness, Cajal realized that focusing his attention exclusively on the microscope was not healthy. "It was necessary to allot to each cell its rations and to each reasonable instinct a convenient opportunity for exercise," he recalled. He needed to balance his mind. From time to time, he played chess with the local champion—we can assume that he beat Cajal; otherwise, Cajal would have bragged about the outcome. In addition to the Agricultural Society, Cajal joined the Valencia Athenaeum, an organization founded in the wake of the 1868 Revolution. Its elegant salons hosted speeches, debates, conferences, expositions, and even art classes, and men of all ages came to discuss social issues, literary theory, and scientific discoveries. Cajal participated in the "Section of Exact, Physical, and Natural Sciences," which attracted experimentalists who discussed new ideas like Darwinism and microbiology. At the Valencia Athenaeum, he attended a conference in 1885 called the Cell Theory and was later elected director of Biological Studies and Medical History and Philosophy.

In the Chimney Salon at the Athenaeum, named for its fireplace, Cajal befriended a local pharmacist named Narciso Loras, who decided to organize his own *tertulia*, held at his Mulberry Tree drugstore, about five hundred yards from Cajal's home. Regulars included a climatologist, a naturalist, a chemist, an experimental pharmacologist, and the future curator of the National Science Museum. The Mulberry Tree *tertulia* seeded an offshoot called the Gaster-Club, a "gastronomic-recreative society" devoted to Sunday excursions among the carob, palm, pine, and rosebay trees on the outskirts of the city and to savoring the famous paella Valenciana, with rice, chicken, sausage, and red peppers. Cajal was in charge of drafting the constitution, which prohibited all talk of politics, religion, or philosophy; science and art were the only permissible topics.

With his trusty camera, he also served as the group's official photographer. Of the more than two thousand photographs contained in Cajal's archive, most of the nonscientific images are self-portraits, family portraits, or depictions of landscapes, but the pictures from his time in Valencia are notably social—many depicting a group of bearded and mustachioed men in black suits and vests gathered together, sometimes accompanied by their families, clearly enjoying themselves. In one photo, the men are seen at Albufera—the pristine blue lake that was home to flamingos—boarding a sailboat while raising their hats in the air, like seamen before a long voyage. In the woods, they lounge in the grass, posing affectionately with one another. And in front of the ancient ruins of Sagunto, Cajal appears vibrant and self-assured, comfortably surrounded, for the first time since childhood, by a group of friends.

THE AMPHITHEATER IN THE GARDEN OF THE PROVINCIAL HOSPITAL was packed for Cajal's first anatomy class in Valencia. "No one skipped," a student recalls, "because word already spread in Valencia that the 'new one,' as the boys called him, was a man of great talent." The students were in awe of the timbre of his voice, a resonant, nasally drawl, which reminded an Aragonese student of the *jota*, the traditional highland folk song evolved from the sounds of men and women calling to each other in the fields. All the men in Cajal's family were blessed with that talent for folksinging, according to Pedro, who recalled his teenage brother crooning while working at the shoemaker's. While lecturing in class, though, Cajal opted for something closer to a drone. His lectures, recalled another student, were set in a minor key. "It was as though he were in a *tertulia*," said a student, "calling bread bread and wine wine."

In 1884, Juan Bartual Moret, a twenty-one-year-old graduate of the University of Valencia medical school, had returned from studying abroad in search of a teacher who could prepare him for the doctoral examinations. His former mentor recommended the professor that they called "the Aragonese." Bartual went to the designated address and climbed the stairs to the top floor. "Why not be honest?" he said. "They sent me to a man who lived in an attic, for no other name suited the room where he found himself settled in with his family." The room was practically empty, with only a small cabinet inside, which seemed almost "magical." Bartual did not expect the new star of the faculty to look so disheveled, even rustic. Baldness broadened Cajal's already broad forehead, and slight hollows, the faded stigma of past illnesses, still scarred his hands. He was not one for pleasantries. The two men exchanged perfunctory greetings, then immediately got down to work. "Soon, very soon," Bartual recalled, "I realized whom I was with, captivated by that potent intelligence and his love of scientific work bordering on obsession." Cajal's true home was wherever he set up his microscope. "There, in his laboratory, he lived only for his science."

According to Bartual, Cajal's teaching determined the path of his career. "[Cajal's] aspect, his ideas, his vigorous speech produced in me the effect of a great city as seen from the point of view of a bird," Bartual said. The preparatory course was "so surprising, so unusual, and so impressive" that, at the end of the year, Bartual and some of the students at the university wrote Cajal a tribute, which Bartual read to him out loud:

> Dear Professor: It is so sad to be separated from the person who has made us think, to leave someone who spread the truths of science one by one to nourish our intelligence. A veil of sadness seizes our spirit. Must we leave you? It is necessary to resign ourselves and continue, but know that our hearts will never forget you; when in the future we hear a scientific principle repeated, a truth that is already known, let us say, we already knew this, do you not remember? Our dear teacher taught it to us.

"You are more than a teacher," the tribute ended. "You are a father."

One day, Cajal was in the laboratory when he noticed a man on the opposite balcony who appeared to be spying on him. Day after day, the man continued to watch him. One afternoon, Cajal heard a knock at the door. It was the man from across the street; he introduced himself and handed Cajal a business card. He judged his neighbor to be either petit bourgeois or working class, not a man of culture or sophistication. Cajal imagined what his neighbor must have been thinking as he stood

in the threshold of the laboratory, scanning the room: "Was he a watchmaker? A sewing machine repairman? An engraver? A fabric dyer? Or just your average madman?" When he caught sight of the microscope, he begged Cajal to let him look. Cajal chose to show him a simple preparation, so as not to confuse him: a sample of kidney tissue, stained pinkish red with carmine. The structure reminded Cajal of fruit hanging from a tree or an elaborate bird's nest. His neighbor stared long and hard through the eyepiece. Finally, he spoke. "What a beautiful drawing for a piece of waistcoat!" he said. Cajal looked down at the man's card: it said "Tailor."

CAJAL NEVER FORGOT THE ABSENCE OF HIS COUNTRYMEN FROM THE science textbooks of the day. In Valencia, he decided to write his own; he considered it an act of "scientific emancipation." "My motto was always this," he later explained: "to demonstrate that the inhabitants of the Iberian Peninsula . . . are capable of creating pure science and that even the most modest people can emulate the prestigious representatives of foreign science if they determine it with strong will and unwavering perseverance." He devoted his first two years in Valencia to the project, which he released in installments whenever he felt that he had made a technical improvement or discovered an unreported anatomical detail. He thought of Spanish science as "the Dulcinea of [our] dreams." "I took the role of Don Quixote seriously," he said.

Cajal's *Manual of Histology and Micrographic Technique* took the same form as other contemporary textbooks, only his was written for Spaniards by a Spaniard. He shared the secrets of microscopy, including advice about the best table ("large, solid and black, so as not to tire out the vision of the observer"), slides ("cut from glass perfectly free of striations, bubbles, and deformations"), and light ("diffuse, proceeding from a white wall or a well-illuminated cloud"). He reviewed different models of microtomes (machines for cutting tissue), micrometers (devices for measuring small distances), and *camera lucidas* (optical devices used as drawing aids), as well as different formulae for stains. Cajal visited the press, located five minutes from his house, to oversee the printing process himself; well into old age, he would have nightmares about discovering typographical errors while correcting proofs. The first fascicle of his *Manual*, totaling 192 pages of small print, was published on May 1, his thirty-second birthday.

Valencia was the first city in Spain to boast a printing industry. Since ancient times, it had been known as a center for science publishing, and while Cajal lived in Valencia, the industry was in the midst of a renaissance. His *Manual* was published by Pascual Aguilar y Lara, who had produced many Castilian translations of foreign texts after the 1868 Revolution, including Virchow's *Cellular Pathology*, which had inspired Cajal in medical school, as well as *Essays on Cellular Psychology* by Ernst Haeckel, the famed zoologist who named thousands of new species, and *Cholera* by Robert Koch, the famous German pathologist. Likewise, the faculty at the University of Valencia was famous for promoting laboratory medicine. While Cajal was arguing against his vitalist professors back in medical school, the faculty in Valencia had been protesting to the central authorities that they should have the right to teach cell theory too. Students in Valencia revered Virchow as a god. Before histology was on the national curriculum, the Valencian faculty supported it as an official course and brought histologists on staff. The dean of the medical school, the pathologist Enrique Ferrer y Viñerta, belonged to Maestre's histological society and performed microscopic studies of lesions and tumors. Amalio Gimeno, the chair of therapeutics, was an experimental pharmacologist who wrote a textbook inspired by Virchow and Claude Bernard, the giant of French physiology. Peregrín Casanova, Cajal's fellow anatomy chair, was an early champion of cell theory and corresponded with Haeckel, whom he considered his master, for years. Casanova maintained that anatomical studies should no longer be descriptive but *predictive*, based on evolutionary theory.

In *On the Origin of Species*, Darwin observes that species of a phylogenetic class, regardless of their life experiences or circumstances, are built according to the same anatomical plan. "What can be more curious," he wrote, "than that the hand of a man, formed for grasping, that of a mole for digging, the leg of a horse, the paddle of a porpoise and the wing of a bat, should all be constructed on the same pattern and should include similar bones, in the same relative positions." Darwin referred to this study of form as *morphology*, a word coined from the Greek (meaning "the study of change") by Johann Wolfgang von Goethe in his 1790 essay "The Metamorphosis of Plants." Goethe, in his studies of botany, mineralogy, and anatomy, sought common patterns that went deeper than the doctrine of classification by the Swedish botanist Carl Linnaeus, which was based on external appearance. All organisms, Goethe believed, had an inner "drive to formation." "When," he said in his 1817 work *Zur Morphologie* (*On Morphology*), "having something before me that has grown, I inquire after its genesis and measure the process as far back as I can, I become aware of a series of stages, which, though I cannot actually see them in succession, I can present to myself in memory as a kind of ideal whole." Morphology was rooted in a fundamentally Romantic worldview, portraying nature as the expression of a universal creative process.

In Valencia, surrounded by stimulating new ideas, Cajal would come to understand how classical anatomy, the body of knowledge transmitted by his father, was "pure abstraction," insufficient as an explanation of the living world. He would focus on how structures change. He would become a morphologist.

CHAPTER 13

"The Religion of the Cell"

In 1884, an epidemic of cholera—the fifth of the nineteenth century—was decimating populations around the globe. At first, the symptoms of cholera mimic simple food poisoning or stomach flu, until victims expel torrents of fishy smelling, rice-water stool, sprinkled with bits of intestinal lining. (*Cholera* is named for the ancient Greek word meaning "roof gutter.") The skin turns dark blue and leathery. Racked with cramps, the body depletes itself entirely. If cholera is left untreated, death can come in four to twelve hours, so quickly that masqueraders infected at a ball in 1832, the poet Heinrich Heine recalled, were buried still in costume. The real horror, doctors reported, was consciousness. Victims were doomed to watch themselves transform into corpses as their minds remained clear until the end. All that doctors could do was administer opium.

The ancient theory of miasma, which persisted into the nineteenth century, held that disease was caused by "bad air." In the 1860s, Louis Pasteur discovered that many diseases are caused by microscopic organisms, which he called germs. He invented vaccines to inoculate animals against epidemics of anthrax and swine flu. Older doctors, resistant to new ideas, clung to the miasma theory, while younger doctors tended to believe in microbes. Robert Koch went to India and Egypt to investigate the cause of the cholera outbreaks there and was able to isolate the cholera bacterium from the intestinal walls of the deceased. When examined under a microscope, the bacillus looked "a little bent, like a comma." Koch found the *comma bacillus* in all of the almost one hundred autopsies that he conducted.

When Cajal arrived in Valencia, Spain was under quarantine. At the time, Valencians were not aware of the concept of water contamination or of innovations such as drain traps and air ventilation systems, which improved sanitation in cities. Butchers handled meat without gloves; there was no such thing as refrigeration. In Burjassot, four miles from the city, waste was heaped on the banks of the Turia River, where Valencians washed their clothes. Every house had a water closet, but they all emptied into the same drain. Filth and garbage turned into black, stinking mud when it rained.

In April 1885, Valencians received the chilling news that they had long feared: the disease had penetrated the city. The irrigation canals, Valencia's circulatory system, carried the infection to all corners of the city, tainting every element of civic life like carmine flushed through the bloodstream of a frog. The city erupted in panic. The wealthy fled so fast that their beds remained unmade and food sat uneaten on their tables. Public officials abandoned their posts, and armed men stood outside cordoned-off homes. Bonfires were set in the streets, burning greenwood treated with sulfur, the smell of which was meant to reassure passersby of the neutralization of bad air. Newspapers piled up outside houses whose residents were too terrified to crack open the door. Placards in offices and advertisements in papers warned "Do not shake hands." A rumor spread that even holy water had been contaminated. The "cholera wagon," gray with a conspicuous red top, stalked the streets for victims, alive or dead. Death registration offices were open twenty-four hours a day, seven days a week, but many people did not file reports, fearing the stigma associated with the disease. Corpses rotted in homes. Many Valencians refused to even say the word *cholera*; the Spanish government, to avoid economic disruption, propagated the euphemism "the suspicious illness." Mobs stoned doctors who failed to cure patients. Men from the south who begged for food and slept in the fields, calling themselves "faith-healers," rubbed patients' stomachs and urged them to pray to God. In June, seven hundred cases of cholera were reported per day. The epidemic spiked in early July, when 217 people died in one day. A woman who lived a few blocks away from Cajal fell ill. Then someone became sick in Cajal's building, in the apartment almost next door. Silveria was pregnant with their fourth child. Meanwhile, Cajal was in the kitchen, growing bacteria.

The provincial council of Zaragoza, where the epidemic also spread, had commissioned Cajal and another doctor to study the origins of cholera and the efficacy of a new vaccine that Jaime Ferrán i Clúa, a country doctor from Catalonia, claimed to have invented. After injecting the bacterium in its purest form, suspended in beef tea or chicken broth, into guinea pigs, he reported that those that survived became immune to further exposure. The efficacy of the vaccine—"the Ferrán Question"—was debated in academies and athenaeums, in articles in medical journals, and in the daily newspapers. "There is not a tittle of evidence to show that the fluid used for the inoculation is capable either of producing cholera or of conferring immunity from it," *The Times* of London stated. Ferrán insisted that the vaccine was simple and harmless, and that of the thousands he had treated, no one had experienced ill effects. No one could verify the statistics, which Ferrán and his assistants compiled themselves. The word *ferranista* became synonymous with *liberal* or *progressive*, while *anti-ferranista* meant *conservative* or *immobilist*. People defended their theories with vehemence, even violence.

Long lines of fearful people extended outside a clinic—men with bared arms and women with holes cut in dress sleeves, where the hypodermic needle would be inserted approximately four inches above the elbow. Three doctors manned the clinic at a time from eight in the morning until eleven at night; the flow of patients was unceasing. Ferrán administered the first few thousand inoculations for free, but he advised that, since the duration of protection was short, patients should receive shots every two months. Each injection cost the equivalent of between two and seven dollars. Ferrán was responsible for vaccinating more than fifty thousand people between April and June. "I have frequent news of Ferrán," Cajal wrote in a letter to a former student. "He has discovered the cholera vaccine and vaccinated himself. His work is making a lot of noise. He has sent me some microbes with which I have performed some experiments."

For months, Cajal abandoned his beloved tissue cells and turned his attention to bacteria. In Cajal's story of the "life of the infinitely small," microbes played the villains, "the invisible enemies," a "traitorous gang" terrorizing good cellular citizens of the body. Fighting germs was an "all-out war," and bacteriology was the new battlefield. Stopping a national epidemic constituted the purest form of patriotism, Cajal thought. He ordered expensive incubators and sterilizers for the laboratory and consulted as many books as he could find on the subject. Koch had postulated that the same microorganism was present in every case of the disease, meaning that any sample could offer the researcher the material that he needed. In order to observe cholera directly, Cajal went to the hospital and asked for a sample of the virus, an odorless, light-pink serum sprinkled with floating white flakes, which he carried in carefully sealed test tubes on the half-mile walk back home. He removed a speck from the vial with a fine platinum needle that he had sterilized in a flame and stained the sample with dahlia violet, highlighting the bacillus against the background of clear glass. Under the microscope, the comma shape was obvious.

The international scientific community could not believe that Spain would be the country to which they turned for a medical breakthrough. "Those times were very difficult for Spaniards fond of investigation," Cajal recalled. "We had to fight with universal prejudice of our lack of culture and our extreme indifference toward the great problems of biology. One could admit that Spain produces some genius artist, such as a long-haired poet or gesticulating dancer of either sex; but the idea that a true man of science would emerge from there was considered absurd." Doctors from France, England, Belgium, Portugal,

Italy, Brazil, Russia, and the United States traveled to Valencia to investigate Ferrán and his methods, flocking to the house where he prepared the vaccines in an improvised kitchen laboratory. A French commission asked Ferrán for information. "I have to keep my secret," Ferrán reportedly said. "I see what I give up by surrendering it, but I do not see what collateral you give." His renown would not last for long. After vaccinating seventy nuns at a convent, it was observed that fourteen died and forty became ill, while all the nuns who did not receive the vaccine survived. Public opinion turned on Ferrán. People began preventing him from entering their towns.

THE RESULTS OF CAJAL'S STUDIES WERE PUBLISHED IN A MONOGRAPH in July, titled *Studies on the Vibrion Microbe of Cholera and Prophylactic Inoculations* and illustrated with eight lithographs, printed in color, which the council published at its own expense. Cajal affirmed that the epidemic was indeed caused by cholera, a conclusion that opponents of bacteriology still rejected, and he expressed doubt about the efficacy of the vaccine. "With regard to the principal point, namely prophylaxis," the report said, "I declared myself not to be favorably inclined toward the procedure of Ferrán." Silveria was deeply superstitious, but Cajal promised her that if they followed the recommendations of bacteriologists, boiling their drinking water and cooking their food thoroughly, then their family would be safe.

The epidemic also offered Cajal a chance to hone his microscopy skills. In an 1885 article on bacteriology titled "The Safest and Easiest Way to Color Microbes," he announced that he had produced slides that are "even more beautiful than those obtained by the method of Dr. Ferrán."

As a reward for Cajal's work, the provincial council gave him a new microscope, the model that he had wanted: a Zeiss Stativ with a 1.18 objective, the best in optical amplification at that time, to replace his old Verick, which had become like "a rickety door bolt." Now no microscopist in Europe had better equipment than he did. A new path of research lay open for Cajal. Bacteriology offered the prospect of fame and financial reward; Koch and Pasteur were celebrated as heroes. But studying microbes was also expensive; bacteriologists were constantly buying stoves and autoclaves for incubation and sterilization and paying for gas to run them. Histologists lived lives of relative poverty and obscurity. If they were lucky, perhaps two or three dozen experts might read and appreciate their work. And Cajal, after all, was not interested in applying scientific research to solve medical problems. "I am going to leave the vibrion [the microbacterium], whose terrain I judge to be very sterile," Cajal wrote to Ferrán, "and I am going to embark in other directions."

EVEN AS HE REJECTED CLINICAL MEDICINE, A LIFE OF RESEARCH AND teaching was not enough to satisfy Cajal. Just as it had when he was young, his imagination began "impatiently champing at the bit," and so he spent what little free time he had writing science fiction stories again. In "Life in the Year 6000"—a story salvaged from his notebooks and published after his death—a doctor is dried up like a piece of jerky and reanimated thousands of years later, a plot inspired by Claude Bernard's research on rotifers, microscopic marine animals that, by desiccating themselves, can resist death. In Cajal's vision of the distant future, life is totally automatized, "no more than the application of a mechanical rationale and advanced calculations to the phenomena of life." Idealists, poets, dreamers, mystics, philosophers, and romantics are eliminated from society. Darwinism and positivism are laws of the land, and vitalism is relegated to its rightful place, treated as a form of insanity. Romantic love—the most dangerous force known to humanity—is suppressed by a vaccine. There is no such thing as fiction; every "novel" is a work of natural history. "Life in the Year 6000" showcases Cajal's sardonic sense of humor; the most famous musician in the world is named Cicada, whose greatest hits are "The Endocarditic

Waltz" and "A Cough with Noisy Fifths." Professors are replaced by giant phonographs, which prove to be infinitely more effective teachers than human beings.

Cajal wrote thirteen short stories over the course of his life, seven of which he claimed were lost (including "Life in the Year 6000"), probably during his military service. Five were published in 1905, of which four had been written in Valencia, around 1885 or 1886, while one had been written much later, in Madrid. Cajal edited the stories before publishing them, and only the revised versions remain. The four stories hint at Cajal's inner life at the time of their original composition, before he became a famous researcher and public figure, and while he was still forming his identity as a scientist and questioning his own ideas. He likened the writing process to an "unbosoming," a "dynamic compensation" for his overworked self. The stories were published under the title *Vacation Stories*, alluding to a departure from his everyday life, and signed with the pseudonym Doctor Bacteria, a characteristically dark joke given that the story was written at the height of the cholera epidemic.

One story, "For a Secret Offense, a Secret Revenge," is about an esteemed bacteriologist named Max von Forschung ("Max Research" in German). Like his author, Forschung spends all his time "working in isolation in his laboratory," living "as happily as scientists can live who are disquieted and kept awake at night by the devouring fever of investigation and the desire to emulate glorious reputations." Having become "inoculated with the terrible toxin of love," Forschung marries an American doctor named Emma Sanderson. After discovering that his wife is having an affair with his assistant, Forschung infects them with bovine tuberculosis. Cajal speculates about how the science of bacteriology, touted for its potential to save lives, might also augur human destruction. Forschung develops a cure for his wife's tuberculosis, reconciles with her, and returns to work, developing a serum to cure adultery by speeding up the aging process of women's bodies. He calls the drug "senilina."

The protagonist of another story, "The Corrected Pessimist," most closely resembles the author himself: Juan Fernández is a young doctor, "not lacking in talent, but utterly pessimistic, with a misanthropic streak," convalescing from a serious illness, who competes in an *oposicion* for a chair "unsuccessfully, but with honor." Fernández is about to commit suicide when "the spirit of science" visits him and gives him microscopic vision for a year. "There was a whole world of beauty buried and hidden in the infinitely small!" Fernández says. Yet he is also forced to see that the world is covered in germs, spoiling his romantic enchantment. When the year of enhanced vision is up, Fernández wins a hospital post, gains patients, is surrounded by friends, and marries well. The story provides an origin myth for Cajal as a scientist; the microscope saved his life after illness, failure, and rejection.

CAJAL FOUND ANOTHER DIVERSION IN THE STUDY OF PSYCHOLOGY. HIS fourth Valencian story, "The Fabricator of Honor," is a tale about the power of suggestion. When he was a student, the psychology textbook used in Spanish secondary schools declared that psychology is "a science that deals with the soul," which "is not, and cannot be, material." The French philosopher Auguste Comte, whose ideas were popular among Spanish intellectuals in the second half of the nineteenth century, believed that psychology could never be practiced objectively because the mind conducting the experiment is inseparable from the mind being investigated. "The interior observation gives birth to almost as many theories as there are observers," he lamented. In the 1870s, another French philosopher, Théodule Ribot, insisted that psychology be separated from philosophy, metaphysics, and religion and grounded in the methods of empirical science. Ribot proposed that psychologists study mental disorders, "the anomalies, the monsters of psychological order," which he called "experiments prepared by Nature." He proposed a new school of psychology—*psychologie nouvelle*—which treated occult phenomena such as human automatism, multiple personalities, double consciousness, demonic possession, fugue states, faith cures, and mediumship as legitimate topics of scientific investigation. Also included was the phenomenon of mental suggestion, which had fascinated Cajal since his days at Panticosa.

The most prominent figure of *psychologie nouvelle* was the neurologist Jean-Martin Charcot, nicknamed "the Napoleon of Neuroses," who conducted scientific investigations of anomalistic mental phenomena. In 1882, Charcot read his findings on hypnotism at the Academy of Sciences, and hypnotism became a frequent and controversial topic in medical publications. Charcot and his "Salpêtrière school" argued that the capacity to be hypnotized was a symptom of hysteria, while Hippolyte Bernheim and his fellow physicians from the "Nancy school," after hypnotizing over one thousand patients, concluded that the capacity to be hypnotized was just another form of suggestibility, which every person possessed.

In the 1880s, Cajal became acquainted with hypnosis and the Salpêtrière-Nancy debate through books and articles in Spanish translation. He convinced friends from the Casino de la Agricultura to form the Committee for Psychological Investigation to settle the Salpêtrière-Nancy debate on hypnosis. They rounded up hysterics, neurasthenics, maniacs, and spiritualistic mediums as subjects. Cajal even attended a séance, easily dispelling the effects and explaining them away. The findings of the committee supported Bernheim; by practicing his methods, Cajal was able to hypnotize lawyers, physicians, and other subjects who were not neurotic. One patient changed the routine of his life for a week based on a special program of absurd behaviors that Cajal suggested to him. "I succeeded in performing wonders which would be envied by the most skillful of the miracleworkers," Cajal boasted. He also said that he was able to suggest to subjects that they were experiencing hemorrhages.

One of Cajal's patients was a thirty-five-year-old pregnant woman, whom he described using the psychological classifications of the day: her temperament was not nervous but "sanguine-lymphatic"—

language borrowed from phrenology. She was pregnant for the fifth time and, having experienced extreme pain during each previous birth, was afraid that she would suffer with the same intensity. For a few months, Cajal employed suggestion on the woman, using Bernheim's method: "Your lids are closing, you cannot open them again," he told her. "Your arms feel heavy, so do your legs. You cannot feel anything. Your hands are motionless. You see nothing, you are going to sleep." He told her, "You will have consciousness of the most strenuous pains, but they will be so mild that you will not be able to differentiate them from the lightest ones." Remarkably, Cajal claimed that the treatment worked. During childbirth, the woman remained hypnotized. It was as though her nerves had been paralyzed—she felt no pain. Half an hour after giving birth, she woke up with a vigorous appetite, and after only five days she resumed ordinary life. A few years later, Cajal published a report on the subject of this experiment, the first study of hypnosis during childbirth to appear in Spain. The woman was, in fact, Cajal's wife, Silveria, and the child was their fifth, a son named Jorge.

Cajal's reputation as a healer spread across Valencia, he claimed. Crowds of disturbed people flocked to his small apartment. A lucrative new path lay open to him. But once again, Cajal remained faithful to his first love; he was devoted to "the religion of the cell."

CHAPTER 14

"Moved by Faith"

Cajal published six fascicles of his histology manual during his time in Valencia, from 1884 through 1887. Together, they offered a survey of nearly every type of tissue in the human body: epithelial, crystalline, corneal, blood, conjunctive, adipose, cartilaginous, bone, muscle, and dental. All that was left to explore was nervous tissue, the most difficult to study. Because nervous tissue does not produce its own glucose, it cannot survive without oxygen, and so, when an organism dies, the nervous tissue deteriorates first. It is so delicate that one careless touch will destroy its structure; for this reason, Cajal sometimes sectioned the brain while it was still in the skull. He consulted manuals and textbooks to help guide his research, but he could not find a clear and accurate image of nervous tissue anywhere. One of those books, hailed as "the fullest and most systematic ever undertaken" by contemporary reviewers, contained only six illustrations. The images in German textbooks, though sophisticated, seemed "perfectly irreconcilable" with one another. In 1850, the Swiss researcher Albert von Kölliker, the leading authority in the field, had declared the nervous system the only unknowable structure in the human body. Little progress had been made since then.

The anatomical study of organic tissue could progress only as far as technique and technology would allow. In the mid-seventeenth century, Marcello Malpighi, the first scientist to attempt a microscopic investigation of nervous tissue, boiled the brain in hot water to separate the nerves into layers, which he stained with ink. Malpighi saw spheres and clusters, which he assumed were glands secreting "animal spirits." The objects turned out to be artifacts. Indeed, many microscopists who studied nervous tissue in the seventeenth and eighteenth centuries reported the presence of granules or globules. The most common histological technique involved macerating tissue in water, and what they saw were probably droplets left on slides and flattened by coverslips. ("I have already told thee," said Don Quixote to Sancho Panza, debating whether or not a mill in a river is really a fortress, "that all things are transformed and changed by the power of attachment.")

The first anatomist to accurately observe nervous tissue under a microscope was Christian Ehrenberg, who studied leeches that he had skimmed from a pond in 1833. He saw "ganglionic corpuscles," roundish forms resembling cell bodies, and branching fibers, or "projections," that might have been attached to them. Anatomists asked if and how the bodies and fibers were connected to each other. Though cell theory had established that all life is composed of anatomically independent cells, nervous tissue was thought to be the sole exception. Ehrenberg also assumed that the structure of the nervous system was like that of the vascular system, which had been discovered in the previous century: an interconnected network of capillaries, arteries, and veins. Almost every researcher in the middle of the nineteenth century believed that the fibers of the nervous system were connected. When Joseph von Gerlach examined nervous tissue with his carmine staining technique, he also saw a network, and his illustrated descriptions, which appeared in a widely distributed textbook, further entrenched the idea. Gerlach held that the nervous system is composed not of independent cells but of a single, continuous fiber. He referred to this structure as a *reticulum*—a fine, intricate net. He even claimed that he had observed a reticulum in the most exalted nervous tissue of all: in the human brain.

Absent definitive evidence, some researchers refrained from conclusions, while others fashioned theories explaining the unknown. Everyone was arguing about something that no one could clearly see. Cajal tried the leading staining techniques—carmine and hematoxylin— but they colored only the bodies of nerve cells, not their fibers. Under the microscope, the tissue resembled "a tangled thicket," Cajal recalled, "where sight may stare and grope ever fruitlessly, baffled in its efforts to unravel confusion, and lost forever in twilit doubt." A single sample of nervous tissue, as small as the size of a fingernail, contained so many elements that all he could see was an indistinguishable mass of color. Nothing was revealed. He also experimented with an innovative technique known as microdissection, which required fixing tissue in osmium tetroxide, an expensive and dangerous chemical contained in white crystals, sold in flame-sealed vials because the odor was so foul—and if any chemical vapor reached the researcher's eyes, it would begin to fix the cornea. After staining, the sample was mounted on a slide and placed onto the stage of the microscope against a black or white background for contrast, while the researcher painstakingly teased apart the tissue with a tiny needle to see if it was possible to isolate a single cell.

The German anatomist Otto Karl Friedrich von Deiters, pioneer of the microdissection technique, managed to distinguish between two distinctive kinds of fibers projecting from the bodies of nerve elements. Under the microscope, he saw that one kind—which he called "protoplasmic prolongations"—tapered and branched out repeatedly; they were later termed "dendrites," from the Greek root *dendr*, meaning "trees." The other kind of fiber—which Deiters called "axis-cylinders"—was thinner, longer, branched out only at right angles, and did not taper; the term was later shortened to "axons." Even though he examined hundreds of samples from all areas of the nervous system using the microdissection technique, Deiters could find no evidence of anastomoses, the connections between different fibers. No network could be seen to exist.

And yet, despite his own findings, Deiters could not abandon the idea of a reticulum. "Every modest, sensible working researcher without preconceived notions must come to the same conviction," he concluded. When he articulated a theory of nervous system function, he could not escape the possibility of anastomoses once again—"one falls back upon the fine nerve fibres," he said, "which ramify and therefore may well unite."

Cajal soon abandoned microdissection; the technique required "the patience of a Benedictine," and he was eager for answers. To be a microdissectionist, he wrote, was to be a "tormented soul." He wanted a more dramatic confrontation with an individual cell.

IN THE MIDST OF HIS RESEARCH INTO THE NERVOUS SYSTEM, IN January 1887, Cajal was summoned to another *oposicion*, only this time as a judge. Near the Plaza del Sol, the bustling center of Madrid, was an unofficial "Biological Institute," where young researchers gathered to share their work. There, Cajal met Luis Simarro, a practicing psychiatrist who had just returned after five years in Paris, where he studied neurology and psychiatry with masters such as Charcot and Louis-Antoine Ranvier, whose textbook served as Cajal's "technical bible." French neuropathologists, following Ranvier's program, used microscopes to study the effects of illness on brain tissue. Simarro had learned the latest techniques and brought slides of preparations home with him to show his fellow Spaniards.

Another address that Cajal would never forget is 41 Arco del Santa María Street, Madrid, where Simarro had his private laboratory. The laboratory was even depicted in a painting, *An Investigation*, by the Valencian impressionist Joaquín Sorolla. Wearing a white laboratory coat, Simarro is seated at his worktable, under the light of a shaded gas lamp, holding tweezers between his thumb and forefinger and focusing his dark eyes through his thin, gilded spectacles. In the foreground is a microtome, its blade like a guillotine waiting to make its cut. On the table is a large glass bottle of a bright red-orange solution labeled potassium bichromate. Crowding behind Simarro are six men in white dress collars, having come straight from work, watching over his shoulder with expressions of inquisitiveness and amazement, as though he were performing a miracle. Cajal would have been among such spectators in real life.

One new technique in particular caught Cajal's eye. In 1873, an Italian anatomist named Camillo Golgi was experimenting in the supply closet of the mental hospital that he directed in a town outside Milan. Golgi had become frustrated by the available techniques for staining nervous tissue, and he devoted himself to finding more effective ones. "I have obtained magnificent results and hope to do even better in the future," Golgi wrote in a letter to a friend, touting his method as so powerful and clear that it could reveal the structure of nervous tissue "even to the blind." He called it the black reaction. One of Golgi's students recognized "the marvelous beauty of the black reaction . . . [which] allows even the layman to appreciate the images in which the cell silhouette stands out as if it had been drawn by Leonardo."

Simarro showed Cajal slides that had been prepared using Golgi's method, which Cajal had never seen before. Unlike the technique using carmine and hematoxylin, this method stained only a small fraction of the nervous tissue to which it was applied, effectively pruning "the inextricable thicket." "On the perfectly yellow background," Cajal recalled,

> *sparse black filaments appeared that were smooth and thin or thorny and thick, as well as black triangular stellate or fusiform bodies! One would have thought that they were designs in Chinese ink on transparent Japanese paper . . . The eye was disconcerted, accustomed as it was to the inextricable network in the sections stained with carmine and hematoxylin where the indecision of the mind has to be reinforced by its capacity to crit-*

icize and interpret. Here everything was simple, clear, and unconfused . . . The amazed eye could not be removed from this contemplation. The dream technique is a reality!

As soon as he returned to Valencia, Cajal began using the Golgi method himself. Nervous tissue was hardened for several days in potassium bichromate—the bright-red solution in the foreground of Sorolla's painting—then immersed in a bath of silver nitrate, provoking a chemical reaction that precipitated microcrystals of silver chromate, which, unlike synthetic dyes such as carmine and hematoxylin, clung to both the bodies *and* fibers of nervous elements. Cajal followed the protocol, placing slices—no thicker than a cubic centimeter—in different beakers of potassium, where, after fifteen minutes, they began to darken like photographic prints. He removed sections at different intervals, of between one and four days, and dried them with blotting paper before placing them in beakers with silver nitrate, where a brownish cloud immediately formed. The tissue remained immersed in silver nitrate for at least twenty-four hours. Cajal then dehydrated, embedded in paraffin, sliced, and mounted the samples on slides, and then rushed to the microscope to examine his handiwork.

And yet, for all Cajal's enthusiasm, the scientific community was skeptical of the Golgi method, which was by then fifteen years old. The 1870s were a time when many journals publicized new histological methods, which often proved unreliable. Though reviews appeared immediately in Italy and abroad, for years only Golgi and his students used the method. Some researchers even refused to call the Golgi technique a "method" because good results seemed purely accidental. If two sections were prepared in exactly the same way, one might be excellent and the other worthless; a hundred preparations might be necessary to produce a single workable sample. Sometimes the stain yielded stunning preparations, Ranvier admitted, but more often than not irregular silver deposits left black rings on the sample. Some claimed that what the black reaction stained were not nerve elements themselves but vessels that carry lymph. Fibers were often incompletely stained, precluding observers from tracing them.

By the time Cajal encountered the black reaction, most histologists had abandoned the technique. Critics characterized it as "capricious and highly uncertain"; Cajal was the lone researcher who believed in the method's potential. "For Cajal," one of his disciples later said, "technique and science were the natural, spontaneous result of a secret activity moved by faith." Taming the reaction meant mastering the relationship between potassium dichromate and silver nitrate—the two reagents—by optimizing the volume, concentration, and duration of chemical baths. Golgi's method called for immersing tissue sections in potassium for fifteen to forty days (the length of time would depend on the temperature of the surroundings). Cajal understood that the task of refining the stain would require great patience. "Ideas do not show themselves productive with those who suggest them or apply them for the first time," he wrote, "but with those persevering workers who feel them strongly and put all their faith and love in their efficacy." Simarro himself dismissed the technique for his own research purposes; the black reaction "*suggests* rather than demonstrates," he said. To Cajal, the implication was that Simarro thought that the results were illusory. Cajal was sure that they were real.

Cajal began to believe that, with the help of the black reaction, he could answer the most important question facing anatomists of his generation: How are the elements of the nervous system related? Are there anastomoses between the fibers, as the reticular theorists said? Do they form a singular, continuous network? If Cajal could prove that the fibers were not linked, he would establish that the nervous system is made up of cells, like all the other systems in the body. The nervous system would be the final territory conquered by cell theory. He thought that, like Cervantes or Velázquez, in order to advance Spanish culture he needed to somehow become a celebrity. In a letter to a fellow anatomist, in 1888, Cajal announced that he had abandoned his study of other tissues to focus solely on the nervous system.

IN JANUARY 1887, CAJAL DECIDED TO APPLY FOR A CHAIR IN THE NEW field of normal and pathological histology. Two vacant chairs became available to him: one in Zaragoza and one in Barcelona. At first he leaned toward Zaragoza, wanting to be near his family; by this, he meant his mother, his sisters, and Pedro, who had followed in Justo's footsteps and worked as the town surgeon in Fuendejalón, about thirty miles outside the city. But Cajal knew that his old friendships in Zaragoza would demand energy and time, and that that kind of social life was incompatible with his scientific aims. This, too, was the problem with Valencia—between teaching, short-story writing, psychology experiments, and *tertulias*, Cajal complained in a letter to a friend, he had "no time to work" on his own research. His black reaction studies of the nervous system would now become his only priority. Barcelona offered better resources, including a fully equipped modern laboratory and more opportunities to publish articles—and, perhaps most importantly, it was farther from his father.

CHAPTER 15

"Free Endings"

In *Don Quixote*, Cervantes celebrates Barcelona as "the seat of courtesy, haven of strangers, refuge of the distressed, mother of the valiant, champion of the wronged, abode of true friendship, unique in both beauty and situation." Barcelona was where Justo's journey had ended; the rebellion in the city nearly destroyed him. Barcelona is also where Don Quixote is finally vanquished by the White Knight, forced to renounce his fantasy before returning to his native village to die. In the decades since the departure of Cajal's father in 1848, the population of Barcelona had increased to five hundred thousand. Though far less populous than London, the city of Barcelona was twice as dense. For decades, unskilled laborers and journeymen had been relocating to Barcelona because of drought, plague, war, or other misfortunes, bringing their relatives with them, huddling in single-room shacks called *barracas*. A housing report from around that time expressed shock that human beings could even live in such squalor. When Cajal arrived, Barcelona was in the midst of a transformation in advance of the 1888 Universal Exhibition, an opportunity for the newly industrialized Catalan capital—and not its rival Madrid—to show itself on the world stage.

Despite Cajal's new position, his salary did not increase. He, Silveria, and their now five children moved to the Raval, the most crowded neighborhood in the city, home to working-class and migrant families where developers divided five-story buildings into tenements, and industrialists built slaughterhouses, tanneries, and other factories deemed too dirty and dangerous for wealthier areas. Dark, twisting streets became narrower and narrower before suddenly opening into squares where residents hung bedsheets from iron window grills, the universal banner of the proletariat. Cajal's apartment was in an old wedge-shaped building on the corner of Lluna and Riera Alta Streets. In the Middle Ages the plot was home to a Capuchin monastery, where monks had taken vows of poverty. The apartment was miserable and narrow. "It was difficult to find a chair in which to seat yourself," a colleague of Cajal's recalled. "They were few, and they were occupied by books and magazines."

Having never taught pathological anatomy, Cajal needed to familiarize himself with the subject. Down the street from his apartment was the Santa Cruz Hospital, a medieval stone fortress that housed the medical school, where he spent hours each day in the decrepit dissection room autopsying corpses stricken with tumors and infections, which were given to him by local hospitals and veterinarians. He made important contributions to the study of cancer, and his research served as material for his second textbook, *Manual on Pathological Anatomy*, which appeared in installments between 1889 and 1890. Less ambitious than his first manual—"without pretensions," he wrote, and "dedicated to students"—the new one relied largely on the research of others, though many of the descriptions and illustrations were Cajal's own. Lack of time kept him from producing a more original work, he said; his research on the nervous system consumed him. The dean of the medical school threatened to punish Cajal for sneaking department microscopes home with him at night. One colleague who lived near him used to chat with him before class. "I have four children," he recalled Cajal once saying, "and my fifth is the microscope." (Actually, Cajal had five children at the time, which either he forgot or his friend misremembered.)

Cajal performed experiment after experiment, altering a single variable each time and carefully observing the effect. His instinct was to intensify the reaction. He stained the tissue with silver nitrate twice, even three times, for longer periods of time and in more highly concentrated formulae. The goal was to trace nerve fibers from their origin to their endings, to see if they touched. Though his "double impregnation" technique increased silver deposits, which risked further obscuring the image, it allowed the stain to penetrate thicker sections of tissue, which showed more of the fibers. After several weeks of practice, his results started to improve. After several months, the most elusive cells appeared, splendidly stained. Cajal consulted Golgi's masterpiece, *On the Fine Anatomy of the Nervous System*, published the year before, which Simarro had given him. (Anyone with a basic education in Spanish could read Italian, said Cajal.) In his book, Golgi described parts of the nervous system either for the first time or in unprecedented, revelatory detail. "There is certainly to be found a very widespread network of filaments anastomosing one with the other," Golgi claimed, bolstering the reticular theory. He also acknowledged that some of the thinnest fibers had yet to be seen.

Initially, Cajal's results confirmed Golgi's own, but he was not content with an incomplete picture. Why did he not abandon the black reaction, like his colleagues, and simply accept the reticular theory? Cajal did not want to make the same mistake as he had with muscle fibers in 1880, seeing a network where there was none. According to Golgi, fibers that join the reticulum "lose their individuality," an assertion that Cajal found unconvincing, perhaps even offensive. He considered his search for the independent nerve cell to be "an act of rebellion."

THE RAMBLA IS LIKE THE CENTRAL NERVE OF BARCELONA, CONSISTING of five distinct sections extending all the way to the sea, its broad promenades lined with plane trees. Every social class could be found there, from fishermen in slouching red caps, their jackets slung over their shoulders, to ladies in brightly colored skirts, their faces half veiled by *mantillas*. The streets were so crowded that women had no room to fan themselves, and the flow of pedestrians continued from morning until late into the night. Paperboys ran around hollering the names of the dailies, while flower girls sold fragrant bouquets. Men in long, dark cloaks hawked birds in stacked cages. The coats of the mules pulling tramcars were clipped in patterns so fine that they almost looked embroidered. The year Cajal arrived in Barcelona, the sumptuous Liceu opera house, modeled after La Scala in Milan, was staging its first production of *Carmen*; a young architect named Antoni Gaudí was touching up the spectacular facade of the Palau Güell; and the Socialist Workers' Party held its first congress, promising "constant and rude war" against the bourgeoisie. Catalan separatists plotted secession, whispering about "cutting ties."

"Modern winds that bring something of the future" was how the poet Rubén Darío described the atmosphere of the Rambla at that very moment. Cajal barely looked up from his dead tissue to feel the breeze. He disparaged the kind of man who "needs at all hours, in order to feel alive, the tumult of the streets, the emotion of the theater, or the mumbling of the *tertulia*." In Barcelona, Cajal adopted a program of "severe abstention and renunciation" reminiscent of his father's ascetic approach to life. Careful not to slacken "the creative tension of the mind," he avoided gossiping and reading newspapers, ceased writing short stories, abandoned the study of hypnotism, and even quit playing chess. He exercised his will not because he was uninterested in the world around him but precisely because he knew himself to be so distractible.

"To bring scientific investigation to a happy end once appropriate methods have been determined," Cajal counseled in a scientific advice book written a decade later, "we must hold firmly in mind the goal of the project." When he was not looking through the microscope, he devoted every waking moment to focusing on the images that he had observed. "One must achieve total absorption," he said. "Expectation and focused attention are not enough." "When he worked," a colleague claimed, "he did not think about anything else other than what he was doing; the rest of the world, for him, was indifferent."

The only leisure activity that Cajal allowed himself was the *tertulia*. "The laboratory man requires the invigorating atmosphere of the club to keep him from intellectual isolation or from falling into eccentricity," he said. His primary haunt was the Café Pelayo, which, while not luxurious, had ample space for many *peñas*—or circles—to meet at once. Men played *ombre*, a fast-paced trick-taking card game, and shot billiards in the basement. Among the *peñas* sat Gaudí, before he turned to religion and designed one of the most famous cathedrals in the world. (Cajal and Gaudí were born in the same year, had acquaintances in common, and may have met, though Cajal never mentioned him.) Cajal's *peña* was made up of fellow university faculty, specialists from different fields, most of whom taught in the sciences. When the Pelayo closed, Cajal's *tertulia* migrated to a nearby café, the brand-new Great Café of the Nineteenth Century—it boasted a large open room surrounded by windows, commonly known as "the Aviary," featuring a small stage with a grand piano for daily concerts under a sparkling chandelier. Silverware shone on white tablecloths. Patrons signaled waiters by clapping their hands. Both cafés were located near the Plaça de Catalunya, where, as part of an attraction for the Universal Exhibition, a panorama showcased newfangled "moving images" of the Battle of Waterloo for only a cent.

A *peña* was like an organism, Cajal thought, with its own specialized, adaptive roles. One man might talk about science, another business, and another current events; one man might demonstrate a passion for art, while another might supply levity and humor. From time to time, someone might even pretend to be a conservative, defending religion, the monarchy, or social norms, just for the sake of variety. To avoid pedantry, experts were forbidden from discussing their own fields. The *peña* had to be kept small, Cajal said, since men expressed honest opinions only in an intimate setting—while in the presence of an audience they tended to perform. The most important role was played by a silent arbiter, to whom the debaters instinctively turned for affirmation. *Contertulios*—*tertulia* participants—had to possess complementary temperaments; Cajal considered anyone who disturbed the balance at the table to be more dangerous than a gossip or a bully. *Tertulias* were as sensitive as chemical reactions. "To conserve the cordial atmosphere of a *peña*," Cajal explained, "one must never escalate arguments nor try to be right in every difficult situation." At the end of each *tertulia*, strictly limited to one hour, he asked himself: "Have I learned something noble, useful, or agreeable? Do I leave the gathering better or worse than when I entered? Will the mental exhaustion and emotion provoked by irritating discussions hinder or interfere with my daily work?"

Though his research consumed him, Cajal continued his teaching duties, priding himself on never missing a class. One of his private students, Josep Maria Roca, later wrote a memoir of his time with Cajal during this phase. "Withdrawn, isolated from the world," Roca wrote, "[Cajal] saw only through the objective of the microscope." While he worked, he always looked surly, often mumbling to himself, sometimes interjecting "a distinctly Aragonese word or phrase." It was as though he did not realize there was anyone else in the room. "He was so absentminded," Roca explained, "so obsessed with his fixed idea, that he was completely withdrawn from the world around him, isolated inside the diving suit of his constant preoccupation." Some of his techniques seemed crude; nervous about placing a good preparation on a slide, he would clean the glass first by hand, removing any grease, hair, or dust, then breathe on it and dry it with the left tail of his jacket, the same one that he wore each day. Once, Cajal accidentally took a student's hat from the entrance hall and wore it for an entire day. Another time, when Cajal and his students planned a trip to Badalona to catch a species of newt unique to Spain for their experiments, the professor was hours late and, since he had not eaten yet, insisted on stopping for breakfast. At the restaurant, he left his handkerchief on the table and took the napkin home instead.

SINCE ANCIENT TIMES, THE CITY OF BARCELONA HAD BEEN ENCIRCLED by walls, around which a militarized zone had been established to protect the city. The Romans built the walls when they founded the city, the Visigoths reinforced them centuries later, and, in the nineteenth century, so did the French. For Barcelonans, the walls were a symbol of occupation. Finally, in the second half of the nineteenth century, the government decided to demolish the walls and expand the city. The Catalan engineer Ildefons Cerdà was selected to lead the project. Cerdà, who studied modern hygiene and technology, conceived of a healthier and more efficient city, including streets that were wider and straighter, a mathematically equidistant length apart, with chamfered corners to increase visibility for drivers making turns. Cerdà was primarily concerned with the well-being of the individual citizen, and his plan was inspired by the utopian socialist Étienne Cabet, whose ideal city consisted of equal cells.

The new area of Barcelona, called l'Eixample, or "the Extension," connected the old city to the surrounding towns, allowing for much-needed population dispersion. Once buildings were constructed, land prices soared. Developers bought up as many plots as possible; after they had developed a block or two, they would sell off the rest to other developers. Lots were halved and halved again, becoming smaller and smaller, a Zeno's paradox of real estate. The first plots to be developed were those closest to a natural water source, as on Bruc Street, which was on the petit bourgeois side of the new railway tracks that bisected the Eixample. This was where Cajal and his family moved.

Buildings in the Eixample were shaped like octagons, with two sides open, allowing for maximum passage of sunlight and air. In the middle of every block, a space was left for a communal garden. Cajal raised and kept animals there for use in his experiments. Bruc Street apartments were "sunny, spacious, and salubrious." By Cajal's standards, his new home was luxurious—it even had an extra room that served as his most comfortable laboratory yet.

THOUGH THE BLACK REACTION DRAMATICALLY REDUCED THE NUMBER of nerve elements visible on a microscopic slide, those elements were still so densely packed that their fibers appeared inextricable from one another. Traditionally, researchers studied nervous tissue from adult humans who had died naturally, after a normal life span. The problem was that in the adult nervous system, the fibers were already fully grown and, therefore, the most structurally complex. Looking for a solution to this problem, Cajal turned to embryology—also known as ontogeny—which he had first read about in a college textbook. "If we view the natural sequence in reverse," Cajal explained, "we should hardly be surprised to find that many structural complexities of the nervous system gradually disappear." In the nervous systems of younger specimens, cell bodies would in theory be simpler, fibers shorter and less numerous, and the relationships between them easier to discern. The nervous system was also well suited to the embryological method, because as axons grow, they develop myelin sheaths—insulating layers of fat and protein—which repel the silver microcrystals, preventing the enclosed fibers from being stained. Younger axons, without thick sheaths, more fully absorb the stain. In addition, mature axons, which sometimes grow to be a few feet long, were more likely to get chopped off during sectioning. "Since the full grown forest turns out to be impenetrable and indefinable," he wrote, "why not revert to the study of the young wood, in the nursery stage, as we might say?"

The first structure that Cajal chose to investigate was the densely wrinkled part of the brain that sits at the back of the skulls of vertebrates, which Aristotle, thousands of years ago, named the cerebellum, or "little brain." By lesioning the brains of animals, nineteenth-century researchers had shown that the cerebellum controls the coordination of movement. Compared with other brain regions, the cerebellum was relatively easy to study, with its uniformly organized structure divided into regular layers and populated with distinctly classifiable cell types. The cerebellum contains some of the largest cells in the nervous system, twenty microns wide and thirty-six long, named Purkinje cells after the Czech researcher Jan Evangelista Purkinje, who discovered them. Their bodies are shaped like teardrops, some researchers observed, and their fibers are tangled in a "complex and bizarre manner." In other words, under a microscope, a Purkinje cell was difficult to miss. Were Purkinje cells a special exception to cell theory, connected to other cells by their fibers, or were they completely independent?

Golgi's first black reaction studies had focused on the cerebellum, and so that was the territory on which Cajal sought to challenge him. Whereas Golgi had studied mammals, Cajal chose to work with birds, a classic model for embryologists because of the ease of incubating eggs. On the evolutionary scale of vertebrates, birds were considered lower than mammals, implying that the structure of the bird nervous system would be less complex. Cajal figured that knowledge about more basic nervous systems would help him understand those of more advanced vertebrates—and eventually, and most importantly, human beings.

At the age of thirty-six, his beard sprouting wayward fibers and his head prematurely balding, Cajal found himself incubating eggs, just as he had loved doing when he was a child. This time, instead of waiting to witness "the metamorphosis of the newly born," Cajal cut into the eggshell after a few days and removed the embryo. Embryonic tissue was too delicate to withstand pressure from the clasp of the microtome. So holding the block of tissue between the thumb and forefinger of his left hand, he cut sections with a razor blade, applying his training as a barber in a fashion that he could never have foreseen. A private student of Cajal's in Barcelona, who worked in the laboratory with him, attested that his hand-cut sections—often between fifteen and twenty microns thick—were as perfect as those cut with any machine.

In April 1888, Cajal prepared samples from the cerebellum of a three-day-old pigeon embryo. Through the microscope, he fixed his gaze on a clear, fine axon as it arced downward from its base—a soft, conical bulge on the cell body—and followed the black line, transfixed, as if he were still a boy following the course of a river. The axon curved, running alongside the layer of cells below it, until it started to branch, with some of the branches turning back in the opposite direction. In Cajal's eyes, the Purkinje cell, stained with the black reaction, resembled the "most elegant and leafy tree." He traced one branch all the way to its end, where it approached the "pear-like" bodies of other Purkinje cells, forming a kind of "basket." Though intimately and inextricably related, the "pear" of one cell and the "basket" of another never touched. Cajal sensed a "new truth" arising in his mind: nerve cells ended freely. They were distinct individuals.

Golgi's illustration of the cerebellum—artistically stunning in its own right—depicted the axons of Purkinje cells running their course and then simply fading away, reflecting his uncertainty about their endings. In Cajal's depiction of a similar tissue sample, he rendered the endings of fibers more boldly, a clear statement of their independence. Golgi positioned the elements more deliberately, with almost equal distances between them, organizing the image in a way that appeared schematic rather than realistic. Cajal's illustration, on the other hand, appears more chaotic, more true to life.

The job of an anatomist is to restore lost depth from a compressed image. Tissue samples were sliced consecutively, stored in numbered wells, then mounted in order on the same slide, so that the researcher could observe one after the other, visualizing their relative positions in space. If Cajal had a special talent, it was his visual memory; the same person who reproduced byzantine maps was able to hold the images of six or seven sections in his mind at once and synthesize them into a full-scale model.

But Cajal could not yet be sure of his results. More preparations were necessary to complete the anatomical picture of the tissue. He used other stains to effect a comparison with his black reaction results. Furthermore, to corroborate his findings, he had to prepare even more samples than he would have with any other stain, because of the black reaction's reputation for producing artifacts. He was overturning the established view, and so others were sure to challenge his findings.

In search of further proof against the reticular theory, Cajal examined the retina, the membrane that lines the back of the eyeball and sends impulses via the optic nerves to the brain. Like the cerebellum, the retina is layered and relatively uniform in its composition. In the 1860s, after comparing the retinas of diurnal and nocturnal animals, one anatomist suggested that certain cells shaped like cones and others shaped like rods might be responsible for vision under different light conditions. Underneath the rods and cones, four distinct cell types were identified: horizontal, bipolar, amacrine, and ganglion. Well into the late nineteenth century, the understanding of the internal connections and the pathways between them was still vague. As in all areas of the nervous system, most researchers presumed the existence of anastomoses.

Cajal applied his improved black reaction to the retinas of mammals, birds, and amphibians. "We have never been able to see an anastomosis between ramifications of distinct [axons], nor the filaments emanating from [axons]," he reported. "The fibers are interlaced in a very complicated manner, engendering an intricate and dense plexus, but never a network . . . Every time the impregnation has been fine and complete, we have seen free endings." Cajal published forty-three papers during his four years in Barcelona. He wrote about the structure of the retina in birds, the nerve endings of the muscle spindles in frogs, the structure of the cerebellum, the texture of the muscle fibers of the heart, the structure of the optic lobe and of the optic nerves of birds, and the spinal medulla of birds and mammals. Independent nerve cells were everywhere. "By this time," he said, "I nourished the certainty that I was not mistaken."

THE UNIVERSAL EXHIBITION IN BARCELONA WAS OFFICIALLY INAUGU-rated on May 20. Five thousand people attended the ceremony, including Maria Cristina, the queen regent of Spain, who sat on the podium with her two-year-old son, the future king Alfonso XIII, in her lap. Twenty-seven countries participated in the exhibition; over the course of about 250 days, one million people would attend from around the world. A hunger artist, his stomach numbed by a magical elixir, performed feats of strength during a thirty-day fast; a woman shot herself from a cannon; a three-hundred-foot-tall captive balloon hovered in the air. There were concerts, fireworks, and roller coasters; poetry competitions, floral-arrangement contests, and bullfights. There were pavilions dedicated to the fine arts, industry, agriculture, and the sciences, with a natural history exhibit, a geology museum, and a botanical garden. Not all visitors were impressed. "The Science Pavilion in itself is capable of disappointing the pretensions of the most exaggerated," one commentator wrote in a Barcelona newspaper. "Four pots of pills and drugs, a desert of dynamite, and some basic school exercises constitute Spanish science in its entirety."

In the circular hall of science, among the cameras, telegraphs, telephones, medicinal plants, chemicals, and experimental instruments, Cajal stood at a table, exhausted and disheveled, with a microscope trained on his best slides, for spectators to come and marvel at a sight more wonderful than any other: the infinitely small world inside the brain. The organizers of the exhibition must have agreed—they awarded him a gold medal.

Cajal was eager to establish his priority in making discoveries about the nervous system. He wrote up his findings and sent the papers to local medical journals, but the perpetual delays in their publication schedules frustrated him. He decided to found his own journal, *The Trimonthly Review* ("*Revista*") *of Normal and Pathological Histology*, to serve as a vehicle for his work. His publisher was not an established house but the Casa de Caritat, a long-standing almshouse where the neediest and most unfortunate people in Barcelona learned trades to support themselves, including printing. Cajal was his own editor. He insisted on using the highest-quality paper and ink to make a good impression on his readers. The expenses "entirely swallowed up" his income, he wrote.

Most petit bourgeois families, including professors like Cajal, hired maids to help with the housework. Maids typically woke up as early as 5:00 A.M., lit the stove, cleaned the house, shopped and ran errands, cooked breakfast, dressed the children, served lunch and dinner, and went to and from the river to wash clothes. But Silveria did all these chores to save a couple hundred pesetas a week. She also educated their young children by herself—a sixth child, Pilar, was born in 1890, and a seventh, Luis, in 1892—and took them to mass on Sundays, without their father. "She divined no doubt that there was maturing in my brain something unusual and of decisive importance for the future of the family," said Cajal. He referred to Silveria as the family's "spiritual director."

Though "not sullen or disagreeable," according to her son Luis, Silveria was an "authoritarian" mother, and her first priority was always maintaining a stable and tranquil home, shielding Cajal from the stress and anxiety of domestic life. Only the most serious questions about the children were brought to his attention. "Preoccupied with my studies," Cajal admitted, "I hardly concern myself with affairs of the house, which run under the stewardship of my wife." Every night, the family had dinner together, a ritual that Luis recalled as happy and normal, and Cajal was warm and engaging at the table. Other than that, the children never saw him. He was not an absentee father; he was working in his laboratory, right inside the house. Their mother told them never to disturb him. "He walked around like a god," Luis said.

THE FIRST ISSUE OF THE *REVISTA*, WHICH APPEARED ON MAY 1, 1888, Cajal's thirty-sixth birthday, contains the first unequivocal evidence that nerve fibers end freely—indisputable proof that the nervous system comprises independent cells. Cajal could afford to print only sixty copies of his *Revista*, which he sent to leading scientists abroad, free of charge, accompanied by letters promoting his discoveries. He published two more issues of his *Revista*, in August 1888 and March 1889, authoring every article and etching each lithographic plate himself. He feared being exposed as "deluded" or as a "pretender." Would his story end like Don Quixote's? Would the dream be revealed as an illusion? He reviewed his findings again and again, plagued by doubt, trying to root out any errors. With his work in the hands of the masters, Cajal lived in "a state of restlessness and suspicion." None of the scientists responded. "I was a little alarmed by the silence," he admitted.

Though reticular theory predominated in the 1870s and 1880s, a few researchers had provided tempting counterevidence. In 1877, the British physiologist Edward Sharpey-Schafer, using the gold chloride staining technique, found that nerve fibers in jellyfish "do not come into anatomical continuity." They may bend toward each other, veer off course, hook around each other, run in proximity or even in parallel for a distance, and entangle themselves—yet the fibers "never actually coalesce . . . In the nervous network, the constituent fibres, although frequently in contact, never actually unite." Schafer found the new fact "as startling as it is novel" and "at first sight almost incredible." He submitted his findings to the Royal Society, and on the outside of the envelope, he wrote: "So far as I know, this paper contains the first account of a nervous system being formed of separate nerve units without anatomical continuity." The committee, doubting his findings, asked him to withdraw them.

When Cajal began his black reaction studies in 1886, the Swiss researcher Wilhelm His, a professor of anatomy at the University of Leipzig, also employed the embryological method to see if nerve fibers were continuous. His collection of twelve human embryo samples aged two to eight and a half weeks had been sent to him from laboratories throughout Europe. He examined the samples in chronological succession to see how far nerves had grown and noted the pattern of their development. "Connections of nerve cells among one another have long been recognized as illusions," he wrote in an 1886 paper, "but probably most of us believe silently that in some way the central nerve fibers serve for the mutual connections of cells." His found no proof of a reticulum. "I believe that one arrives at simplified notions," he wrote, ". . . when one gives up the idea that a nerve fiber . . . must necessarily be in continuity with it." He doubted his own observations, however. Though it was clear that embryonic nerve cells were independent, His assumed that their fibers must eventually fuse together at a later stage of development.

That same year, another Swiss researcher, Auguste Forel, a professor of psychiatry at the University of Zurich, arrived at the same conclusion about the discontinuity of nerve fibers, albeit by a different approach: the degeneration method, pioneered decades earlier by a British physiologist who severed the nerves of a frog's tongue and observed the effects of the injury over the course of three weeks. The fibers that were still attached to the bodies of nerve cells survived the injury, while those separated from the bodies decomposed. "By means of these alterations," the inventor of the degeneration method, Augustus Waller, had concluded, "we can most exactly determine the course and distribution of the whole nerve." If nerve fibers were continuous, Forel reasoned, then damage to any fiber should pervade the entire system. In experiments with the degeneration method, he severed fibers and saw that the damage was always limited to either one cell or a group of cells at most. Distant fibers were never harmed, which contradicted the prediction of the reticular theory. "It was as though scales had fallen from my eyes," Forel recalled. "I asked myself the question: 'But why do we always look for anastomoses?'" He had disproved the logic of the reticular theory, but he did not demonstrate the existence of the independent nerve cell.

For his entire life, Forel maintained that he and Wilhelm His were the real founders of neuron theory. "Well, our two papers were simply ignored," Forel wrote in his autobiography. Cajal swears that he did not know of the work of His or Forel at the time of his own discoveries, a claim further legitimized by the fact that Spain lay beyond the bounds of the mainstream scientific community, and that His and Forel themselves, despite being countrymen, were not even aware of each other.

Still, that same year, Fridtjof Nansen, a young Norwegian zoology student, launched a comparative study of the nervous system of invertebrates and primitive vertebrates, namely sea creatures, which were easily accessible in Bergen, the coastal city where he studied. As dissatisfied as any histologist was with the efficacy of available stains, Nansen visited foreign laboratories to learn new techniques, including a trip to the city of Pavia, where Golgi taught at the university. He became the first researcher to apply the black reaction to the nervous system of invertebrates, and among the first outside Italy to publish results. Nansen found no fibers that anastomosed. He wrote his dissertation in English, suggesting that he wanted to disseminate his findings widely. Though Nansen was fluent in the language, his prose is riddled with minor grammatical and typographical errors. He was busy planning to become the first person to ski across the interior of Greenland, and as he wrote his paper on the nervous system, two rival explorers had landed on its eastern coast. Four days after defending his dissertation, he set off for Greenland, abandoning histology for good.

When his foreign journals arrived, Cajal was dismayed to find that his work was either unmentioned, dismissed, or misinterpreted in support of Golgi's. "The facts described by Cajal in his first publications proved so strange," one foreign colleague later attested, "that the histologists of the era . . . greeted them with the greatest skepticism."

Most scientists did not even read Cajal's work. The *Revista* was written in Spanish, which almost no scientists could read, and its author, with his unusual name, was unknown to them.

Nonetheless, Cajal called 1888 his "Golden Year," echoing the Spanish "Golden Age." Like the explorers of legend, Cajal rushed to map the terrain and claim land. "During the sacred fever," he explained, "when one feels himself in a vein to produce, it is advisable to force one's luck, monopolizing, if possible, all the tickets of the lottery." Hour after hour, he examined sample after sample, taking notes, sketching his observations, correcting his galleys, and consulting the works of other authors. "I work when the machine calls for activity," he wrote. "I rest when fatigue notifies me. My nerves are my clock . . . Are there Sundays in Nature?" Cajal recalled barely sleeping for two weeks. "Ideas boiled up and jostled each other in my mind." His rate of discovery was dizzying. "There is no rest for the fitful obsessive in the midst of discovery," he later wrote.

Cajal forged on into the spinal cord, the long, tubular structure encased by the vertebrae, extending from the brain stem to the base of the spine. Herophilus was the first to identify that the spinal cord contains "the *neura* that make voluntary motion possible." Galen later described the structure as "a river rising from its source, extended from the brain, continuously sending forth a nerve channel to each of the parts it meets, through which both sensation and motion are conveyed." He referred to the spinal cord as a second brain.

During the Battle of Trafalgar in 1805, while urging his men on, the British naval hero Vice-Admiral Horatio Nelson was shot through the chest by a sniper. His men carried him below deck and called for the surgeon. "Ah, Mr. Beatty!" Nelson is alleged to have said. "I have sent for you to say what I forgot to tell you before, that all power of motion and feeling below my chest are gone." The bullet had struck his spine.

Studying the spinal cord was thought to be key to understanding the function of the nervous system as a whole. Sensory impulses—such as information about pain and temperature—travel along ascending pathways of the spinal cord to the brain, which coordinates a response and sends the motor signal along descending pathways, eventually stimulating muscles. Sensory fibers were known to enter the spinal cord and motor fibers were known to exit, but what happened in between was a mystery.

A transversal section—or cross section—of the spinal cord resembles a butterfly. The area of the spinal cord outside the butterfly shape is known as white matter because it consists mainly of myelinated axons, which appear white under the microscope. Contents inside the butterfly shape, made up of numerous cell bodies and dendrites, are known as gray matter. Gray matter is so dense that even under the strongest microscope it looks like an "impenetrable thicket," while white matter, due to its myelination, cannot be stained well. Golgi thought that the gray matter was an area where input and output fibers met in an immensely rich reticulum. Most anatomists agreed. From previous findings, Cajal suspected that fibers in the spinal cord ended freely, as in the cerebellum and the retina. He studied the torpedo fish, a relatively simple species, whose spinal nerves were thick, separable, and easy to study. He found independent nerve cells wherever he looked.

IN JULY 1889, CAJAL'S FOUR-YEAR-OLD DAUGHTER, ENRIQUETA, WHOM the family called "Kety," was recovering from a case of the measles when she contracted meningitis, an inflammation of the membranes of the brain and spinal cord. The child's sickbed was in the room next to Cajal's laboratory. He spent many sleepless nights with his microscope by his daughter's bedside. As a father, he was distraught; as a physician, he understood that the disease was fatal. "Continually awake," Cajal recalled, "exhausted with fatigue and distress, I developed the habit of drowning sorrows during the small hours of the night in the light of the microscope, so as to lull my cruel tortures." In the middle of the night, Cajal was sequestered in the laboratory when Silveria cried out, "Santiago! Santiago!" but received no response. She even pounded on the door, disturbing him at work for perhaps the first and only time. In the early morning, when he finally exited the laboratory, Cajal found Silveria seated on a chair with the body of the dead child in her lap, covering her own mouth to stifle her sobs.

The episode may be apocryphal. Kety's death certificate gives her time of death as 4:00 P.M., which contradicts the claim that she died at night. The story appeared in a biography of Cajal written by Francisco Alonso Burón and Garcia Duran Muñoz, the latter of whom was married to Cajal's granddaughter and had close access to the family. The two authors were also loyal Francoists, and they tailored Cajal's image to suit a Catholic, nationalist ideology. "He loved his children very much," the biographers said, "but infinitely more, without possible comparison, he loved Spain, and if placing one's country ahead of the family means being selfish, then Cajal was selfish, and we can only be left to bless such a defect, thanks to which humanity as a whole can now benefit from the deprivation imposed on his family."

Cajal's children Fe and Luis denied that the episode happened and were outraged at the defamation of their father's character. "What ignorance to the sensibility of my father!" Luis said. "He wasn't even able to work on account of other unfortunate events, much less when one of his children was ill."

The night of her death, Cajal recalled, he was chasing a particularly elegant discovery, following an axon to its very end. For the rest of his life, he associated the microscope image of the blackened nerve cell with his poor daughter's "pale and suffering" face. At the microscope, he wept.

CHAPTER 16

"Doubting Certain Facts"

Twenty-five hundred years ago, the Greek philosopher Praxagoras asked his pupil Herophilus the question from which the study of the physiology of the nervous system can be said to derive: How is the intention of movement transmitted to the extremities of the body? How are decisions communicated so as to be realized in actions? Herophilus addressed the question empirically, by dissection, and his work constitutes the first known study of anatomy. What Herophilus uncovered were cylindrical bundles of fibers, which he called *poroi*, or pathways. When he followed the *poroi*, they seemed to lead to the brain, which he assumed was the seat of intelligence.

Though he never observed them directly, Praxagoras had imagined similar structures—like arteries whose walls had collapsed—which, he theorized, might serve as conduits for the senses. He called these theoretical structures *neurons*, an ancient Greek word that appears in *The Iliad* to describe "sinews" or "tendons," whether those exposed by a battle wound or those used to make twinelike cords for sandal thongs, bows, and lyre strings. ("Then," *The Iliad* reads, "gripping the notches and ox-gut string [*neura*] together . . .") Marcus Aurelius employed the term to mean puppet strings, referring to the machinations of our impulses pulling us in every direction.

In the late eighteenth century, the Italian physiologist Luigi Galvani noticed that the legs of a recently killed frog twitched when sparked by an electrical machine. Galvani believed that electricity was a property inherent to animal bodies; his colleague and rival, Alessandro Volta, demonstrated that the body conducts electricity. What moves from the brain to the muscles? What is transmitted through the *poroi*? Galvani and Volta proved that the substance was not a substance at all—not *pneuma*, animal spirits, magnetic fluid, or ether—but rather common electricity.

Anatomists were left asking what course electrical impulses took through the nervous system, and what happened between the stimulus and the response. The greatest scientists of the era eagerly awaited images of the endings of nerve fibers in the hope that they would provide answers. "The possibility of our establishing an accurate theory of the action of the nerves," said the German physiologist Johannes Müller in 1840, "consequently rests wholly on the question of whether the primitive nervous fibers anastomose, or not." "All we are really talking about here are connections," summarized Claude Bernard. Cajal spent almost all his time alone in the laboratory trying to determine the nature of these connections, if there were any connections at all. In each of the forty-five different papers he published between 1888 and 1891, he used the word *connections* in the title or a subtitle.

WHEN GOLGI CONSIDERED THE STRUCTURE OF THE NERVOUS SYSTEM, he began with a question: How does the system function as a whole? Within a continuous network of interconnected pathways, reticularists argued, impulses could reach anywhere, traveling in every direction at once, allowing every part of the system to communicate with every other. According to Golgi, the reticulum was "clearly meant to permit the greatest variety and the greatest complexity of [the fibers'] mutual relationships." Any sensory receptor could affect an "infinite number" of cells in the central nervous system, and a cell in the central nervous system could affect an infinite number of cells on the periphery. Golgi did not believe that the actions of individual cells, conducting impulses to a limited number of other cells along discrete pathways—which he called "isolated transmission"—could account for such complexity. In Golgi's view of the nervous system, the individual cell was not the protagonist. Cell bodies served a purely nutritive function and were not active in the conduction process at all. They were passive nodes, placed at the intersections of the continuous fiber network.

Some recognized an inherent problem with this model. "If there were an anastomosis between the nerves," wrote the French Enlightenment philosopher Denis Diderot in the eighteenth century, long before the articulation of the reticular theory, "there would no longer be any order [*regle*] in the brain, the animal would be mad." There is, after all, a basic order to the functioning of the nervous system: receiving stimuli and coordinating responses. Like Diderot, Cajal objected to the reticular view of brain function because it invited chaos; he referred to the reticulum as an "unfathomable physiological sea," fed by streams of impulses from the sense organs, with motor activity flowing out. "What is the direction of the nervous impulse within the neuron?" Cajal asked himself. "Does it spread in all directions, like sound or light, or does it pass constantly in one direction, like water in the watermill?" Cajal was sure that electrical impulses followed discrete pathways, like the courses of rivers that so obsessed him as a boy.

While observing nervous tissue, Cajal tried to ground himself in the anatomical facts alone, but such questions inevitably arose in his mind. He set out to articulate a general theory of impulse conduction in vertebrates and invertebrates, both in adults and embryos, "applicable to all cases without exception." In 1889 and 1890, while studying the olfactory bulb and the retina, he noticed that dendrites were always oriented toward the external world, while axons faced the nervous organs inside the body. This seemed like common sense: the olfactory bulb and the retina are sense-gathering organs, receiving information from outside and relaying it to the brain's decision-making centers. Cajal inferred that the path of conduction within the nerve cell must lead from dendrites to axons, or else the fibers of the cells would not be oriented that way. Nervous impulses travel in one direction only, he hypothesized, from dendrites to the cell body to axons. He called this direction of impulses the law of dynamic polarization. (Interestingly, Cajal used a similar term to describe the workings of his own mind— "cerebral polarization," by which he meant the orientation of his thoughts in one direction only, toward a solution or goal.) He saw his theory as one of "simplicity and elegance," creating "perfect harmony" within a complex system and offering researchers an exit from the labyrinth.

IN THE SECOND HALF OF THE NINETEENTH CENTURY, MOST HISTO-logical research took place at universities within a circular region in central Europe, encompassing Berlin to the north, Bern to the south, Bonn to the west, and Breslau to the east. But with the increasing prominence of the telegraph and railroad, European science became more interconnected, and innovations in printing technology allowed for better illustrations, including some in color, in science periodicals. "Nothing convinces like things actually seen," said Cajal, speaking from a lifetime of experience. In order to relay his findings, he decided to apply for membership in the German Anatomical Society, the most important organization of anatomists and histologists in Europe, which met each year in a different city. As a young nation, Germany wanted to hold up science and intellectual prowess as evidence of its modernity. Meetings provided a forum for members to debate problems, explain methods, and demonstrate findings. "[Cajal] went to the mountain," his student Roca said, "seeing that the mountain would not come to him."

Unable to secure a grant, Cajal marshaled all his savings, which amounted to five hundred pesetas, money he had earmarked to fund more research. He sold the remaining first-edition copies of his histology manual at a loss and the rights to the second edition to his publisher.

Before leaving for Berlin, Cajal published a review article in the Barcelonan journal *Practical Medicine* titled "General Connections of the Nervous System," a *pronunciamiento* for the record, declaring the independence of the nerve cell and summarizing his interpretations of his findings. One of the colleagues to whom Cajal sent his work was Golgi himself, whose findings Cajal was striving to invalidate. Golgi had not performed histology research for years; his focus was now on infectious disease. When Cajal had sent Golgi the March 1889 issue of his *Revista*, Golgi sent back a paper on malaria. He did not take the insignificant Spaniard seriously.

Meanwhile, Cajal challenged Golgi's claim to every micrometer of territory in the nervous system. On August 27, 1889, he wrote to Golgi to inform him, "My discoveries in the embryonic spinal cord are so surprising that I have the absolute necessity of showing my preparations . . . Fortunately my preparations are so clear, so analytical, that doubting certain facts is absurd." Those "certain facts"—the free endings of nerve fibers—threatened to demolish Golgi's view. In an attempt to be chivalrous, Cajal addressed Golgi as "Sir and dear colleague," while at the same time dismissing his opponent's position as "absurd." Golgi, known to be a timid and humble man, found Cajal's behavior aggressive and pretentious. "I intend to visit you in Pavia and show you the preparations realized with your admirable method," Cajal continued in the summer of 1889. "Also I would desire to examine [your preparations] and hear your sage advice . . . I am thinking of going to Berlin the month of next October."

By the late nineteenth century, international scientific conferences had become routine. And yet most of Cajal's colleagues had never met a Spanish scientist before. His dark hair and olive skin made him stand out as exotic to them; once again, he was seen as the *forano* (foreigner), as he had been in the town square of Ayerbe as a child. He arrived at the hall at the University of Berlin hours early to set up his station. On the table, beneath the giant windows, were a few state-of-the-art microscopes—Cajal brought his Zeiss, the only instrument he trusted.

Other researchers gave their demonstrations, but he did not listen because he was consumed with anxiety about his own. He placed the slides that most clearly revealed the free ends of nerve cells on the stages of the microscopes. He even focused the images in advance so that all an observer had to do was bring his eyes to the ocular. His slides, lacking cover glass and smeared with the blood of the specimens, appeared crude. The other researchers either ignored him or flashed a dismissive smile. Some were suspicious that any important discovery could come from "such a poor environment" as Spain, to quote one skeptic present at the conference. Cajal stood alone in a corner of the hall, his heart pounding, calling out to passersby in the halting, primitive French that his father had taught him in the cave in Valpalmas. He had worked for countless hours alone with no support, scraped together his last savings, and traveled all the way to a foreign country just for a chance to show his discoveries to the world's leading authorities. But what if he could not get their attention? His scientific future was at stake.

The man whom every aspiring researcher at the conference wanted to impress was Albert Kölliker. Then in his seventies, with pure white hair and a mustache, he was a "handsome old man everyone rises to their feet for," according to one colleague. Kölliker, it was said at the time, "knew more by direct personal observation of the microscopic structure of animals than anyone else who has ever lived." His textbook had been among the sources from which Cajal procured illustrations for the "biological novel" that he wrote in college. Kölliker was a worldly man who spoke English, French, and Italian, and published in all three languages. He did not have a favorable opinion of Spanish science. Forty years earlier, he had visited Madrid and toured the Museum of Natural Sciences, where he noticed a brand-new French microscope in a glass display case, still in its mahogany box. He asked the museum director if anyone there practiced microscopy. The microscope was just an object of curiosity, the director said—no one knew how to use it. A few years before the conference, Kölliker, the most respected anatomist in the world, had criticized Cajal by name for his 1881 paper promoting the reticular theory of muscle fibers—a paper that Cajal had long since disavowed.

In 1886, Kölliker had traveled to Pavia to learn the black reaction from Golgi himself. Kölliker began experimenting with the technique, and the following year he published a paper stating that he found no evidence of a reticulum. Previous claims about nerve fiber connections, he said, were "based upon deception." And yet Kölliker only *believed* that nerve cells were independent—considering himself the arbiter of intellectual disputes, he was too judicious to offer a new theory. For decades, like so many histologists, he had awaited a clear picture of nervous tissue to settle the issue. Cajal practically dragged Kölliker to his corner of the demonstration room. As soon as he looked through Cajal's lens, Kölliker was "enchanted." The proof was undeniable: fibers ended freely, showing that the reticular theory was indeed false. "I have discovered you," Kölliker told Cajal somewhat patronizingly, "and I wish to make my discovery known in Germany."

Following Kölliker's lead, prominent scientists gathered to view Cajal's "extremely excellent" slides. Many had tried and failed to clarify the endings of nerve fibers themselves, and all were excited by the new revelation. Those who had struggled with the black reaction were eager to learn Cajal's secrets, and Cajal unreservedly shared every detail of his technique. Attendees left Berlin determined to explore the possibilities of the black reaction with the help of Cajal's improved method. The obscure Spaniard had finally made his name.

ONE PROMINENT RESEARCHER HAD BEEN ABSENT FROM THE BERLIN conference, however: Golgi. On the way home, feeling like a conquering hero, Cajal stopped in Pavia to visit him as promised. According to some accounts, Golgi was there but refused to receive him. But the Italian school year had not started, so Golgi was likely still at his country house in Varese, fifty miles away. Meeting Cajal was not worth curtailing a vacation. He still did not take the Spaniard's challenge seriously. In recent years, Golgi had welcomed many visitors to his laboratory, but none had come to ask for his intellectual surrender. Cajal lamented that he did not encounter Golgi during that trip in the fall of 1889, for had they met, he maintained, their ensuing rivalry—which would become among the most legendary in the history of science—might have been avoided. Cajal claimed that, at the time of his visit, Golgi was away in Rome, fulfilling his duties as a senator. But Golgi did not become a senator until 1900; Cajal either confused the date or sought to excuse Golgi's impoliteness.

Cajal's findings revolutionized the study of the anatomy of the nervous system. "The laboratories of Europe were in a ferment," according to a Belgian colleague, Arthur Van Gehuchten, who helped to spread news of Cajal's work in Francophone countries. "We all wished to carry our stones for the new edifice, which, under the brilliant direction of Cajal, became so magnificent." The most famous histologist in Paris, Mathias Duval, turned Cajal's drawings of the nervous system into giant wall charts and used them for his lectures. "This time," the Frenchman would tell his students, "light comes to us from the south, from Spain, the country of the sun." In August 1890, Cajal received a letter from Wilhelm His, recently returned from another international congress, which Cajal did not attend. "I have shown your beautiful preparations in the Anatomical Institute," His told Cajal. "In general, your absence has been much lamented. It had been a very brilliant section." Golgi was one of the conference's presidents, and the discussion about the nervous system, reported His, was "rather animated."

When he returned from Berlin, Cajal intensified his attacks on the reticular theory. His study of the olfactory bulb, published in three parts in *The Barcelona Sanitary Gazette* in 1890, contained the first illustration of olfactory receptor cells, whose endings he declared to be "perfectly free." He attributed Golgi's error to "a typical case of the crippling influence of theoretical prejudice." He also published two papers, translated into German, in *Anatomischer Anzeiger*, among the most prestigious scientific journals in the world. In one paper, Cajal discussed his discovery, in the spinal cord of chicken embryos, of delicate fibers branching off from axons, which he termed "collaterals." Golgi was furious; ten years earlier, he had made the same finding. "Even if this neglect might be attributed in part to the limited diffusion of the Italian language and to the unsatisfactory conditions of scientific publishing in our country," Golgi later wrote in a letter, "still I cannot believe that such a neglect is completely justifiable, nor that I should continue to bear it." Neither Cajal nor any other researcher—including Kölliker, always the most informed—was aware of Golgi's discovery, which he reported in a small provincial journal but did not include in his comprehensive 1885 tome. Cajal had also mentioned axon collaterals in a March 1889 *Revista* article that he had sent to Golgi, who, having never read the issue, raised no concerns. Now, Golgi heard his name constantly associated with Cajal's and was bewildered. Cajal was almost ten years younger, and he had been working on the nervous system for only two years. All that the upstart had done, Golgi thought, was apply *his* method.

Cajal referred to the years 1890 and 1891 as his "Palm Sunday," alluding to Jesus' triumphal entrance into Jerusalem. In 1890 alone, he published nineteen monographs, four of which were translated into either French or German. Only three years after beginning his studies of the nervous system, he had more or less conquered the cerebellum, revealing its structure completely and even beginning to probe the mechanisms behind its functioning. He continued his work, expanding its scope to include the spinal cord, olfactory bulb, retina, optic nerve, and structures in the peripheral nervous system, such as nerve fibers in the wings and feet of insects. At the same time, he published three articles denouncing Ferrán and his cholera vaccine, putting to rest what he considered to be a national disgrace in a paper titled "Truth Against Error."

In the laboratory, Cajal let the chemistry of the black reaction guide his research. Since the technique worked better on embryos than on mature organisms, he asked himself a simple question: "Why should I not explore how the nerve cell develops its form and complexity by degrees?" The developmental process of nerve cells was poorly understood; before Cajal's studies, researchers could not accurately see embryonic nerve fibers. His research had demonstrated that, during the earliest stage of development, the bodies of embryonic nerve cells were rounded, with no projections. Later, fully mature axons were found to extend and branch outward in myriad directions, seeking out other cells.

How did nerve cells grow? In the absence of definitive evidence, no matter the topic of study, reticularists continually challenged the conclusions of Cajal and his supporters. "No one has seen in the embryo the free ending of a nerve in the course of its growth," they maintained. The reticularists argued that nerve cells grew by fusing together with elements from other cells, which would then form part of a reticulum, meaning that young nerve cells were not independent for long.

Cajal prepared almost four hundred sections from the spinal cords of three- and four-day-old chicken embryos. In commissural cells, whose young axons grew across the midline of the nervous system, contacting cells on the opposite side, Cajal observed "a conical lump with a peripheral base" on the ends of axons. The structure remained independent of other cells during the course of its development, Cajal showed. He dubbed the structure the "growth cone." If the opposing theory—known as the polygenic or catenary theory—were true, and nerve fibers eventually fused together, then how could the nervous system ever change? Cajal thought. If the material of the mind became fixed after childhood, then how could a person like him have learned?

Examining growth cones at varying stages of development, Cajal observed that no two looked alike. They were not only distinct from each other but also distinct in their appearances. The developmental study of nerve cells presented Cajal with a new scientific challenge. Anatomical images are static; the technology to follow growth in vivo did not yet exist. What he needed to see was a succession of images depicting a growing axon as it approached its target in order to prove that the neuron retained its independence every step of the way. This required him to do what he had practiced as a child, animating fixed forms in his mind, bringing dead structures to life.

What Cajal saw were young nerve cells "relentlessly forging a path." He described the growth cone as exhibiting a mixture of belligerence and sensitivity, discipline and waywardness: he compared it to "a living battering-ram . . . pushing aside mechanically the obstacles which it finds in its way." He compared growing nerve cells to soldiers—"breach[ing] the membrane" and "assaulting the interior." It was as though the life of the infinitely small recapitulated his own experience. Cajal's first drawing of growth cones shows probing axons, with bulbous forms at their tips, having neared the midline, boldly prepared to cross.

The leading theory of nerve growth held that axons travel along preexisting paths. In a widely cited paper in the August 10, 1890, issue of *The Barcelona Sanitary Gazette*, Cajal presented axons in the retina as a counterexample. "If the route across the retina was traced in advance, with the axons gliding there as inevitably as a train on its rails," he postulated, "then the growth cones would follow their path without error or hesitation." But many axons, after penetrating a new layer of tissue, seemed to remain in a fixed position, "leaning" against a structure for a while, as though loitering. "They apparently have not succeeded in finding their way," Cajal concluded. Sometimes he came across fibers that seemed to follow a route contrary to that which they should take, away from "the region of the future." "Must we accept . . . the existence of pre-established aberrant and useless pathways?" Cajal asked. "And if so, how can we explain how the strayed fibers abandon the wrong road and reach their destination?"

At crucial moments, it seemed as though young nerve cells had to "choose" their own directions. This was when, Cajal observed, their structures often underwent dramatic changes—he could discern the journeys of these cells by examining their adult morphologies. "Thus the total arborization of a neuron represents the graphic history of conflicts suffered during its embryonic life," Cajal wrote—just as his own childhood conflicts bore on his identity as an adult.

How do axons navigate through the tissue? How do they manage to find their targets? It seemed to Cajal that they proceeded "without deviation or error, as if guided by an intelligent force." He attributed this force, a preoccupation throughout his career, to the growth cone's "exquisite chemical sensitivity." His idea was probably influenced by contemporary research on the movement of leukocytes toward bacteria, which followed gradients of chemicals in the environment, a phenomenon known as "chemotactic ameboidism." Cajal proposed that growth cones were similarly oriented toward stimulating substances produced by their targets, which he termed the "chemotactic" hypothesis, later revised and renamed the "neurotropic theory." Certain axons, at the beginning of the nerve cell's life, appear "disoriented and often tortuous at the outset" but "progressively become oriented" and straighten their course. It is as though, "after hesitation and groping, they end up by feeling the effects" of some mysterious influence.

CHAPTER 17

"The Only Opinions That Matter to Me"

In June 1890, Cajal's mentor Maestre died, leaving an open chair in histology at the University of Madrid. The University of Madrid was the "Central University" of Spain; all other universities were considered to be peripheral. Every academic in the country sought a pathway to the capital. Cajal knew that the *oposicion* would distract him from his research, but financial reasons alone were enough to convince him that he needed to compete. His time in Barcelona had depleted his resources. He was not independently wealthy, nor did he earn income from clinical practice, like many of his peers. Textbook sales were the biggest source of additional income, and his two existing volumes were not selling well—provincial researchers were not considered authorities, while the textbooks of professors in the capital became part of the national medical curriculum.

The gamesmanship began soon after the *oposicion* was announced. Cajal's friend Simarro wrote to ask him about his plans; Simarro knew that he would lose if Cajal competed. "If by an implausible chance I was able to win the chair against you it would be difficult for me in good conscience to accept the post, since I consider you the only histologist in our country," he wrote. Most histology professors in Spain had been Cajal's students, either at universities or at his private academy. Rival candidates and their supporters clamored for judges to recuse themselves, causing the *oposicion* to be suspended multiple times. Cajal was irritated by the long delays; he could not afford to stay in the capital indefinitely, whereas Simarro, his main rival, who lived there, incurred no extra cost. One night, in the Puerta del Sol, the busiest square in Madrid, Simarro and a friend literally bumped into Cajal, who was walking with his head down, reading a German textbook. "How can you deal with this man Cajal?" Simarro's friend asked, exasperated. Cajal held up his book and said, "These are the only opinions that matter to me."

In the end, Cajal triumphed in the *oposicion* by a unanimous vote, the first in history in the medical field at the central university. He was also the first Aragonese to hold a position on the faculty of medicine. His colleagues abroad were incredulous that Cajal's status should ever have been in doubt. "Who in Spain could compete with you?" Kölliker asked.

While Cajal was consumed by the *oposicion* process near the end of 1891, a German anatomist named Heinrich Wilhelm Gottfried von Waldeyer-Hartz published a six-part review article in the *German Medical Weekly*, a popular, seventy-year-old journal dedicated to community medicine, public health, and clinical techniques. In the final section, printed in a different font, Waldeyer comprehensively summarized the work of Golgi, Cajal, and others, deferring to Kölliker for judgment on controversial issues. Waldeyer termed the independent nerve cell the "neuron." In later years, though Cajal used the word, he generally preferred *nerve cell* so as not to bring attention to Waldeyer's role, since Cajal resented that the German was mistakenly credited with the underlying discoveries. He was reminded "more fully, and somewhat bitterly, [of] the impotence of his country's language as the medium for holding, let alone catching, the scientific ear," according to a colleague. But some people have a gift for naming things, and Waldeyer was a brilliant synthesizer and communicator of scientific information (he also introduced the term *chromosome*).

• • •

IN APRIL 1892, CAJAL MOVED PERMANENTLY TO MADRID WITH Silveria and their six children. At forty years old, he was the "prince of the academic world," as one colleague liked to tease. The child of Petilla de Aragon suddenly found himself in the cosmopolitan Spanish capital, at the apex of Spanish society and culture. His apartment at 131A Atocha Street was just a five-minute walk from San Carlos Hospital, where the medical faculty was located. An independently rich faculty member, impressed by Cajal's *oposicion* performance, funded the construction of a new laboratory for him at the university, with long marble tables, stools for hundreds of students, and a library full of the latest international journals. Another wealthy colleague offered him free accommodations for life, but Cajal did not accept, because the house was too far from his laboratory.

At first, the excitement of Madrid "demagnetized and bewildered" Cajal, weakening his powers of concentration. Though his salary had risen to six thousand pesetas a month and his laboratory expenses were now paid for, he rented as modest a flat as ever, costing only eighty pesetas a month. He no longer had to teach private lessons. Instead, he became a student himself, auditing history, philosophy, and law classes taught by Spain's most eminent scholars. Twenty-five years after nearly failing secondary school, he now became a model student, never missing a class, taking copious notes, and poring over the assigned texts.

Hall Four, where Cajal taught his undergraduate anatomy and histology classes, was at the top of a steep stone staircase. The front door opened onto Santa Inés Street, where the rattling of old carriages over the cobbles was the only sound to disturb the silence in the room. Students packed the wooden benches for Cajal's first class. In Barcelona, he taught six or seven students; capacity in Madrid was four hundred. They expected a religious experience, one student recalled. Cajal entered the room without fanfare, looking youthful and robust, with an air of absentmindedness and distinction. Draped across his broad shoulders was a traditional Spanish cape, customary among intellectuals in the capital. He stripped off his cloak, revealing an old, long-tailed, poorly fitting frock coat. His old brown suit jacket was always stained. One day, when he took off his overcoat, he was in shirtsleeves—he had forgotten to wear his jacket. "The man's appearance, frankly, was not very pleasant," recalled another student. Cajal sometimes wiped the classroom chalkboard with his handkerchief, the same one that he used in the laboratory to clean excess wax from blocks of tissue.

Cajal greeted the class with a few formal opening remarks, then reached into his jacket pocket for the small, leather-bound attendance book and called the roll. When he came to a long and difficult surname—Van Baumbergen, Idogaja, Gazteizgoycasca—he would simply give up and say, "Etc." For the majority of medical students, Cajal did not live up to the hype. A group of students seated in the uppermost row waited anxiously for him to finish calling roll so that they could escape the class, slipping out through the door or hanging from the windows and lowering themselves down to the patio. Cajal thought of this exodus as a human "diapedesis," referring to the passage of blood cells through capillary walls during inflammation. Instead of the typical minus sign to mark absences, he drew a scribble next to missing students' names—the spiral shape of the cholera bacterium, whose tail he drew longer and longer as more absences accrued. But Cajal never scolded or punished students who left early. He simply waited for the racket to die down and the deserters to vanish. "Let's see if those emigrant leukocytes leave soon!" he would say, smiling. Those who fled had to cross the patio, known as the Horse Yard, where the department kept corpses for dissection. When the weather was warm, the smell was extremely unpleasant. "The bodies were piled up, I will never forget," a former student said. "There were children, young and old, whose eyes nobody had closed."

Professors usually stood at a podium to lecture. Cajal sat at the reddish slab used for demonstrations. "[He] is not an orator," a reporter who profiled him wrote. "He does not aspire to enshrine his ideas in the filigree of eloquence. His concise, serious explanation goes right to the heart of things." During each class, he told the first-year students about his and other scientists' latest findings. "The majority, recently departed from institutes, did not understand his elevated meditations," said one student. Cajal's diction still retained "a certain Aragonese aftertaste." Most agreed that his style was uninviting and that his material was too esoteric for the unversed. The majority of students found him boring. Some decided that it would be better to learn from his books.

One former student described Cajal's freshman class as "a collection of donkeys escaped from the woods." He was no breaker of colts—or donkeys, for that matter. Cajal's tests were legendary pieces of cake. He asked only basic questions and expected clear and direct responses, not elaborate explanations. He hated when students launched into discursive detours to avoid answering a question head-on. If a student did not know an answer in class, all that Cajal did was add twists to his trademark spiral. In his "pedagogical dictionary," one student recalled, "the word *fail* did not exist . . . he appeared to have a phobia of zeroes." He often threatened massive tests but never followed through. "Both my father and my professors would have been wiser if they had employed with me the methods of persuasion and kindness," Cajal wrote of his own educational experience, "instead of inflicting upon me corrections which were sometimes excessive and always exasperating."

Cajal had a soft spot for students who reminded him of himself, the "somewhat indomitable, highly scornful ones, unmoved by superficial motivation," according to one alumnus. Cajal's favorites were the students who spent all their time not on medicine but on writing, art, and philosophy. Some were poets; all were Romantics. At first glance, they seemed lost or distracted, but deep down, he knew that they harbored passion and intensity. They were the ones no one expected would become scientists.

• • •

AFTER CLASS, CAJAL SPENT THE REST OF THE MORNING EITHER attending lectures or working in the laboratory. At around two-thirty, he went home for lunch, prepared for him—as every meal was—by Silveria. Typical Spanish meals include five or six courses; Cajal ate only two plates, almost always with the same food, to facilitate his digestion. With each meal he drank a bit of white wine, a habit that, as he wrote sheepishly in an editorial called "The Uses of Wine," had nothing to do with his health. He classified this as a vice, perhaps overly harshly, since he served himself wine with the lowest alcohol concentration possible and never more than a small glass. He prided himself on moderation, claiming that he never overate and never snacked. "I do not abuse bread," he stated. But his secretary would later say that he had a sweet tooth, especially for German chocolate cake.

It was vital that Cajal, now in a strange city, find another *tertulia*, a respite from his daily work for an hour after lunch. Luckily, by the turn of the century, Spain had become "an archipelago" of them, as one journalist put it. He gravitated toward the Café Levante on Arenal Street, near Puerta del Sol, a refuge for wanderers and eccentrics, who whiled away their days reclining on plush sofas, keeping a keen eye on the tables, eager to seize a chair from anyone naive enough to abandon his seat. As far as cafés went, the Levante was neither the fanciest nor the seediest. The mirrors were tall and the couches were wide. There was a platform with a grand piano in the center of the room and a billiards table in the basement. Cajal's *peña* at the Levante consisted of acquaintances from Cuba, mostly military physicians. Conversation was lively and frank. Flanked by fellow veterans, Cajal could relive his wartime past, reclaiming a part of his youthful identity. But the thrill was gone. By that time, the Spanish military had begun thinking of itself as a class apart from the rest of society, a self-renewing caste. With no official policy for promotion, soldiers and officers relied on favoritism to secure their futures. Cajal's old comrades were obsessed with petty politics, and all that they read were medical gazettes. He was looking for more diverse company, men who shared his breadth of intellectual interests, especially in the arts and sciences. Nor did his political ideology align with theirs; military men were notoriously traditional, monarchist, and conformist, while Cajal was a free thinker, a religious skeptic, a liberal republican, and a staunch individualist.

Cajal left the *peña* at the Levante, seemingly leaving his past behind. Sudden exits like these, which violated social norms, exposed him to accusations of rudeness and arrogance. "Cajal, man, say goodbye," a politician later told him. "Does it hurt to be polite?" But all he wanted was to stimulate his mind. "Pure, but sanctified egoism," he wrote, "for without that no serious work is possible!" Cajal was most comfortable around men who, like himself, cared only about teaching and research. Though renowned among histologists, both in Spain and around the world, Cajal was not yet recognized by the general public. It thrilled him to be anonymous. "Certainly, no one thinks of me," he recalled saying to himself, "but, in return, what freedom of thought and work do I enjoy!"

Tertulias in Madrid were invitation-only. A colleague of Cajal's formally introduced him to the *peña* at the Café Suizo, located on the corner of Alcalá Street, one of the busiest streets in Madrid. With slender columns, marble tabletops, elegant chandeliers, imported wood floors, gilded bronze walls, space for five hundred patrons, and large windows facing the street, the café was "the axis of social life" for the intellectuals of Madrid. It was said that at the Suizo, news was discussed before the events occurred. Financiers and scroungers, politicians and artists, writers and unemployed actors, and bullfighters and peripatetics alike met there daily for coffee and conversation.

Cajal's *peña* consisted mostly of doctors but also included lawyers, professors, business owners, and other professionals, some of whom went on to become senators, university rectors, and even government ministers. Philosophy was their topic of choice. Once, a new *contertulio* wrongly claimed that Socrates did not believe in immortality, and the refutation from the rest of the *peña* was so forceful that the poor man never returned. Evolutionary theory was preached as gospel, with

Darwin, Haeckel, and Herbert Spencer, the social Darwinist, venerated as high priests. Cajal's speeches could be so ingenious and scathing that some nearby *tertulias* of politicians, artists, writers, or stockbrokers would divert their attention and listen.

A dream of Cajal's was fulfilled when a friend brought him to the *tertulia* of his boyhood political idol, Emilio Castelar. Short and obese, with a walrus mustache, Castelar looked more like an aristocrat than the man of the people that he claimed to be. In the presence of his "torrent of sonorous words," full of "unimagined splendors," Cajal was "in ecstasy." He was introduced as a devotee of the world of the infinitely small. "You do well," Cajal recalled Castelar telling him. "Life is a secret and the cell deserves our attention all the more in that we carry it within us and that it often influences our actions." Cajal considered replying to Castelar: "No, those tiny cells, which keep hidden in the minuteness the mystery of life, are *the whole man*, in his two aspects, rational and physiological." On the way home, the friend punctured Cajal's idealism, informing him that Castelar was in massive debt and had used his influence to let a creditor off for murder. Perhaps, in the end, cells were less complicated.

CHAPTER 18

"The Absolute Unsearchability of the Soul"

Cajal's move to Madrid came in the midst of his studies of the retina, which he now resumed with a broadened scope that included all vertebrate species. He had begun with birds, then moved on to amphibians and reptiles such as adders and green lizards, then on to mammals like rabbits, dogs, cows, sheep, and horses, and finally to fish. In 1892, he published an article titled "The Vertebrate Retina," the result of seven months of research and featuring seven large lithographic plates, in the French journal *La cellule*. A German translation appeared two years later as a book. Cajal's research, his German translator informed the reader, has "advanced our knowledge of the structure and function of the nervous system beyond all hopes and expectations" and should "be regarded as proven and acknowledged by the most eminent scientists." In Madrid, Cajal also began working on the hippocampus, the brain structure responsible for consolidating memories, named for its resemblance to a seahorse. What drew Cajal to the hippocampus was not its function but rather its beauty: "[Its] pyramidal cells," he wrote, "like the plants in a garden—as it were, a series of hyacinths—are lined up in hedges which describe graceful curves." His paper on the hippocampus was translated by Kölliker himself. "I have learned your language rather well," he wrote to Cajal, "for the necessity of studying your work." The only problem, he said, is that it was impossible to read Cajal's handwriting.

• • •

SOON AFTER WALDEYER'S REVIEW ARTICLE, HUNDREDS OF RESEARCHERS around the world adopted neuron theory as the basis for their understanding of the brain's structure. By the mid-1890s, the neuron had even come to serve as the foundation for theories of mind. Advances in histology, led by Cajal, inspired researchers in other fields to incorporate brain anatomy into their research. "Every new work of yours sheds light on the great dark questions," one colleague told him. Psychologists investigated mysterious phenomena such as memory, hallucination, and sleep, searching for material correlates to satisfy an empirical standard. This new field was known as objective psychology, or rational psychology—Cajal called it "mental histology." "It is well known that psychologists immersed in contemplating the mind overlook the brain," Cajal declared. "When Cajal opened up the path of elementary structures," one German psychologist wrote, "an invaluable foundation was established for all scientific study of the psychic faculties, for a Psycho-Physiology that is not subject to the mystical element." "Cajal has given us the key that opens the mysterious cave of the brain," his colleague Van Gehuchten said, "opening with it a new world, the world of the thinker."

Ideas about the brain might be older than human history. *Australopithecus* skulls from three million years ago show cranial fractures that are close together yet distinct, and the imprints of the wounds roughly match the outline of an antelope bone, indicating that the australopithecines were clubbed to death. Our distant ancestors may have realized that brain damage could be deadly. Similarly, specimens of *Homo erectus* were found murdered by blows to the head. Skulls from the late Stone Age, around sixty thousand years old, contain holes so perfectly round that they could only have been bored by specialized tools, suggesting that Neolithic humans may have tried to treat illness by relieving the brain of evil spirits. The word *brain* first appears in hieroglyphics in the Edwin Smith Papyrus, dating back to the third millennium B.C.E. But the ancient Egyptians believed that the heart was the seat of the soul, later weighed in the afterlife against a feather to see if a person was free of sin. In the process of mummification, the brain was dealt with unceremoniously, churned up with a long metal rod and spilled out through the nostrils.

At different moments in the history of ideas, "the mind" or "the soul" has been believed to inhabit the heart, the bloodstream, the liver, the opening between the stomach and the small intestine (called the pylorus), and the four empty cavities between our flesh (known as ventricles), to name a few locations. But the heart was by far the most popular site. The Epic of Gilgamesh located feelings in the heart, and the Rigveda presented the heart as the home of thought, as did the records of the Aztec and the Maya. Neither the Quran nor the Bible mentions the brain. As pointed out by the Stoic philosopher Chrysippus, we indicate *I*, naturally and instinctively, by pointing to ourselves at the place in which thought appears to be: the chest. (We still say that we remember things "by heart.") It was Hippocrates, the father of medicine, who first articulated the brain's special role. "Men ought to know," he declared, "that from the brain, and from the brain alone, arise our pleasures, joys, laughter, and jests, our sorrows, pains, griefs, and tears . . ." He called the brain "the interpreter of consciousness."

Many philosophers and scientists have tried to break down the brain into functional parts, a trend known as localization. For twelve hundred years, physicians and philosophers believed that thought is produced by ventricles. Despite his many anatomical discoveries, Andreas Vesalius still held this view. In the seventeenth century, René Descartes identified the pineal gland as the seat of intelligence, because there was only one of them, not two, and it was surrounded by cerebrospinal fluid, believed to be a reservoir for animal spirits. A few of his contemporaries placed intelligence in the corpus callosum, a bundle of fibers that connects the two hemispheres of the brain. A generation later, the great British neurologist Thomas Willis located the mind in the *matter* of the brain, not the spaces of the ventricles, identifying the cerebral gyri, the ridges on the brain's surface, as the sources of memory and willpower. Willis had faith that the anatomy of the nervous system could "unlock the secret places of Man's Mind and [look] into the living and breathing Chapel of the Deity." But his contemporary Nicolas Steno, a Danish scientist who later became a priest, lamented that "when you look into its inner Substance you are utterly in the dark."

"Where is the soul?" asked Immanuel Kant. "Has it taken up a little spot from which it directs the whole body?" Kant, who sought to understand the nature of thought, lamented that he could be only a "mere spectator" to its representations since he did not "know the nerves and fibers of the brain." Before visions of angels called him away to mystical realms, Emanuel Swedenborg, a major influence on American transcendentalism, focused on brain anatomy "solely for the purpose of discovering the soul," he said. Though he never performed an experiment, after studying contemporary anatomy, Swedenborg became the first person to correlate thinking, remembering, imagining, judging, willing, and understanding with the cerebral cortex, the thin, two-millimeter layer of tissue cloaking the outside of the brains of humans and other mammals. Cortical localization did not come to prominence until the nineteenth century, with Franz Joseph Gall's theory of phrenology, which held that the surface of the skull reflected the relative size of underlying regions of the brain. "Where there is variation in function," Gall assumed, "there must be variation in controlling structures." Phrenology was wildly popular in the nineteenth century, but the theory was complete nonsense, and some used it to justify European superiority over other races. Still, Gall was grasping at the idea of localization. He was looking at the skull, but he needed to be looking inside it.

The reticular theory conceived of the brain as a unitary whole, and scientists likely would not have searched for functional areas in the cortex if they had continued to think of the brain in that way. On the other hand, neuron theory encouraged researchers to apply localization to the microscopic level, assigning functions to different kinds of neurons. Emboldened by his discoveries, Cajal joined in the speculations in the 1890s. "It is a rule of wisdom . . . not to theorize before completing the observation of facts," he wrote. "But who is such a master of himself as to be able to wait calmly in the midst of darkness until the break of dawn?" Correlating brain structures with movement or speech was difficult enough; the complexity of psychological functions seemed nearly irreducible. The neuron, however, seemed like a candidate for a long-theorized part of the nervous system: an individual cell that might serve as a "storage unit"—one for one—for individual mental impressions. The materialist doctrine, insistent on physiochemical explanations for every natural phenomenon, called for an "exact science of the fundamental relationship between body and mind," and Cajal's work seemed to offer a way forward. In the black reaction, Cajal had recognized not only a solution to a technical problem of visualizing the nervous system but also, so it was believed, a method of illuminating "the utter darkness of the inner mechanism of psychic acts."

This was not the first time Cajal's neuroscientific investigations had ventured into psychology. In March 1892, before leaving Barcelona, Cajal had delivered a series of lectures at the Academy and Laboratory of Medical Sciences of Catalonia, aided by giant color diagrams in which he synthesized all his findings on the structure of the nervous system from the previous four years. Students diligently took notes. The majority of the audience, however, did not consist of histologists. He sensed that his listeners wanted him to speculate about the psychological implications of his findings. And so Cajal announced his discovery of a certain neuron, with a triangular cell body and multiple dendrites, which he named the pyramidal cell. The pyramidal cell, he explained, appears in all vertebrate species and becomes larger and more complicated in more highly evolved ones. He suggested—albeit "with some reservations"—that mental function must be related to pyramidal cells, which he took to calling "psychic cells."

Descartes believed that the mind and body are distinct and made up of different substances. Mental functions, Descartes argued, are separate from the physical matter of the brain but could somehow influence it. Cajal rejected this age-old "dualism." As a stalwart materialist—"with a certain amount of petulance, unshakeable in his views"—he insisted that all bodily functions, no matter how advanced, from spinal cord reflexes to conscious thought, "cannot be mediated other than by the axonal and dendritic processes." Every neuron is built according to the same model, he said, with essentially the same axons, dendrites, and mechanisms of transmission. But "no hypothesis can elucidate" the mechanisms behind that "very special activity," he admitted. "Neither materialism nor spiritualism explains to us how a phenomenon of excitation arriving at the first cerebral layer is converted there into such a different thing as an act of consciousness."

"The Spaniard of Barcelona, of such world-wide reputation, has with perfect right ventured over the line of strict anatomy into the provinces of psychology," one American psychologist wrote in a review. "The invasion is a welcome one." Shortly thereafter, Cajal's students transcribed their notes from Cajal's Barcelona address into a series of articles, which Cajal amended and revised. A German translation of Cajal's series appeared the next year, followed by one in French, which in two months sold through three printings. The success of his book would later inspire Cajal to compose a larger, more definitive work, one that would take him ten years to complete, which would become his masterpiece.

IN FEBRUARY 1894, CAJAL RECEIVED AN INVITATION FROM THE ROYAL Society of London to deliver the Croonian Lecture, a 156-year-old tradition recognizing the greatest discoveries in biology from the previous year. Cajal was stunned. The Royal Society, founded in 1660, was the oldest and most prestigious scientific association in the world, whose first president was none other than Sir Isaac Newton. Cajal felt intimidated by the list of past speakers, giants in the field, many of whom he counted as his personal heroes. He hesitated, unsure whether to accept. "I feared that I should not measure up suitably to the honor," he recalled. One of Cajal's daughters was sick, and he did not want to leave her, having already lost Kety. But the Royal Society offered a prize of 50 pounds sterling, equivalent to 250 pesetas. To further offset his travel expenses, the Ministry of the Interior, which had denied him money for Berlin for the anatomical congress, provided him with 2,000 pesetas—his first grant. No other Spaniard had ever received an invitation to deliver the Croonian Lecture.

What convinced him, Cajal said, was not the financial incentives but rather a kind and welcoming letter from Charles Sherrington, a professor at the University of London. Sherrington, a thirty-six-year-old British physiologist, had received recognition for his studies of the reflex arc in animals, the most basic sensory-motor response. The previous year, in a paper on the "knee jerk" in dogs, he had observed that the reflex response was not instantaneous, which suggested a delay in transmission of the nervous impulse. Sherrington had been reading Cajal's papers; he knew some Spanish and had spent time in Spain researching cholera (even meeting with Ferrán, about whose vaccine he had had similar reservations). Cajal's findings of independent nerve cells implied the presence of gaps between them, providing an anatomical explanation for the delay that Sherrington had recorded. It was Sherrington who had convinced the Royal Society's secretary, his mentor, Michael Foster, to invite the Spaniard.

Cajal stayed at Sherrington's home for two weeks that March. "Bristling with oddities," Sherrington later told a friend—"but a *great* man." Though not born an aristocrat, Sherrington, educated at one of the finest private schools in London, was a member of the elite, a connoisseur of art and culture and a famously warm host. Neither he nor his wife had ever met anyone like Cajal. One dinner, Mrs. Sherrington served whitebait, a piscine delicacy meant to be eaten whole. Cajal

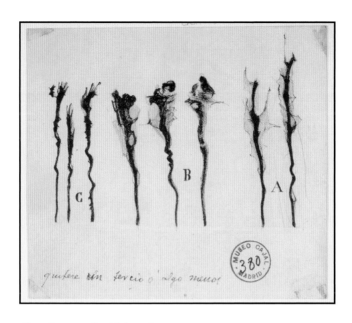

Growth cones in chick embryo marrow

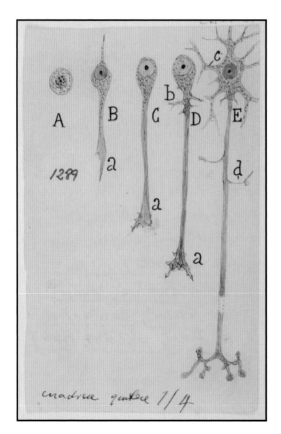

Phases of
development
of the neuron

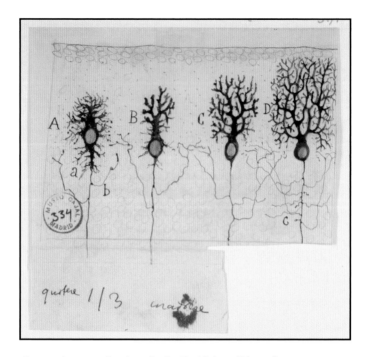

Successive complications in the Purkinje cell branch

Purkinje cell in
a human brain

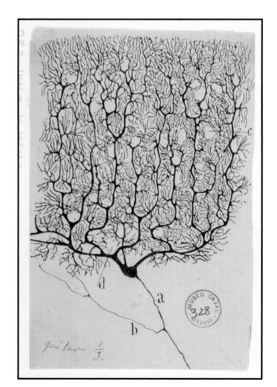

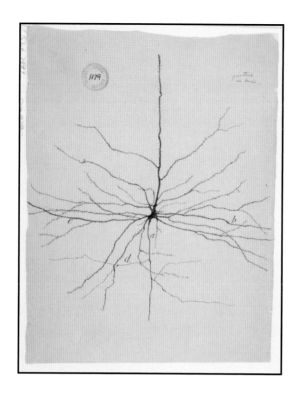

Giant pyramidal
cell of the
motor region

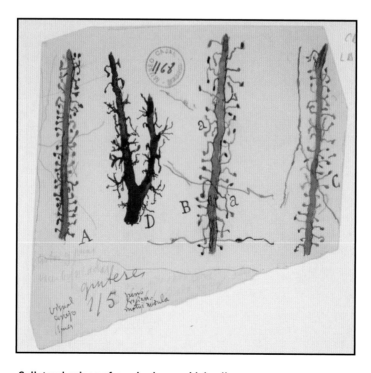

Collateral spines of cerebral pyramidal cells

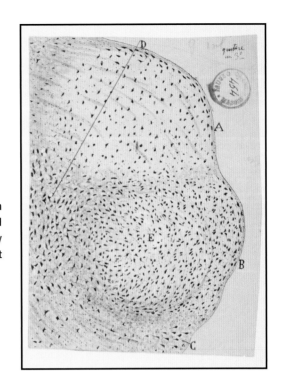

Frontal section
of the medial
geniculate body
of a cat

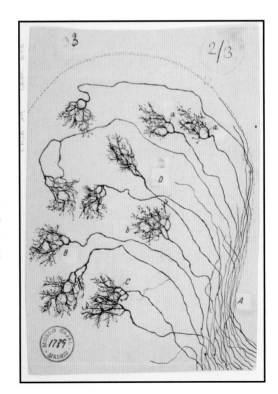

Nerve endings from
the lateral nucleus of
a twenty-four-day-
old mouse thalamus

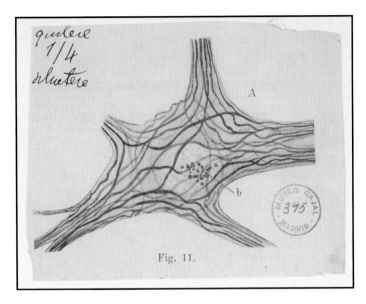

Hypertrophy and simplification of the neurofibrils of the spinal cord of a rabid animal (giant funicular cell)

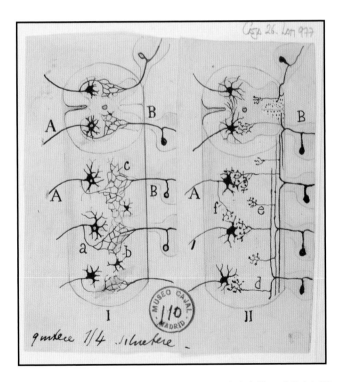

Sensory-motor communications according to Golgi (I) and Cajal (II); note how the fibers in I are continuous, while the fibers in II never touch

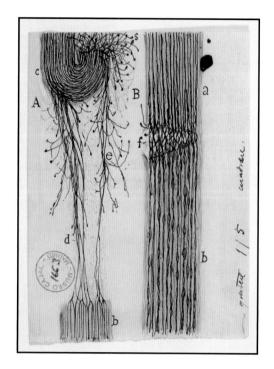

Innervation of the peripheral stump in a section and hemisection of a nerve

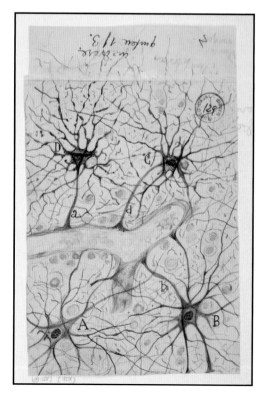

Neuroglial cells in the white matter of the brain

this point, and we will limit ourselves to indicating only those connections that appear to be the best determined or the most probable." His findings contradicted the popular belief, promoted by phrenologists, that intelligence is related to the brain's size and weight. Cajal instead focused on the number of neurons and the richness of their connections. As a way of increasing connections, he cited the example of "men devoted to deep continuous mental exercise," having witnessed such effects in himself, and called this mental exercise "cerebral gymnastics." He again attacked the reticular theory, this time on psychological grounds. To say that "everything communicates with everything else," he declared, "is to assert the absolute unsearchability of the soul."

There was no doubt in Sherrington's mind that Cajal was "the greatest anatomist of the nervous system ever known." After centuries, even millennia, of ambitious but frustrating attempts, the obscure Spaniard had revolutionized humanity's understanding of the brain in three short years. At first, when Cajal talked about his research, Sherrington was startled, even disturbed. "Cajal treated the microscopic scene as though it were alive and were inhabited by beings which felt and did and hoped and tried even as we do," Sherrington recalled. To compare the nervous systems of different species, Cajal said, was to follow the "dramatic history lived by the neuron in its millennial progress through the animal series." In Cajal's words, a growing nerve fiber "groped to find another," motivated by passion and rivalry. Their eventual connection he called "protoplasmic kisses, the intercellular articulations, which seem to constitute the final ecstasy of an epic love story."

"If we would enter adequately into Cajal's thought in this field," Sherrington continued, "we must suppose his entrance, through the microscope, into a world populated by tiny beings actuated by motives and striving and satisfactions not very remotely different from our own." Cajal's perspective seemed almost childish to Sherrington—unbefitting of a great scientist, to say the least. "Listening to him I asked myself how far this capacity for anthropomorphizing might not contribute to his success as an investigator," Sherrington said. "I never met anyone else in whom it was so marked."

CAJAL HAD ALSO BEEN INVITED TO DELIVER A PLENARY ADDRESS later that month in Rome at the eleventh international Medical Congress—the largest gathering of its kind in history—with over seven thousand doctors from ten countries expected to attend. Rudolf Virchow was among the luminaries chosen to speak. Cajal declined the invitation. The congress occurred almost immediately after the Croonian Lecture, and he either could not miss more classes or could not afford the cost of the trip, despite the fact that the organizers had arranged for members to receive special discounts from railway companies. Instead, he sent a copy of his lecture, titled "General Considerations on the Morphology of the Nerve Cell: On the Mechanism of Ideation, Association and Attention," which was read aloud at the proceedings and later published in two Spanish journals and translated into German. Cajal's Rome address ushered in an unprecedentedly speculative period in his scientific thinking.

Since the dawn of the study of the nervous system in ancient times, researchers have tended to compare its structure to contemporary technologies. The ancient Egyptians saw in the exterior casing of the brain, with its fissures and convolutions, the corrugated slag left over from smelting ore. The ancient Greeks thought that the brain functioned like a catapult. René Descartes believed that animal spirits flowed from the brain through hollow nerves and inflated the muscles, just as hydraulic fluid traveled through machines in the royal gardens at Saint-Germain. In the nineteenth century, a new era of transportation, Otto Deiters, among many others, had conceived of the nervous system as a railroad, with junctions at which traffic could be routed, similar to the lines crisscrossing Europe.

In the 1870s and 1880s, the railway metaphor for the nervous system gave way to another transformative technological advance: the telegraph. The German biophysical school, headed by Hermann Helmholtz and Emil du Bois-Reymond, led the charge. "The wonder of our time, electrical telegraphy, was long ago modeled in the animal," said du Bois-Reymond in an 1850 speech. He argued that the similarity between the nervous system and the electrical telegraph ran far deeper. "It is more than similarity," he wrote. "It is a kinship between the two, and agreement not merely of the effects, but also of perhaps of the causes." In turn, engineers who designed telegraph networks, like Samuel Morse and Werner von Siemens, looked to the biological nervous system as a model of centralization and organization. With people traveling across countries for the first time and communicating with each other across the world, interconnectedness became a social ideal. When Germany finally unified in 1871, its telegraph network, centered in Berlin and reaching all its territories, became both a symbol and an instrument of imperial power. The following year, perhaps influenced by the predominant metaphor, Gerlach looked at nervous tissue through his microscope and saw a reticulum.

Cajal, who grew up in the preindustrial countryside, saw in the nervous system the natural images of his childhood. "Is there in our parks any tree more elegant and leafy than the Purkinje corpuscle of the cerebellum or the psychic cell, in other words, the famous cerebral pyramid?" he asked. He observed branchlets of axons "in the manner of moss or brambles on a wall," oftentimes supported by "a short, delicate stem like a flower"; a year later, he settled on the term "mossy fibers." These mossy fibers, he found, end in "knotty thickenings"— which he called "rosettes"—that approach the dendrites of other cells but, again, do not touch them. There are "nest endings" and "climbing fibers," which cling "like ivy or vines to the trunk of a tree."

Above all, the cells seemed to connect like "a forest of outstretched trees." Gray matter was an "orchard"; pyramidal cells were packed into an "inextricable grove." Cajal hit upon the embryological method for studying the nervous system, he said, while reflecting on the difference in complexity between the "full grown forest" and the "young wood."

The cerebral cortex, impenetrable and wild, was a "terrifying jungle," as intimidating as the one in Cuba. By force of will, Cajal believed, human beings can transform "the tangled jungle of nerve cells" into "an orderly and delightful garden." Cajal always feared that the backwardness of his environment had stunted his intellectual growth. "I regret that I did not first see the light in a great city," he wrote in his autobiography. But the undeveloped landscape of his childhood became the rich ground that nourished an understanding that was distinct from that of his contemporaries.

Though he evoked the telegraph from time to time, in his Rome address Cajal fundamentally rejected the metaphor. His opposition was rooted in both his anatomical findings and his observations of his own mind. "A continuous pre-established net—like the lattice of telegraphic wires in which no new stations or new lines can be created—somehow rigid, immutable, incapable of being modified," he said, "goes against the concept that we all hold of the organ of thought, that within certain limits, it is malleable and capable of being perfected by means of well-directed mental gymnastics." He knew, in other words, that he could change his own mind. That was why he could not tolerate the reticular theory's assertion that the structure of the nervous system was fixed. The nervous system must have the capacity to change, and that capacity, he argued, is crucial to an organism's survival. Cajal relied on a variety of terms to express this concept: "dynamism," "force of internal differentiation," "adaptation [of neurons] to the conditions of the environment"—and, most consequentially, "plasticity."

Cajal was not the first to use the term *plasticity*, though his 1894 Rome address, delivered before a broad international audience, was probably responsible for its popularization. In *The Principles of Psychology*, William James discussed the relationship between anatomical changes and habits acquired by learning. He identified plasticity as a property of organic material, built into the fiber of living beings. James provided no anatomical evidence for this idea, nor did he imply that learning changes the structure of the nervous system itself. The effects of plasticity, according to James, may be "invisible and molecular, as when an iron rod becomes magnetic." James cited Golgi but was unaware of Cajal's work, which was not translated into English until 1891, one year after *The Principles of Psychology* was published. Neither was Cajal aware of James; his work was not translated into Spanish until 1900. And in his typical fashion, Cajal did not like the term *plasticity* and hardly used it after 1894, probably because he did not coin it himself.

CHAPTER 19

"Grand Passion in Service"

Skeptics had long dismissed Cajal's results on account of the black reaction's tendency to produce artifacts. The work of neuron theorists was "long regarded with a certain suspicion," according to the Swedish anatomist Gustaf Retzius, who recalled receiving letters "from celebrated histologists abroad, who were otherwise favorably disposed toward me and my work, exhorting me in the most serious and moving terms not to go on experimenting with that wretched Golgian method which only resulted in artifact." The best way for Cajal to eliminate doubt was to demonstrate the same results with a different technique. He felt "the determination, the desire, nay more, the urgent necessity" to confirm his previous findings, validating his reputation in the process. He turned to another histological method, known as methylene blue, invented by Paul Ehrlich in the 1880s to stain microbes. When injected into a live specimen, the synthetic dye could color neural tissue without damaging its structure, eliminating the need for fixation and hardening. Like the black reaction, the methylene blue method stained axons and dendrites selectively, but methylene blue samples were seen as more accurate than those stained with the black reaction. Cajal was familiar with the method from his previous studies of cholera, but he had preferred the black reaction for nervous tissue. In 1896 and 1897, he employed this new technique almost exclusively, once again studying the spinal cord, cerebellum, hippocampus, and other areas of the brain in different animals, recapitulating a decade's worth of work. Now that his stains were more colorful, Cajal introduced aquarelles, a type of watercolor, and little pens (*plumillas*) into his arsenal of artistic media.

The methylene blue method confirmed all Cajal's black reaction findings unequivocally, including one of his earliest and most bizarre-seeming claims. In his 1888 studies of Purkinje cells in hens, Cajal had observed that "the surface [of their branches] . . . appears bristling with tips or short spines," reminding him of rose stems. Initially, he assumed that the minuscule structures were artifacts. In section after section, in birds and in mammals, and on the dendrites of other types of neurons, the spines continued to appear. The tips of these spines were not pointed, Cajal noticed, but instead exhibited a ball-like shape. He began to incorporate the spines into his drawings. Not only was Cajal convinced that these spines were real, but he also insisted that they played a role in transmitting impulses from one neuron to another. Proving their existence became a personal crusade, and his methylene blue studies slowly began converting his peers.

BY THE TURN OF THE TWENTIETH CENTURY, MOST SCIENTISTS HAD accepted the neuron doctrine. The "onus of proof . . . lies with those who deny the nervous function," a contemporary textbook author wrote, "not with those who maintain it." The mid-1890s to early 1900s represented the most ambitious and speculative period of Cajal's career. Along with his contemporaries, Cajal was swept away by the hope and belief that, with new scientific discoveries, humankind could finally solve the ancient mysteries of mind and self. "To discover the brain," Cajal said, "was the equivalent to ascertaining the material course of thought and will."

Theories of consciousness abounded. According to Mathias Duval, a French colleague of Cajal's who wrote the prologue to one of his books, consciousness and unconsciousness were mediated by the extension and retraction of axons and dendrites into the gaps between neurons, facilitating or obstructing the transmission of impulses. The movement of neurons, Duval believed, was like the movement of another type of cell: the amoeba. He based his conclusions on observations of neurons exposed to different environmental conditions, including experiments in fatigued mice and in the retinas of fish, which shrink when exposed to light. Cajal, however, doubted Duval's conclusions. He suspected that the chloroform that Duval used to kill his specimens may have influenced his results. Cajal's theory was that altered mental states, such as those during anesthesia, hypnosis, or hysterical paralysis, alter the structure of neurons themselves, as do stimulants such as caffeine and normal functions such as imagination, memory, and the association of ideas.

Still, Cajal had to account for this with his model of independent cells. How could isolated pathways of conduction—impulses transmitted from one neuron to the next—account for myriad coordinated responses? In 1890, he had formulated what he called the "avalanche principle": sensory neurons record external impressions and relay signals toward the "nerve centers"—the nineteenth-century catch-all term for functional areas like the hippocampus, optic lobe, or thalamus—and as those impulses rush across conduction pathways, they trigger adjacent neurons as well. Thus, no matter how small an initial stimulus might be, a "simple movement" would expand and amplify itself, as the total number of neurons engaged increased, like snow rushing down a mountain slope, fanning outward as it descends.

ON PRADO STREET, A NARROW, UNASSUMING SIDE STREET A FEW minutes' walk from the Café Suizo, was a slender facade with a filigreed black iron gate, watched from above by busts of Velázquez, Cervantes, and King Alfonso X, "the Wise." A frieze above the entrance displayed, in Grecian font, the words ATENO—Athenaeum. Inside, up an elegant staircase, was the most prestigious and influential intellectual institution in Spain, consisting of an academy, a library, and a conference room, where concerts and university classes were held. For ten pesetas a month, men could access the magnificent library, read newspapers in comfortable lounges, and, during the harsh Madrid winters, enjoy the modern heating. The atmosphere could be as zany as it was staid; in a salon known as the Crock Shop, a reference to its abundance of both Greek vases and utter nonsense, everyone thought himself an expert—unemployed men bloviated about the economy, doctors about literature, ministers about poetry, and poets about politics. Genteel conversations devolved into belligerent disputes, which occasionally turned violent.

The Spanish Athenaeum was founded by a cadre of liberal, Francophile intellectuals in 1820, during a respite from Ferdinand VII's suffocating absolutist regime, and then abandoned a mere three years later, when the constitution failed and the king returned. Like a fugitive on the run, the Athenaeum moved from location to location, always vulnerable to changes in Spain's volatile political climate, before settling in its own building on Prado Street in 1884. During the nineteenth century, a class of men affiliated with universities and with access to the political elite emerged as a select minority—an "aristocracy of the intellectuals," as one journalist wrote—especially prominent in contrast to the illiterate masses. In the second half of the century, as Spain nationalized its education, railroad, press, and bureaucracy, the Athenaeum served as a vital instrument for the creation and dissemination of national culture. The Athenaeum became, according to the newspaper *El Liberal*, "the brain of Spain."

The intellectual class in Spain had developed a reputation for elitism, cloistered in the comforting warmth of the Athenaeum. Toward the end of the nineteenth century, members devised strategies to disseminate knowledge to a broader audience. The Athenaeum began to focus on influencing public opinion by increasing the number and diversity of its members and shifting its attention toward the Spanish people via essays, manifestos, and public meetings. "Life of the

Athenaeum-Goers" became a popular recurring newspaper column. Cajal directed the Athenaeum's natural sciences department, organizing conferences on the discoveries of Pasteur, proposing a series of education reforms, and editing the biology section of the Athenaeum's magazine.

One of the Athenaeum's initiatives was an extension of the university called the School of Higher Studies, intended to "agitate philosophical inquiry in Spain." For the first term, during the academic year 1896–1897, twenty-eight subjects were chosen, ranging from literature to law to evolutionary theory. The Athenaeum paid its professors fifty thousand pesetas, a rate of "more than a peseta per minute," for teaching. Expectations for the curriculum were high; more than three thousand students attended classes. Cajal's course was titled "Structure and Activity of the Nervous System." His teaching style, unlike that of the other professors, was not polemical. His presentations were characteristically straightforward, displaying a critical attitude even toward his own results. Every Monday, from October through April, except during winter break, Cajal walked to the Athenaeum after dinner to teach in the splendid great hall decorated with neoclassical paintings under a golden dome. Cajal's class enrolled 221 students, the second highest total in the school.

In his course, Cajal defined the brain as an "organ of equilibrium and coordination of voluntary movements" that "works automatically." Many of his lectures addressed the challenge of publicly promoting scientific discoveries; at the time, the Catholic Church was fighting a trend it labeled "Scientificomania." *El Heraldo de Madrid*, which covered the School of Higher Studies in a regular newspaper column, noted that Cajal's course generated an "extraordinary interest by a numerous audience." *El Liberal* and other newspapers extolled his ability to parry conservative attacks and his "amazing faculties for making the most tangled chapters of microscopic anatomy clearly comprehensible." In a country that did not believe in its potential for developing a scientific culture, Cajal, rigorous and empirical, became its public representative. "How can he be anything less than someone who brilliantly cooperates in our scientific rehabilitation?" one student said.

On December 5, 1897, Cajal delivered his induction speech before the Royal Academy of Sciences. Titled "Rational Foundations and Technical Conditions of Biological Investigation," his address, reflecting twenty-five years of scientific research, was intended "to awaken in our disinterested young teachers the passion and taste for scientific investigation." For years, Cajal says, he had thought about writing a guidebook for new researchers. "I got no such advice from relatives or teachers when forming the reckless desire to devote myself to the religion of the laboratory," he said.

> *At the time, I was usually doing things that did not work, was lost, and more than once was hopelessly discouraged about my ability to pursue laboratory science . . . And there were so many times—for lack of discipline and more than anything for living so far from the kind of scientific atmosphere that stimulates and energizes the young investigator—that I thought about abandoning my work, tired and disgusted (as much from the work as from my sad and exhausting isolation).*

Great discoveries, Cajal said, are not inspired by books but by "the living logic that man possesses in his spirit." A scientist relying on methodological texts was like an orator studying the anatomy of the larynx. With each experiment, scientists must evolve like a species across generations, forming and correcting their hypotheses, patiently and with grit, until fortune finally smiled upon them. The most important quality in a scientist is "grand passion in service of a great idea." One either has the spark of investigation, Cajal said, or one does not.

Among the attributes that Cajal emphasized are intellectual independence, love of science, perseverance in work, desire for glory, and fervent patriotism. Cajal acknowledged that, when "fervor extends to chauvinism," patriotism can be an odious and destructive force. But for the man of science, he said, patriotism "has an entirely positive connotation," making him "eager to enhance the prestige of his country, but without destroying the reputation of his contemporaries." He insisted on proving to foreigners that Spain could produce not only warriors, explorers, artists, and writers but also scientists, who could contribute to global civilization and progress.

Praise for the speech appeared in leading newspapers. "Aside from its determinism and materialist tendencies," the monarchist newspaper *El Día* declared, almost grudgingly, "the work of the new academician is the notable and highly original work of a great thinker." *El Globo* named Cajal "the great Spanish savant" at a time when, it reminded its readers, "Spanish savants do not exist." Some saw his research as a way to bridge the gap between progressive science and traditional Spanish values. Appraising the significance of his research, at least one commentator recognized Cajal's findings not just as a collection of esoteric details but as "the sketch of an anatomical doctrine of intelligence." The first journalists who profiled Cajal made sure to emphasize his unlikely personal journey. "Before he was an illustrious academic," one article reminded readers, "he was the noblest son of Petilla de Aragón." Thus began the legend of Cajal.

CAJAL'S ADDRESS AT THE NATIONAL ACADEMY OF SCIENCES CAME AS Spain was in the midst of a crisis. Two years earlier, exiled Cuban rebels had returned to the island, inciting a rash of uprisings as they declared their independence for a second time. Spain's official response was "to fight war with war." The Liberal prime minister delivered a speech in the Cortes, the Spanish legislature, vowing to sacrifice "the last peseta of its treasure and the last drop of blood of the last Spaniard before consenting that anyone snatch from it even one piece of sacred territory." Spain's wartime motto became "to the last peseta and the last Spaniard." The army marshaled sixteen thousand troops to join the twenty thousand already stationed on the island. Within their first two months in Cuba, an estimated half of all men contracted a disease; for every battlefield fatality, an estimated ten deaths were caused by illness. Hundreds of military officers on the mainland applied for early retirement to avoid deployment.

The Spanish economy, reliant on import and export duties, could not afford to lose its income from Cuba. If Spain abandoned Cuba, the government feared riots, sparking a revolution at home to match the one abroad. But the end of the empire was inevitable, and the government saw war as the only way to lose the territory without losing the regime. The new governor-general, Valeriano Weyler, was granted carte blanche to destroy the rebellion. At four foot ten inches tall, Weyler was even shorter than Napoleon and twice as brutal. The Spanish army forced four hundred thousand Cubans into concentration camps, the first in history, where an estimated one hundred and fifty thousand died of illness and starvation. A Spanish journalist, on assignment to the island for a liberal newspaper, compared the conditions there to those in Dante's *Inferno*. Cajal despised Weyler's "war of extermination."

Spain was not the only country with interests in Cuba. Since its founding in 1776, the United States—youthful, wealthy, and eager to expand—had coveted the island, which was "almost in sight of [its] shores." Some considered Cuba to be a "natural appendage" of America, like a fiber attached to a cell body. There was both a humanitarian and a commercial interest in the island. Millions of U.S. dollars were invested in Cuba, and the war with Spain was bad for business, as rebels slashed and burned sugar and tobacco crops. The United States, concerned by reports of brutality in Cuba, pressured Spain to pass reforms. "It was not civilized warfare," President William McKinley said of the Spanish campaign, "it was extermination." Newspaper publishers Joseph Pulitzer and William Randolph Hearst, vying for the highest circulation, stoked populist anger to incite the government to intervene, pioneering the tactics now known as yellow journalism. The

U.S. press published cartoons depicting Spain as a savage rapist and Cuba as a damsel in distress.

Three months after Cajal's patriotic speech, in February 1898, the USS *Maine*, an American warship stationed in Cuba, mysteriously exploded. The Americans blamed the Spanish; "the cause must be removed," said the American envoy to Spain, "before the disease can be cured." The Spanish president, in an April speech to the Cortes, declared that war was "inevitable." Six weeks later, the Spanish-American War began.

CHAPTER 20

"From Catastrophe to Catastrophe"

In the summer of 1898, Cajal and his family traveled thirty miles north of Madrid to Miraflores de la Sierra, a scenic mountain town that was a popular destination in the late nineteenth century for the haute bourgeoisie. He had won two prizes in the previous two years—one thousand pesetas from the Royal Academy of Medicine in Spain and fifteen hundred francs from the Biological Society of Paris—and could now afford to vacation with his family. He rented a small villa next to that of his colleague and friend Federico Olóriz, whose son also served as Cajal's laboratory assistant. Cajal and Olóriz had met twenty years earlier while competing against each other in an *oposicion*, "a scientific fight that," according to Olóriz, "engendered mutual esteem." A native of Granada, Olóriz appreciated Cajal's "unpleasant Aragonese sense of humor," characteristically mordant and dry. They spent their leisure time reading, chatting, and walking along the winding road that led up the mountain to Santa Maria del Pablar, a fourteenth-century monastery built by the Carthusians.

At the time, Cajal was in the midst of studying the optic nerve, the "bewildering meshwork of wavy fibres" that transmits impulses from the retina to the brain. Cajal was enthralled by sight; his own revelations always came by way of visual impression, a process that he called a "marvelous alchemy."

Cajal was not the first to take a special interest in sight. More than two thousand years earlier, Alcmaeon had contended that, by following the optic nerve, one would eventually arrive at the soul. When one eye is closed, the medieval physician Mondino de Luzzi thought, the whole soul is transferred to the other. Hippocrates assumed that the purpose of the optic nerve was to shuttle away waste from the brain, secreted in the form of tears. The study of optic nerves was confounded by the fact that, at least in some species, as they tail away from the back of each eye, their fibers appear to cross. Galen, noting that the crossing fibers resembled the Greek letter *chi* (X), named this structure a chiasm. He believed that animal spirits, released from the ventricles of the brain, would meet at the intersection. Razi and Avicenna, pillars of medieval Islamic medicine, among other fields, both suggested that all the fibers of the optic nerves cross—a theory later known as total decussation—while Isaac Newton maintained that only some crossed and that others remained straight.

Strolling through the gardens of Miraflores with Olóriz, Cajal discussed his preliminary findings. After analyzing optic nerves in multiple vertebrate species using different staining methods, he found that some fibers cross—partial decussation—and that the number of crossings varies depending on the species. Ingeniously, Cajal determined that the length of the uncrossed segment is mathematically related to the distance between an animal's eyes. In species with panoramic vision—such as fish, reptiles, amphibians, and birds—eyes are positioned farther apart and angled slightly away from each other. Each eye receives a partial image of the world, which the retina then inverts. In order for the brain to create a unified image of the external world, Cajal reasoned, the brain must correct the distorted and fragmented representations of sensory space. He suggested that the decussation of the optic nerve facilitated a synthesis of visual information from each eye, and that this arrangement proved that, throughout the course of evolution, the retina and the brain had mutually adapted. Cajal had almost finished his paper on the subject, and with the encouragement of Olóriz, he was planning to publish it, despite his reluctance to theorize.

Whenever they grew tired of reading or walking, Cajal and Olóriz would play chess. Though Cajal claimed to have renounced the game ten years earlier, his autobiography includes a photograph of the two men sitting in shirtsleeves at a small table outside a cobblestone house, foreheads sweating in the summer heat, gazes fixed on the board. Olóriz scratches his head as their children look on. One afternoon in mid-July, their game was interrupted by news that the Spanish fleet in Cuba had been destroyed. After a weeks-long blockade in the bay of Santiago, pressure from the Americans had been mounting. In a final, desperate act, the Spanish naval commander had tried to break through on a Sunday, while the Americans attended church. At the same time, American troops, including Theodore Roosevelt and his "Rough Riders," besieged the city of Santiago by land. The Spaniards lost over three hundred men; the Americans, only one. "My bugle sounded the last echo of those that history tells sounded in the great conquest of Granada," a Spanish flagship captain lamented. "It was a signal that the history of four centuries of greatness was coming to an end." *El Imparcial* ran a headline reading "The Great Sin."

In the weeks following the defeat, Cajal noticed a "common cloud of silent sadness" that hovered over daily life. The 1898 war meant the beginning of the end of the empire, "an epilogue to four centuries of history in a world for which Spain pronounced the *fiat* of life almost as much as God," wrote one commentator in the *Heraldo de Madrid*. Despondent and ashamed, politicians, generals, marines, statesmen, economists, orators, comics, bullfighters, and ballerinas fled the country. "Everything is broken in this unhappy country," declared *El Correo Militar*, an outlet rarely critical of the state; "there is no government, no electorate, no political parties; no army, no navy; all is fiction, all decadence, all ruins."

The Spanish press solicited insights from any and all public figures. On the front page of the October 26 edition of *El Liberal*, an editorial by Cajal appeared as part of a series called "The Nation Speaks." "We have learned nothing from past wars," he lamented.

We go from catastrophe to catastrophe. When we seem to have been given the last thump, after which we would not be able to raise ourselves, we are hurled further into the deepest ravine whose end cannot be seen. The only thing that is perceptible, is that each somersault through the air leaves us sillier and with less sensibility, making us forget little by little the first falls.

Cajal, who picked many fights in his youth, now begged his countrymen to "renounce our *thuggery*, our belief that we are the most warlike nation in the world." "The misfortune of a country is not in its weakness," he later wrote, "but in the ignorance thereof of those whose inexcusable duty it is to know it."

Not everyone welcomed the input of "the great savant." Cajal "blunders and derails in his historical, political, and moral speculations," wrote the editor of a far-right, anti-liberal Catholic daily. The editors of *El Correo Militar*, another publication in which Cajal published an editorial, were similarly unkind. "Cajal comes with a poultice of flaxseed meal to cure the ills of the Homeland," one critic parried in an adjoining column titled "Broadsides." "Shoemaker to your shoes . . . and Doctor Cajal, to your syrups."

Cajal was devastated by the crisis in Spain, but he tried to soldier on. "Despite this disgraceful war that Spain sustains," Cajal assured Retzius in September 1898, "I have not lost my enthusiasm for working. Nevertheless I continue doing a little and overseeing the investigations of some outstanding scientists." In 1896 and 1897, Cajal published a total of twenty-three papers; in 1898, he published only eight and, most significantly, made no discoveries. "How can I continue researching science when the country is in 'a death trance?'" he recalled asking himself.

There is another reason why Cajal found it difficult to work, though in his typical fashion he did not address the trauma in writing. His mother, Antonia, died in June 1898, a month before the surrender at Santiago, his namesake. Pabla and Jorja, who remained unmarried and lived with their parents, were her only companions in her later years. Since moving to Madrid, Cajal had made increasingly infrequent visits home to Zaragoza. Pedro, who since 1895 had been living in Cádiz, visited home once in a while. After Antonia became ill, Justo, who was seventy-six but appeared much younger, had an affair with a twenty-six-year-old peasant named Josefa Albesa Arrufat, whom he met during his medical rounds. The woman became pregnant, and after Antonia's death, Jorja enlisted the local priest to convince Justo to marry her. Though he was not religious, Justo obliged. People gossiped at the funeral. Cajal was enraged. Pabla and Jorja moved out of their father's house but continued to see him, and Pedro eventually resumed talking to him as well. Nowhere did Cajal mention his father's affair or their subsequent estrangement. But Cajal never forgave his father. He never spoke to him again.

No trace of Antonia Cajal's own writing remains, but after Pabla died, a letter was found lost among the family's belongings in which Pabla states that her mother was more intelligent than her father— "*considerably* more"—and that it was *she* who most influenced her brother, the great scientist. It was only fitting that history should know him by her name.

"THE DISASTER" LAID BARE A RIFT BETWEEN SPAIN'S DELUSIONAL self-image and reality. Spaniards still thought of themselves as a world power, a belief upon which their national culture was based. The feebleness of economic, cultural, and political life, hidden behind demonstrations of colonial might, was now exposed. Two-thirds of the Spanish economy still relied on agriculture, a statistic unchanged since the beginning of the century. Society was grotesquely unequal, as the old aristocratic hierarchy still prevailed; the largest landowners in the country were descended from nobles, who hoarded potentially productive property as private estates. Most workers were day laborers. The population of Spain had exploded in the final quarter of the nine-

teenth century, but the economy failed to keep up, propelling an estimated million Spaniards overseas to the New World to seek their
fortune. Only half of all children went to primary school, and the per-
centage who attended secondary school was far smaller. Illiteracy remained around 50 percent. In 1898, as much of Europe and the United
States was urbanizing, the vast majority of Spaniards—91 percent—
lived in towns with fewer than five thousand inhabitants, where peasants crowded into market squares, driving their sheep through the
streets while their pigs and chickens rooted about in the sewers.

Intellectuals of every stripe—from Carlists to Republicans—
gathered in clubs, bars, casinos, and cafés, debating the future of the
nation. For the last quarter century, moderate Liberals and Conservatives had peacefully alternated power, a rigged system known as *el turno
pacífico*, or "the peaceful turn." Middle-class Spaniards—who called
themselves the "neutral" class, neither aristocratic nor proletarian—felt

betrayed by the self-dealing Restoration government, having ruled for almost twenty-five years, which they characterized as foolish, neglectful, and vain. Some blamed centuries of misrule, others international capitalism, a Masonic conspiracy, or even divine punishment. Membership in the General Workers' Union, the precursor of the Socialist Party, more than quadrupled over the next two years. Conservatives decried the lack of traditional religious values, while Liberals faulted defects they claimed were intrinsic to the Spanish character. After dozens of governments, multiple failed constitutions, and recurring civil wars, the Spanish people had welcomed stability, even at the expense of their democratic rights. Now election victories by either party seemed illegitimate and out of touch with popular opinion. Protests spread from the barracks to the streets.

In the aftermath, a movement called Regenerationism took hold, calling for a "revolution from below." Calls for reform rose up not from the Cortes floor but from chambers of commerce in cities around Spain, as disenfranchised farmers and shopkeepers grew tired of the indifference of the elites. The leader of the movement was Joaquín Costa, the spokesman for the Chamber of Agriculture of Alto Aragon, who organized farmworkers to lobby the government for radical institutional change. Cajal's *tertulia* embraced Regenerationism. They bandied about the ideas of Proudhon and Marx and discussed a Regenerationist proposal for land redistribution intended to return the country to its agrarian, collectivist past. Cajal had devoured Costa's speeches, books, and pamphlets, replying to the author's ideas by underlining words and making annotations in the margins. Cajal published more editorials in *Vida Nueva*, a pro-Regenerationist magazine founded one month after "the Disaster."

Meanwhile, in the laboratory, Cajal turned his attention back to the cerebral cortex, "[that] enigma of enigmas . . . the subject [that] attracted me with singular force." The singular force of the cortex was its tantalizing potential to explain human psychology. Wrinkled and gray, the cerebral cortex—from the Latin word meaning "bark"—

envelops the outside of the brain, wrapping around the fissures and folds, such that, if laid out flat, its surface area would be tripled. It had taken Cajal only a few years to master the cerebellum, the first area of the nervous system that he studied. When he began to investigate the cerebral cortex, in 1890, he had expected—naively and arrogantly, he freely admitted—to make similarly quick work of it. But the cerebellar cortex has only three layers of cells, while the cerebral cortex has six, each populated with different types, packed together more densely than anywhere else in the brain. Microscopic images of the cortex were so complex that Cajal resorted to a *camera lucida*—as he only sometimes did—to accurately draw them. "In the convoluted brain warp," he said, "it is only possible to advance step by step." Cajal referred to the anatomists who had come before him as "sappers"—military personnel who construct forts, tunnels, or trenches, paving the way for the battle to come.

Spanish law prohibited dissection until twenty-four hours after a person's death, but for the methylene blue and the black reaction to work best, brain tissue had to be as fresh as possible. Cajal's preferred subjects were deceased newborns or infants fifteen to thirty days old, following the embryological method. He befriended the director of the local maternity hospital, a publicly funded charitable organization providing services to pregnant women and children born out of wedlock, who were often abandoned in the doorways of churches. As an epidemic of colitis swept through the city, Cajal gained access to more than two hundred samples from patients who had died of the disease. Nurses at the hospital, who were also nuns, served as his assistants during dissections, wearing cornettes on their heads and scrubs over their habits.

IN JUNE 1899, A LETTER ARRIVED FROM THE UNITED STATES OF AMERica. Cajal recalled being "profoundly surprised and perplexed," as the Spanish-American War had ended less than a year earlier. The letter was from the president of Clark University in Worcester, Massachusetts, which was planning its ten-year anniversary celebration. He invited Cajal to deliver a series of lectures. "It made no sense to me how the United States had become acquainted with a humble Spanish investigator, a professor belonging to the conquered and humiliated race," he recalled. A check for six hundred dollars was also enclosed. In August 1899, twenty-five years after nearly dying in the jungle, Cajal set out for the New World once again.

Researchers were typically accompanied at international conferences by their wives. Silveria had never joined Cajal before, because she had to stay home and take care of the children; and given their strict budget, the expense of a ticket was difficult to justify. Now, for the first time, Silveria accompanied Cajal on his travels. Before arriving in Worcester, they would stop in Paris and New York City. The twelve-day trip was also intended to be a honeymoon, which they could finally afford, twenty years after their wedding. Silveria, however, ended up playing the role of Cajal's nurse.

One of the traits Cajal inherited from his mother's side of the family was a propensity for migraines, a condition that was exacerbated by intense heat, and he arrived in New York City during a heat wave. Temperatures in the city reached as high as 110 degrees, about 20 degrees warmer than in Madrid. Cajal called New York "the stupendous city of skyscrapers, of multimillionaires, of enslaving trusts, and of suffocating heat." He walked the streets carrying his Kodak camera, mostly taking photographs of the architecture. One night, while Cajal and Silveria were sleeping, alarms sounded throughout the hotel when a fire broke out. Other guests left their rooms through the windows and proceeded down the fire escape. Paralyzed by fear, Silveria refused to set foot on the "aerial steps." Luckily, the fire brigade was quick to arrive.

At the train station in Boston, Cajal and Silveria were met by a young professor wearing a chic silk suit. He insisted on carrying their luggage. Full of clothing and laboratory equipment, it was exceedingly heavy. Silveria asked him why he did not call a porter and avoid wrinkling his coat. "We live in America, the home of democracy," the professor said. It was July 4 when Cajal arrived in Worcester, and all through the night he heard shouting and fireworks. Democracy was so loud that he could not fall asleep.

Everywhere he went, Cajal felt that the people of America were looking down on him. Not without irony did his host, a wealthy philanthropist, welcome the presence of "a Spaniard with common sense." "Rarely have I encountered a husband and wife that seemed so made for each other," their host said, an insult disguised as a compliment. The black legend was alive and well. The American press hounded Cajal with questions about Spain, the country that the United States had "known so intimately of late," as one reporter euphemistically put it.

After a week of banquets, sightseeing, and receptions, a large crowd gathered in the lecture room to hear from the group of visiting foreign experts. "Spanish Scientist Will Speak Today," one local newspaper headline read. "Exotic Tongues Speak Wisely," another claimed. Auguste Forel, one of the speakers, was portrayed as the originator of neuron theory, while Cajal, who "blazed new paths in histology and neurology," was said to be the theory's foremost proponent—"no worker has accomplished more." In his talk, "Comparative Study of the Sensory Areas of the Human Cortex," Cajal offered a detailed summary of the microscopic anatomy of the cortex, including his most recent findings, which were still unpublished, and invited the audience to look through his microscope. With a mixture of awe and humility, Cajal summarized his views on the cortex: "The brain is only a savings-bank machine for picking and choosing among external realities," he declared. "It cannot preserve impressions of the external world except by continually simplifying them, by interrupting their serial and continuous flow, and by ignoring all those whose intensities are too great or too small." The audience, full of prominent American scholars, gave Cajal, a Spaniard, a standing ovation. He was shocked.

CHAPTER 21

"The Mysterious Butterflies"

Cajal returned from the United States complaining of heart palpita-
tions and insomnia, symptoms of what was known as "neurasthenia,"
or nervous exhaustion, a vague, catch-all medical condition popularized in the nineteenth century. Americans seemed especially prone to
the complaint, which William James nicknamed "Americanitis," a
phantom "weakness of the nerves." Cajal was an admitted hypochondriac, and his psychological state was especially vulnerable after the
death of his mother. The gloom of his younger days, always latent in
his temperament, returned with a vengeance. Approaching fifty years
old, he was noticing the first signs of old age.

The view through Cajal's window on Atocha Street was thwarted
by chimneys, and Madrid was packed with people. In the afternoons,
after his *tertulia*, he would walk down Alcalá Street, brick-paved and
lined with recently electrified tram tracks, passing telegraph poles and
mansions with impeccable facades lining Retiro Park. He empathized
with the pine trees, looking frail and out of place, "their needles emaciated or reddish, their branches fallen or dry." Pure mountain air was
the tonic that always healed him. He became obsessed with owning a
cottage far from the chaos of the city, somewhere near the mountains,
under the sky and stars, where, in a laboratory surrounded by a garden,
he could seclude himself in the summertime to work in an atmosphere
of peace and calm. "Whatever the preoccupation of the spirit," Cajal

wrote, "we can always find some quiet corner in which the peaceful beauty soothes the pangs of sorrow and opens up a new channel for our thoughts."

When the opportunity arose to buy some land on the outskirts of Madrid, Cajal jumped at the chance. Cuatro Caminos was located between the country and the city, in the hills of the *dehesa*, the thick meadows with hundred-year-old trees and grazing livestock that blanketed the Spanish tableland. The plot was only six miles from his home on Atocha Street, a twenty-minute ride by tram, and he could rent a carriage for half a day to drive him to and from class. In February, Cajal ordered the construction of a house beside the Isabel II Canal, ensuring, despite Madrid's perennial drought, an abundance of water for his twelve-thousand-square-meter garden. Perched atop a slight elevation, the house offered unobstructed views of lush gardens, neatly sown fields, and the majestic Guadarrama mountains. Flocks of people came from the city on Sundays and holidays to picnic and take in the scenery. Cajal could make them out in the distance, but he never had to interact with them.

Cajal felt more comfortable among the working class than the aristocrats and bourgeoisie of downtown Madrid. On Almansa Street, Cajal's street, splendid country houses alternated with ramshackle hovels, and humble shops stood beside large factories. At every step, merchants hawked their wares, their burners set up against the wall to heat *cochifras*, meat fricassees, that they had seasoned at home and brought to work. Every resident knew where the great doctor lived and would point any visitors to his door. Cajal's square, redbrick house was three stories tall and set on an incline. A trained vine draped the entranceway to the garden, and from a simple bench in the gazebo he could sit and take in the panorama. Cajal felt refreshed in his new home, he said, as though he lived in perpetual spring.

"An unequivocal sign of foolishness, frivolity, or ignorance is being bored in the countryside," Cajal wrote:

There, among trees, flowers, and insects, the enchantment of life is revealed to us. And when night turns the starry sky above our heads, the infinite opens itself endlessly before us. Only in the countryside do we take communion with the stars, and our imagination takes pleasure in remembering suns and planets, nebulae and extinguished stars. The Milky Way, which constitutes our small universe, wrapped in hardly conceivable universes. For me, the starry night is the sovereign reactive of the ignorant. One is bored: then one is an imbecile.

In Cuatro Caminos, Cajal began to recover from his ailments. In the mornings, in the light-filled laboratory beside the garden, he studied regions of the human cerebral cortex, and in the afternoons, wearing a straw hat to shield his bald head from the sun, he planted beans, peas, and strawberries, recording planting methods in a little notebook. "Spain is truly the land of melons," Cajal wrote (*melones*, in Spanish slang, is another word for "fools"). He spent hours and hours in his garden, taking photographs, tending the orchard, and planting a botanical garden in a greenhouse to rival the one in Zaragoza from his student days, with heat for exotic species and tropical plants. He would have liked to extend his garden farther, but he was hemmed in by nearby cemeteries. Instead, he explored the endless gardens of the brain. "Like the entomologist in search of colorful butterflies," Cajal wrote, "my attention has chased, in the gardens of the grey matter, cells with delicate and elegant shapes, the mysterious butterflies of the soul, whose beating of wings may one day reveal to us the secrets of the mind."

On August 8, 1900, during a summer of extreme heat, Cajal received a telegram from a colleague. "Congratulations six thousand francs prize awarded today," the telegram read. Cajal had won the first-ever Moscow Prize, awarded at the International Congress of Medicine for outstanding discoveries during the previous three years. The money came "in the nick of time," Cajal noted in a letter, "because

I have the second volume of *Centros*, the second issue of *Revista*, and the third edition of histology completely stuck, for lack of fat, I mean resources and spirits for writing." Cajal was surprised, since he had not submitted work to the International Congress, which was held in Paris. He certainly had not thought to attend.

After a soiree in the Luxembourg Gardens, at which Sarah Bernhardt recited poetry in front of fifteen hundred attendees, the closing ceremony took place inside the assembly hall at the Sorbonne. A French military band performed a rendition of "La Marseillaise," and the president of the conference strode onstage to announce the winner of the inaugural Moscow Prize. The honoree, the president declared, was "a very illustrious man, a very modest man, a great teacher in a poor environment who lives modestly and does not strive for wealth or acclaim." He said someone's name, but the hall was so large that no one could hear it. "It's Spanish!" the Spaniards in the crowd started murmuring. The president went on: this was the man responsible for the neuron, a discovery "of such cardinal importance to anatomy, physiology, pathology, and scientific psychology . . ." "It's Cajal," they whispered, their excitement uncontainable. "The winner of the grand prize is the Spanish savant," said the president, "doctor Cajal." The crowd shot to its feet and applauded. Then came a sound that none of the Spanish attendees ever thought they might hear: a French band playing the Spanish national anthem. One of them recalled bursting into tears. The Spanish press noted the exact time the award was announced— 3:55 P.M.—as though it were a birth.

In the wake of the Disaster of 1898, Cajal's award was treated as a national triumph. Praise and congratulations came pouring in from politicians and from María Cristina II, the queen regent. Accounts of Cajal's travels and achievements soon appeared, stressing the originality and creativity of his work. "Once again the name of Spain is pronounced with enthusiasm in foreign lands," read *El Liberal*. Biologists in France, England, Germany, and the United States already knew his name well. It seemed that Spaniards were the only ones who needed to ask, "Who is D. Santiago Ramón y Cajal?"

The day after the announcement, *El Heraldo de Madrid* sent a reporter to Cajal's new house in Cuatro Caminos. "Look," he told the reporter, ". . . in this little laboratory, despite it facing this window, where the sun shines all day, it has not reached, at most, 24 or 25 in the hottest hours." At night, it was cool enough that he slept with a blanket. Cajal claimed that his intention had been to move his family to the healthiest location in or around Madrid, and after testing the chemical composition of every water source, he had determined that the Isabel II Canal, in Cuatro Caminos, had the lowest pathogen count. Not yet content, he studied the mosquitoes in the area, still haunted by the specter of malaria, and could not find a trace of the disease. In truth, it was most likely the only land he could afford.

When the reporter first arrived, he had found Cajal bent over the microscope, perusing a medulla oblongata, the labyrinthine structure inside the brain stem that is responsible for involuntary functions such as coughing, sneezing, and vomiting. Cajal's house in Cuatro Caminos was technically his vacation home, the reporter explained, "but that is not the right word, if it gives one the sense of resting, because Cajal always works." Another reporter from the popular Sunday magazine *Por Esos Mundos* wrote, "When one thinks of Cajal, it is impossible to separate him, in an imaginary vision, from his microscope, from his disorganized worktable, from his experimental slides, from his tubes and acids, from his whole field, in the end, of scientific operations, which is the site where we find him every day at all hours."

An *El Imparcial* article included several quotes from Silveria, a rarity in the written record of Cajal's life. As she and Cajal offered a tour, she interrupted the conversation to remind the reporter of their humble past. "And now, as you see, he has a little more, because you know that he began with a bad microscope, some tubes and glasses and bottles and a shaving razor, and by force of patience, time, and sacrifices, thanks to God, now he can work in better conditions." Silveria took great pride in her husband's accomplishments, and he, in turn, always credited her contributions and sacrifices. ("It must be advised," the reporter warned, "that Cajal's wife is the primary admirer and enthusiast of her husband's works and discoveries which she helped and stimulated with great care and intelligence.")

Local businessmen organized a banquet in Cajal's honor at his neighbor's home, which was packed with friends, admirers, and strangers who delivered spontaneous, heartfelt toasts. Cajal refused every word of praise. Aragonese folk songs were played in the round, and the crowd sang the kinds of couplets that he had heard as a child, praising legendary heroes like El Cid.

But the honor that Cajal cherished most was a tribute organized by the University of Madrid, attended not only by faculty members in their academic gowns but also by men from his *tertulia* and the Athenaeum. Nothing pleased Cajal more than watching students mingle with some of Spain's most prominent intellectuals. In the assembly hall of San Carlos, ornamented with flowers and fine drapes, Cajal read from prepared remarks in an attempt to control his emotions: "I say to you, the youth, the men of tomorrow in these recent mournful times, the country has shrunk; but you should say 'To a Small Country, a Great Soul.'"

"I am not a genius," Cajal told the crowd, "but rather a patriot." His life, he said, was "the blessed history of an unshakeable will resolved to triumph at all costs."

AFTER THE DISASTER OF 1898, A NUMBER OF MILITARY, POLITICAL, and literary figures published memoirs, and Cajal's friends often asked how he had come to pursue the unusual life of a scientist, especially in his youth. And so, in his "leisure and rest times, though few and far between," Cajal began writing his memoirs, which ran serially in two new pro-Regenerationist publications.

Cajal presented his memoirs as a psychological case study, with himself as the subject, as well as a criticism of the national educational system, which he hoped to reform. He described this work as "the story of a common life, as poor in appeal as it is rich in disillusions and contradictions."

My Childhood and Youth, published on its own in 1901, received a mixed reception from Cajal's peers. Miguel de Unamuno, perhaps the leading voice of the Generation of '98, praised the book as "a teaching of energy and faith and also a teaching of patriotism." The novelist Azorín called it "the rich experience of one of the most powerful contemporary minds," which, despite its limited publicity and sales, was "fundamental to the ideology of [Spain]—in a certain moment—and has constituted one of the factors of its social or literary evolution." Azorín appraised Cajal's prose as "truly literary . . . a clear, precise, clean, enjoyable, suggestive style." Others disagreed. "Cajal is the most important [voice] in Spain," wrote the novelist Pío Baroja, "but his literary style is archaic." Unlike his direct and plainspoken lectures, Cajal's prose was highly affected and bombastic, with long sentences convoluted by extended metaphors and nested clauses—the style of a poor boy from the country desperate to sound cultured and urbane.

Over time, Cajal grew more and more dissatisfied with his memoirs. "There is a lot of muck in the first chapters and even digressions and lyricisms that today I find annoying, inopportune and even pedantic," he wrote in a 1915 letter to Azorín, who had become a friend. He felt self-conscious about what he had divulged and embarrassed by the "lyrical and philosophical" interludes and "monotonous" descriptions of his childhood adventures, yet he gave a free copy of his memoirs to any friend who asked for one. "My arrogance of those days surprised me a little," he wrote in the prologue to the second edition, in 1917. Later, Cajal wrote a second volume, *Mi obra científica* (*My Scientific Work*), which he hoped would be "more interesting and I think less defective." The final edition of his memoirs, released in 1923, still retained his stories of building cannons, stealing fruit, pranking teachers, and escaping prison. The writer Ramón Peréz de Ayala, recognizing Cajal's intentions, appraised both volumes as being "on par with the most successful works of the classic picaresque, for its language, its agreeableness, its humanity, its charm."

IN ADDITION TO WRITING HIS MEMOIRS, CAJAL USED THE MOSCOW Prize as an opportunity to seek additional monetary support for his research, emphasizing to a reporter who visited him that scientific equipment was expensive and that he could never compete with foreign researchers if he continued to finance himself. The cost of printing his work with illustrations was two or three hundred pesetas per month, he said, plus another two thousand pesetas yearly to buy specimens. Foreign students arrived in Madrid hoping to study with the Spanish master, but because his laboratory was so small, he had to turn them away. "Surely there is no savant in Europe who publishes with as much poverty as I," Cajal said.

A proposal to build Cajal a research institute became a rare item of political agreement among the Spanish press. "Spain should do with Ramón y Cajal what Germany did with Virchow, what England did with Darwin, and France with Pasteur," one reporter said. The major liberal newspapers—*Heraldo de Madrid*, *El Imparcial*, and *El País*—called for five-hundred-peseta donations from members of the general public. Conservative outlets concurred. "Spain, not Cajal, has the obligation or incentive to provide this illustrious savant with the independent position that he deserves, with a laboratory to set up as he pleases, with the freedom to spend," proclaimed *La Época*. Councilmen from Zaragoza offered to open a facility themselves, should the politicians in the capital fail to do so.

In October, María Cristina II issued the decree founding the Institute of Biological Investigations, allocating eighty thousand pesetas to set up the facility and cover the cost of instruments, journals, materials, and personnel; by comparison, the same amount was dedicated to new telegraph lines across the country. *El País* called the Institute of Biological Investigations "the only good work of the government in the whole period."

Cajal's laboratory, opened in early 1901, was fully equipped with the latest technology, from microscopes to microtomes to chemical reagents. He no longer had to fund the publication of his own journal; *The Trimonthly Review of Histology* was reincarnated as *Works of the Laboratory of Biological Research*—or *Trabajos*—printed annually. Cajal now felt an even more pressing obligation to publish his work, so as to justify the government's investment. He no longer had the luxury of concentrating on one subject for an extended period of time. His pace of research became frantic. "Fortunately," Cajal wrote in a January 1902 letter, "I am in good health and eager to work, above all since the government has done me the favor of granting me a good laboratory and sufficient funds to finance my publications." Scientists around the world eagerly awaited his findings, and translations appeared quickly. With his own research institute, Cajal could realize his dream of founding a "Spanish school" to compete on the world stage, rivaling the scientific output of the Germans, French, and British. His initial salary was ten thousand pesetas, but he lowered it to six thousand, which he felt was fairer.

Students came from all across Spain and the world to study with Cajal. "[He] makes a personal impression which accords with his works," one researcher who visited the lab at the time recalled, "earnest and forceful, physically strong and vigorous, at the very prime of his powers." His first pupil at the laboratory, Jorge Francisco Tello, hailed from the province of Zaragoza and, like the master, had forsaken a surgical career to pursue research. "When working," Tello recalled, "[Cajal] says nothing, works and works silently, lost in thought, hermetic." Once in a while, when he saw something curious, he would cock his head and talk to himself in Aragonese slang. Tello was convinced that, with his clumsy feet and large hands, the master was bound to break a glass at any moment, but he never saw it happen.

CHAPTER 22

"The Summit of My Inquisitive Activity"

For a brief moment after the Disaster of 1898, the Regenerationists had seemed as though they could overthrow the Restoration government, which retained power despite the national crisis. Then the Spanish stock market went back up, and traditional politicians reconsolidated their power. Fearful of violent uprisings caused by growing inequality, the Regenerationists turned their backs on the working class and were accused of protecting their privileges and benefits. "We washed our hands, modern-day Pilates, and blamed others," one Regenerationist recalled. Regenerationism was reduced to a buzzword, used in corporate advertisements, in politicians' election platforms, and letters from bishops to clergy. Regenerationism "did not move a single fiber of popular sentiment," said the new Conservative prime minister. "We regenerators of '98," Cajal recalled, "were only read by ourselves."

In 1900, Joaquín Costa had formed a political party called the National Union, and Cajal was the first intellectual to appear at his house to sign the manifesto. With the creation of the Institute for Biological Research, Cajal's work now depended on government funding, and as power changed hands abruptly, he would need support from every leader, no matter his party affiliation. The young king, Alfonso XIII, who ascended the throne in 1902, had an interest in science and a high opinion of Cajal, who cultivated a relationship with the new ruler. Costa accused Cajal of abandoning the Republican cause and supporting the monarchy, but for the sake of Spanish science, he could ill afford to be radical. "Ramón y Cajal, who was always on the margins of politics," an acquaintance from the Athenaeum said, "did not have a greater public concern than the scientific development of Spain."

One day a royal emissary appeared at the institute to inform Cajal of the king's desire to meet with him at a date, time, and location of Cajal's choosing. "Is it urgent?" was Cajal's reply. Members of the royal court were shocked; no one kept the king waiting. Fifteen days passed before Cajal appeared at the Royal Palace, unannounced, requesting a royal audience and excusing himself for the delay. On visits to the king, subjects usually trailed impressive entourages; Cajal brought the janitor, the porter, and the doorman from the institute. Alfonso XIII welcomed Cajal immediately. Cajal had inscribed a special edition of one of his books to the king but did not include the customary official seal—either an honest omission by an absentminded scientist or, most likely, the subtlest act of rebellion that a die-hard Republican, in his current conflicted position, could muster.

WITH CAJAL'S NEWFOUND STATUS, HIS CIRCLE OF FRIENDS WIDENED to include Spain's elites. Cajal attended new *tertulias*, including one called "Old People" (he was fifty), which met for meals at the Café Ingles, famous for its chops and steaks. Another highly exclusive *tertulia* that Cajal attended was held at the home of a Valencian sculptor, Mariano Benlliure, whose Sunday paella feasts were so popular that famous writers, ministers, and former prime ministers from all across the political spectrum attended, arriving in shifts. "If those mental tournaments had been copied shorthand," one participant recalled, "they would have been a gift of art and science and an anthology of the most extraordinary and sharpest qualities of genius."

Also at Benlliure's house was an Aragonese poet named Marcos Zapata, who was at work on a book dedicated "to the excellency of the council of your forever heroic city of Zaragoza." He asked his countryman to write the prologue, and Cajal agreed. "You ask me," the prologue begins, addressing the author, "'What is the cause of the fact that I, as so many other literati, live a comedy and write dramas, have happy conversation and sad thoughts?" Cajal presented his theory of behavior as a political struggle among neurons in the brain. "Fallow brain cells, i.e. those responsible for functions not used every day, remind the self of their rights to an active life," Cajal said, alluding to Regenerationist proposals for land redistribution to maximize the agricultural potential of the country. "They loudly demand a turn at the banquet, as soon as the occasion of a general recess arrives."

BETWEEN 1900 AND 1904, OPPONENTS OF THE NEURON THEORY, though by that time in the minority, intensified their attacks. The current area of dispute was not the relationships between nerve fibers but the material inside them. Histologists used to believe that nervous elements were composed of a granular fluid, until 1843, when researchers examined the nerve cells of crayfish and found extremely fine fibrils, bundled like strands of rope. Axons are measured in millionths of meters, about the thickness of a cobweb; neurofibrils are a thousand times thinner, and staining reagents caused them to disintegrate. The black reaction and methylene blue techniques were not capable of staining them. Because of technical limitations, few researchers believed that these neurofibrils existed, and, if they did, they doubted that the fibrils served any function.

In 1897, a Hungarian anatomist named István Apáthy, who had managed to stain fibrils in leeches and worms using the methylene violet technique, claimed that the fibrils formed "a fine elementary network," "threaded" and "continuous," conducting impulses like telegraph lines. In other words, he saw a reticulum among the fibrils *within* nerve cells. He showed his slides at congresses all over Europe, and his drawings were brilliant. Then a German anatomist named Albert Bethe applied gold chloride to leeches and earthworms, rendering their nerve cells so clear that the fibrils inside looked like metal wires wreathing the nucleus. Bethe claimed that the fibrils actually extended beyond and outside the cell and throughout the system, forming a kind of supernetwork, connecting the bodies of other nerve cells as well as muscles and even sense organs. After sectioning the nervous systems of crabs and observing that they continued to transmit impulses, he concluded that neurons were not the agents of nervous transmission—fibrils were. Bethe reduced neurons to mere storage units for cellular energy, which protected and insulated the fibrils. The new neurofibrillary theory stated that, though nerve cells were independent, the neurofibrils formed a reticulum, and that they—not neurons—were the fundamental structure of the nervous system. The neurofibrillists dismissed neuron theory as erroneous and premature. "We must stop looking at the neuron as an anatomical and physiological unit," one neurofibrillist said.

Cajal referred to anti-neuronists like Bethe as "fanatics," moved by "an anarchical and calamitous passion" to fight under the "ancient banner." Reticularism, like a "contagion," became "virulent and widespread." Some of the most ardent neuronists were swayed by the new theory, including Van Gehuchten and Waldeyer, the very man who had formalized the neuron doctrine. Even Retzius, Cajal's close colleague and friend, reacted to the new findings with "astonishment," calling them "great progress not just in Histology but all of cell science."

In his anti-neuronist articles, Apáthy impugned Cajal by name, and Cajal began receiving anonymous, insulting postcards. "He would never say it to me if he were not thousands of kilometers away!" Cajal replied at the time, according to a student. He considered the war already won and was reluctant to waste his time on old research topics. In a letter, despite his respect for the new findings, Retzius appealed to Cajal's competitive instincts. "Now it is necessary to defend the admirable neuron theory against such attacks," he urged Cajal in a letter. ". . . It is you who headed its formulation, who before anyone else must defend the fortress." Cajal, after all, was the same person who never backed down from a rock fight on the streets of Huesca. "I found myself compelled to halt in my path," he recalled, "and descend to the arena."

Cajal was skeptical of the neurofibrillary theory from the beginning. A scientist visiting from Canada at the time reported on his position. "It was interesting to find that [he] not only holds firmly to the doctrine of morphologic units in the nervous system," the visitor wrote, "but is as yet by no means prepared to admit the existence of organic continuity among these units; in other words, he supports the neuron doctrine as strongly as ever." Fibrils can be found in cells throughout the body, Cajal reminded his detractors, yet nowhere else but in the nervous system were they considered pathways for electricity. Cajal saw the neurofibrillary theory as a challenge not only to the independence of neurons but to the independence of all cells.

Bethe's experiments proved impossible for Cajal to replicate, so he sent his German colleague a note requesting the details of his gold chloride staining technique, along with several issues of the *Revista*. Bethe, who was twenty years younger than Cajal, had never heard of the publication. Cajal was insulted. Bethe claimed that he had searched for Cajal's books in the library at the University of Strasbourg, where Bethe taught, but could find only some translated articles in French journals. Condescendingly, he addressed his next reply to Cajal in French rather than German, so that the Spaniard could "understand him better." It was not possible for Bethe to share his methods, he claimed, because they were incomplete. Cajal must have heard echoes of Ferrán; no reputable researcher would withhold his technique unless he had something to hide. Ever since his debut on the world stage at the congress in Berlin, Cajal had gladly shared his tricks of the trade with anyone willing to listen, in the hope that his colleagues would confirm and expand his findings—and doubtless to demonstrate his own prowess.

When Bethe refused to divulge his method, Cajal asked for the slides themselves. Bethe's students had executed the preparations, he told Cajal, who found this odd, as he took pride in performing every step of the staining process himself. Finally, Bethe sent Cajal two preparations—from the cerebellum and spinal cord of a rabbit—which he requested be returned immediately, since they were his only ones. Cajal, who produced thousands of samples and kept hundreds in his drawer, grew even more suspicious. Still, he was nervous; if Bethe was correct, then his life's work would be undermined. His heart racing, he unpacked the preparations and began to examine them. "What a disappointment!" he recalled. "Those invaluable preparations . . . were no better than mine!" (He really meant "What a relief!") He did see the neurofibrils—"pale filaments" stained violet, wrapping around the bodies of neurons "as creepers attach to the trees of tropical forests"— but there was no proof that the structures were not artifacts, since they had yet to be confirmed by a second method. No theoretical edifice could be constructed on such limited evidence, Cajal thought. "I burned with desire to see the aforesaid neurofibrils in irreproachable preparations," he recalled. For two years, he experimented with different stains: ammoniacal oxide of silver, gold chloride with tannin, pyrogallic acid, and haloid silver salts. The "enigmatic system" of neurofibrils still failed to show up.

In the summer of 1903, Cajal went on vacation to Rome with his wife and sisters, to rest his "over-excited brain." He had always admired classical architecture and took many photographs of the Colosseum and other landmarks, but he could not take his mind off the laboratory. "I was seized persistently by hypotheses requiring experimental proof, technical projects apparently crammed full of promises," he recalled. Two years earlier, Luis Simarro, Cajal's old friend and rival, had published in Cajal's *Revista* the protocol for a new staining technique that called for the injection of potassium bromides or iodides into a specimen while it was still alive, concentrating ions inside the neurons so that, if the tissue was later immersed in silver nitrate, photosensitive compounds would form, which blackened upon exposure to light. The inner structure of cells could then be revealed using a photographic developer. This technique was known as the photographic method. Though less susceptible to artifacts, the photographic method was unreliable, and it worked only in larger cells and stained axons, not fibrils. On the train back from Rome, Cajal realized that the bromides that Simarro used were interfering only with the reaction and that the substance staining the neurofibrils was nothing more than "hot, free nitrate of silver."

When Cajal returned home, he rushed into the laboratory to test the new method, which worked spectacularly; fibrils in neurons of the spinal cord, medulla oblongata, ganglia, cerebrum, and cerebellum consistently appeared "splendidly impregnated with a brown, black or brick red colour and perfectly transparent." He published his results in the *Trabajos* later that year. "I did not sleep," Van Gehuchten wrote to him, after successfully applying the method himself, ". . . and neither did many in Germany, Italy, Hungary, France, and Spain." "Nor did I sleep for several days," replied Cajal, who found his "brain throbbing with the outlines of new plans for work." Cajal's improved silver nitrate method proved that neurofibrils form only within cells and *not* between them. It seemed that they were impeded from making direct contact with each other by some substance, a clear barrier between neurons. After the "complicated, time-robbing methods of Apáthy and Bethe," a visiting scientist remarked, Cajal's new technique was "so simple and so widely applicable that he has been able in a very short time to apply it to the examination of the most difficult parts of the nervous system."

Most histologists failed to master a single technique, but Cajal was an expert in many. "The simplification and improvement of methods has been no unimportant result of Cajal's genius," one contemporary said. For the next ten years, Cajal worked almost exclusively with his silver nitrate method, revealing neurons in greater detail than ever before. "The ingenious work of Cajal, his countless discoveries," a pupil later observed, "are the likely result of his genius for creating techniques."

Cajal also diversified his drawing methods to depict the extraordinary new images he uncovered, increasing his use of watercolors to his typical pencil and India ink to highlight the interior of neurons against a gray background. "Just like in painting or sculpture," one of his students later explained,

> *every era is manifested by a style or interpretative manner of the artist, wisely copied or grossly imitated by other more or less commoners, so too in the histological sciences does each important discovery correspond to the invention of a fertile technique . . . One can be a dilettante in music, painting, and sculpture without knowing how to strum an instrument, without having ever painted nor modeled either; but one cannot be a virtuoso in histology without being a performer at the same time.*

"For the histologist," Cajal declared, "every progress in staining technique comes to be something like the acquisition of a new sense open to the unknown."

AS A CONSEQUENCE OF CAJAL'S MOSCOW PRIZE VICTORY, THE 1903 International Congress of Medicine was held in Madrid, and thousands of doctors descended on the Spanish capital. In the lobby of the Institute for Biological Research, Cajal set up microscopes to show the attendees his findings, just as he had fourteen years earlier in Berlin. He and Simarro displayed sections of brain tissue as scientists roamed the hall, chatting and sipping beer. Cajal was determined to prove, beyond any doubt, that his views on neurofibrils were "absolutely objective." His speech at the conference was a polemic against Bethe, whose views amounted to "panreticularism, a kind of ocean where nervous currents pour out and all channels become confused." If Bethe's theory is true, Cajal said, then histologists should hang a sign above their laboratories with the inscription from the entrance to Hell in Dante's *Inferno*: *Lasciate ogni speranza*—"Abandon all hope."

Also at the congress was the Russian physiologist Ivan Pavlov, director of the Imperial Institute of Experimental Medicine in Saint Petersburg, where he was conducting experiments on gastric function in dogs. In the vote for the 1900 Moscow Prize, Pavlov finished second to Cajal, earning six votes to Cajal's fourteen. Pavlov spoke after Cajal, reading from a paper titled "Experimental Method as a New and Extremely Fruitful Method of Physiological Research," written in perfect German, which he pronounced with difficulty. For the first time in public, Pavlov discussed the effects of conditioned reflexes on the psyche. "Essentially one thing in life is of real interest to us—our psychical experience," Pavlov declared. His goal was to "explain the mechanism and vital meaning of that which most occupies man— our consciousness and its torments." Pavlov and Cajal asked the same questions, one by way of psychology and the other anatomy. "Now we must learn Russian," one attendee of the congress said, "just as we have been obliged to learn Spanish in order to follow Ramón y Cajal's papers in Madrid." Four years later, Pavlov won the Nobel Prize for the work that he first presented in Madrid.

CAJAL CALLED THE SECOND HALF OF 1903 "THE SUMMIT OF MY inquisitive activity," a period as exhilarating as his rush of discoveries in 1890 and 1891. Energized by the fight over neurofibrils, Cajal published fourteen monographs in 1903 and fifteen in 1904, some as thick as textbooks. In the midst of his newfound inspiration and productivity, on the morning of September 12, 1903, his father died. Seventy-five percent of Justo's estate, consisting of three houses and some money, was to be divided between Cajal, Pedro, Pabla, and Jorja; 25 percent went to Justo's young son, the child of his second wife. Cajal and Pedro, both tenured professors, made good livings. Pabla and Jorja, who never married and devoted themselves to taking care of their parents, had no income, and so the brothers gave their portions of the inheritance to them.

Cajal never mentioned the death of his father in writing. According to family lore, when Justo was an old man, he came down with bronchopneumonia—and, refusing to accept any medical help, insisted on bleeding himself.

CHAPTER 23

"The Most Highly Organized Structure"

The year after his father died, Cajal finally completed his masterpiece, *The Texture of the Nervous System of Man and the Vertebrates*, known as the *Textura*, synthesizing almost every significant histological finding about the nervous system in all vertebrates. The resulting tome consisted of two thousand pages of text and nearly a thousand illustrations, all drawn by Cajal himself, spread across two volumes. The work distilled eighty papers on two hundred different animals, written over the course of twenty years, into a single testament to the individuality of the cells in the brain.

Cajal's publisher was Nicolás Moya, the first in Madrid to specialize in medical texts, whose bookshop was a short walk from the medical school. The first print run totaled eight hundred, and Cajal funded the publication himself, at a loss of three thousand pesetas, since, as usual, he insisted on printing in color to help the reader distinguish between different nerve cells and their constituent parts. Cajal was eager to win "the respectful praise of critics and the flattering opinions of the most prestigious savants." The *Textura* was intended as a "trophy placed at the feet of the deteriorating national science and the offering of fervent love produced by a Spaniard to his scorned country."

The title is a reference to Andreas Vesalius's 1543 work *On the Fabric of the Human Body*, considered modern anatomy's foundational text. Born in the Hapsburg Netherlands, Vesalius was a subject of the

Spanish Empire, who spent his final years living in Spain, where he served as court physician to Charles V, the Holy Roman emperor. *On the Fabric of the Human Body* was dedicated to Charles's son Philip II, who was Spain's king during its Golden Age. When Cajal first triumphed on the world stage, Kölliker dubbed him "the greatest Spanish anatomist since Vesalius." *On the Fabric of the Human Body* transformed anatomy into a modern science; with his *Textura*, Cajal intended to signify that brain research had entered a new age.

Galen's views on anatomy had ruled as dogma for more than a millennium, despite the fact that, fearing religious taboos, he did not dissect human corpses. A champion of direct observation, Vesalius challenged Galen's theories with concrete evidence, examining the human body in its entirety by himself. Vesalius was also a great artist, who took advantage of advances in woodcut technology to execute his illustrations; *On the Fabric of Human Anatomy* contains anatomical images, created by draftsmen from the studio of the great Renaissance master Titian, that were far superior to those in any other work of the time. Human bodies, stripped of skin, were posed expressively and rendered in exquisite detail, like portraiture subjects. Vesalius's most famous plate depicts a skeleton leaning over a table while contemplating a skull. The table is a tomb. "Genius lives on," the inscription reads, "all else is mortal."

The title also placed Cajal's work in a long-standing metaphorical tradition. Since the earliest microscopic studies in the seventeenth century, organic tissues had been compared to embroidered garments. Before building microscopes to observe invisible life, Van Leeuwenhoek encountered magnifying lenses as a teenager while apprenticing with a draper, as textile purveyors used them to count threads in order to determine the quality of fabrics. When Robert Hooke looked at mushrooms through his microscope, he saw "an infinite company of small filaments, every way cortex'd and woven together, so as to make a kind of cloth . . . Nature in this [is], as it were, expressing her Needlework, or imbroidery."

Of all the "fabrics" in the body, nervous tissue appeared the most knotted and warped. Descriptions in Cajal's day included comparisons such as "thread-like," "an extremely fine interlacement," a "knitted tangle," "a blanket," "a kind of braid-work," "a sheet of muslin," and "a dreadful felt-like maze." These metaphors echo Cajal's first memory of tangling the shuttles and threads at his grandfather's weaving shop.

In the first chapter of the *Textura*, titled "Basic Plan of the Nervous System," Cajal presented "an echo and recapitulation of the dramatic history lived by the nerve cell on its millenary adventure through the animal series." He hypothesized not only about the structure and function of the nervous system but also about its evolution over time, seeking to explain the origin of the brain, as Darwin had the origin of species. Though Cajal did not have access to a fossil record—soft tissue deteriorates over time—he nonetheless managed to deduce the evolution of neurons in the same way that he had inferred their function. He gathered all the evidence that he could, and then he summoned his imagination.

Cajal divided the evolution of the nervous system into four eras. The first was the era of unicellular organisms, and the second that of coelenterates, aquatic animals with hollow bodies. Coelenterates possess two morphologically distinct neurons: one kind receives a stimulus, and the other kind initiates a response. When multicellular organisms first appeared, Cajal argued, they were unstructured, disorganized, and susceptible to the vagaries of the environment. As life became more complex, cells evolved mechanisms to coordinate more complex responses. The bodies of organisms grew larger, and the nervous system had to facilitate communication to and from their different parts.

In "higher" organisms, Cajal observed a greater number of neurons, with more diverse morphologies, coordinating increasingly complex relationships among tissues and organs. Anatomical studies showed that the neurons of vertebrates—and perhaps even higher invertebrates—received stimuli not only from the external world but also from the internal environment, within the organism itself. This third era of the nervous system, according to Cajal, was defined by the appearance of the "interneuron," a cell that has no contact with the outside world at all, transmitting impulses to other neurons only. Cajal found examples of interneurons in worms, which resembled neurons in advanced invertebrates that he called psychomotor cells. Like many scientists of his era, Cajal saw the evolution of the nervous system—"the most highly organized structure in the animal kingdom"—as a linear process, and he held that the brain represented nature's highest achievement. The nervous system was in the midst of an eons-long process of self-perfection.

Almost every animal has a brain, and vestiges of those of reptiles, birds, and nonhuman mammals are still conserved in the human one, which, for all its power and complexity, retains primitive functions from the past. The human brain doesn't just coordinate bodily mechanisms related to nutrition and defense, as is the case with all other life on earth; it also produces psychic phenomena like feeling, thought, and will. In the first edition of his *Textura*, Cajal included a chapter on psychology. In the next edition, he took it out, instead writing that psychic phenomena are "still largely unexplored." The assumption had been that the structure of the brain would reveal the structure of the mind, but in the second edition Cajal admitted that "this ideal is still very distant." "It appears to us truly difficult to admit that processes so complex as memory or personality and voluntary acts could be localized in a given point of the cortex . . . ," he wrote. "But after all this, is the question really answered, or is it just shifted to another level?"

The *Textura* is not only a scientific classic; it deserves to be appreciated as a work of literature in its own right. Cajal's scientific prose style, obsessively edited for clarity, is as artfully deployed as that of any novelist or poet. What is most impressive is that Cajal built his grand, sweeping narrative fact by fact, guided strictly by deductive reasoning. He believed that, just like the nervous system itself, the study of the nervous system was in the process of being perfected, evolving from rough observations and faulty methods toward greater comprehensiveness and accuracy. Literature is a product of human culture, and all human culture is produced, somehow, by the brain. With his *Textura*, Cajal provided a deeper account of our humanity, the story of how our brains became what they are.

IN 1905, IN THE WAKE OF HIS GROWING FAME, CAJAL DECIDED TO print *Vacation Stories* with Fortanet Press, which also published his memoirs as well as *Advice for a Young Investigator*, based on his 1897 National Academy speech. Most people did not understand Cajal's discoveries, but they respected his achievements. According to Cajal, his literary friends had been urging him to publish his short stories for years. "I never dared take [the stories] to press," he said, "both for the oddness of their ideas and the laxity and carelessness of their style." He worried that the stories, which criticized religion and the political establishment while grappling with vexing issues of scientific ethics, might detract from his carefully cultivated reputation for impartiality. In the foreword, he dismissed the stories as "literary trifles," and he did not offer the book for sale, only giving copies to his friends.

That same year, Cajal was named to the Royal Academy of Letters, in recognition of his memoirs, not his stories. Officially, he never took possession of the chair, since he failed to deliver an acceptance speech. He always claimed that he had too much work to do. In reality, he felt insecure about his way with words.

Segismundo Moret, the leader of the Liberal party, whom Cajal had known since his days in Valencia, informed him in 1906 that, in the event that the Liberals regained power, he wanted him to serve as the minister of education. Cajal thanked Moret and politely declined. "The truth is that I neither felt myself to be a politician nor was prepared for the arduous office of Minister," Cajal admitted. After Moret became Spain's prime minister, he reiterated his intention to appoint

Cajal to the post in a March meeting at his home. Again, Cajal declined. "Minister?" he said. "Do I want to be Minister? I have too much work. I assure you that it leaves me no time to waste on silly things."

Moret went to Cajal's house every day for a week to try to convince him. It was time for Cajal to give up the selfish pleasures of the laboratory, Moret said. Scientists must sacrifice for their country too. Cajal finally relented. ("Beware of orators!" he later warned. "Eloquence is often a picklock.") Cajal proceeded to outline a comprehensive, Regenerationist-style plan to reform the Spanish education system, which included improving training and research, granting universities more autonomy, and hiring better professors. He asked Moret how many millions of pesetas he would have at his disposal; Cajal admitted later that he had hoped the plan's high costs would scare the prime minister away. Instead, Moret agreed to every proposed condition.

A few weeks later, while in Lisbon for the 1906 International Medical Congress, Cajal, reflecting on his decision to serve in the government, decided to renege and withdraw. His reputation was founded on the independence of his judgment and single-minded concern for science, he said, and he did not want his colleagues to think that he was ambitious or corrupt. The other ministers were shocked and upset when they learned of his withdrawal; Moret was furious, but the anger soon passed. He ended up choosing a doctor friend of Cajal's, who had introduced Cajal to the *tertulia* at the Suizo. When he heard the news, the friend came barging into Cajal's office and hugged him twice. "Thank you! Thank you!" he said. "You are a marvel!" The next best thing to being Cajal was receiving his imprimatur.

THOUGH HE WAS GENERALLY CIRCUMSPECT ABOUT HIS POLITICAL views, Cajal was a frequent contributor to liberal newspapers, and the Liberals had been trying to convince him to run for president since the Moscow Prize. In 1905, a year before his flirtation with the Ministry of Education, leading figures from the Generation of '98 including Pío Baroja, Ramón María del Valle-Inclán, Benito Peréz Galdós, and José Martínez Ruiz (known as Azorín) invited Cajal to a rally in defense of democratic liberties. "At a recent and solemn occasion," they appealed to him, "you declared with striking energy that a change of attitude and total reform of procedures and customs was necessary for the social and political future of Spain." These activists presumed that Cajal was an ally, but he chose not to join. Baroja, a self-identified anarchist, criticized Cajal especially harshly. Already among the most important novelists in early twentieth-century Spain, he had begun his career studying medicine at the University of Madrid, where, like all medical students, he had taken Cajal's introductory class. Cajal served as the adviser for Baroja's thesis on the physiochemical origins of pain. Baroja accused Cajal of "duplicity"—of betraying his personal beliefs for the sake of his public reputation. Unamuno, however, came to Cajal's defense: "Cajal serves his country, even politically, in another way."

That same year, Cajal wrote the prologue to a book called *Super-Organic Evolution: Nature and the Social Problem* by his friend Enrique Lluria Despau, who also funded the publication of *Advice for a Young Investigator*. Lluria argued that society is rife with inequality, misery, and pain because human beings have forsaken the natural laws of integration and harmony. Cajal blamed this deterioration on the "unhealthy worship of capital." "The world is upside down," he wrote in his scathing critique, "above, vice and laziness sit enthroned and honoured; below, the toilers and helpers struggling with hunger and pain." "The only legitimate capital, anthropologically considered, is the human organism and the forces of nature." The capitalist means of production, Cajal argued, cannot be harmonized according to the laws of evolution unless they are collectively owned and managed. "The earth for all, the natural energies for all, talent for all," Cajal's prologue continued—"this is the fair division of future society."

But Cajal warned of one troubling possibility that he linked with the downfall of capitalism. "Will not that *aurea mediocritas*"—the golden mean—"to which society aspires, enervate the mental faculties, undermining the energy for scientific investigation?" he wrote with his trademark bravado.

> *Will not collective capital be timid, and lack the dash, on ro-mantic and supreme occasions, of individual capital? Will glory, the passion of philosophic and scientific genius, prosper in the grey and subdued atmosphere of the commonweal? When injus-tice is banished, will not the best spring of the mental evolution of humanity perhaps have ceased to act?*

Still, Cajal hated aristocrats, whom he called "parasites," and he maintained that nobles also worked "unconsciously for the degenera-tion of fellow citizens." In his famous 1923 book *Invertebrate Spain*, Jorge Ortega y Gasset wrote that "without aristocrats and egregious minorities there is no society"; in the margins, an elderly Cajal re-sponded, "Pure aristocratic concept!" He insisted that "the economi-cally inferior are the best." The only people whom Cajal excluded from his *tertulia* were unabashed capitalists. He had "notorious leftist ten-dencies," according to one of his peers.

Later that same year, Cajal also wrote the prologue to a speculative book on positive psychology aimed at a popular audience by Tomás Maestre, a forensic medicine professor at the University of Madrid and member of Cajal's *tertulia* at the Suizo. A nation is an organiza-tion of individuals, he claimed in his expansive prologue, just like cells. "We all feel within us the waves of one and the same vital current, sister-cells of the same body."

Like many intellectuals at the time, Cajal had faith that science could bring about an end to the world's suffering. As a positivist, he believed that humanity was on the brink of the final phase of its intel-lectual development: "the subject of science," he wrote elsewhere, "constitutes all that man can know, with respect to himself and the world that surrounds him." In Cajal's utopia, "all can taste the ineffable harmonies and beauties that palpitate in the lap of Nature." And with their newfound leisure time, the result of technical advances, people could devote themselves to science and art, the only worthwhile pursuits in his opinion.

CHAPTER 24

"A Cruel Irony of Fate"

Cajal preferred to tackle great problems in the realm of the infinitely small. In politics, the Regenerationist moment had passed, but Cajal devoted himself to regeneration of a different kind. In the field of medicine, nerve regeneration was a critically important topic; if fibers in the nervous system could regenerate, then damage from brain and spinal cord injuries could be halted or possibly reversed. Cajal began studying nerve regeneration in 1905, not for clinical reasons, but because he recognized a new battlefield on which the neuron theory needed to be defended. Researchers had focused their work on the peripheral nervous system, mainly the nerves that innervate muscles, which were easier to study because they lie closest to the body's surface. Cajal, however, understood that the real promised land was the central nervous system—the spinal cord and the brain. A new theory claimed that axons regenerated not on their own but with direct assistance from other cells. The individuality of the neuron was once again at stake.

As with so many functions of the nervous system, regeneration was an object of curiosity before it was ever a subject of scientific research. In Greek mythology, one of Heracles' epic labors involved slaying the Hydra, a many-headed sea serpent. For every head he severed, two more appeared, as though the Hydra were a monster "made fecund by its wounds," in the words of Ovid. The topic of regeneration fascinated

Aristotle, who, in addition to being a philosopher, was also a biologist. In his *History of Animals*, he noted that if the tails of salamanders, lizards, and snakes are cut, they will grow back, as will the eyes of a swallow chick after being pricked out. Fishermen had long noticed that oysters, their most valuable catch, were preyed on by starfish, which pried the shells open and ate the flesh inside. To fight back, the fishermen caught the starfish, cut them in half, and tossed both parts into the sea, yet the starfish population only increased. It was not resurrection, but it was something close to it.

Researchers in Cajal's time knew that when a young animal's nerve is severed, the portion attached to the cell body (the "central stump") retains its conductivity, whereas, now isolated from its source of nutrients, the other portion (the "distal" or "peripheral stump") degenerates immediately and dies. In samples of damaged tissue, in the scar between severed nerves, histologists observed evidence of new fibers sprouting and growing from one stump toward the other, as though the nerve were reaching for its other half. In some cases, the fibers appeared to reunite, and the function of the nerve, motile and sensitive, was restored. How do damaged nerves regenerate? How do their fibers, despite being wounded, manage to survive and heal? Some scientists believed that the peripheral stump was able to "self-regenerate," an explanation that undermined the fundamental tenet of neuron theory: that fibers emanate from cell bodies and that cell bodies are the functional agents, propelling the fibers' growth.

Before long, Cajal found himself embroiled in what he called "the great controversy," a scientific dispute that had spanned half a century. For his regeneration studies, he injured the nerves of animals in a variety of ways: he transected them with a snip of the scissors; he crushed them with a pair of forceps, dulled so as not to sever nerves accidentally and smudged with carbon to mark the injury site for later dissection; he tied ligatures using sewing thread; and he placed obstacles such as blood clots, fat globules, and other cells between severed nerves to observe how they might change their course. Then, after giving the animal time to heal, he killed and dissected it, staining the nerve fibers to see how they had responded to the injury.

Some nerves, when injured, died almost instantly. Other nerves died "after a more or less prolonged agony," in a "death struggle," as degeneration radiated from the surface of the tissue to deeper layers. When he dissected an animal and stained thick sections of its tissue, he could make out "sprouts" in the process of growing, frozen by death like buds after a cold snap. Some fibers were "completely destroyed" within twenty-four hours; others appeared normal, but a more intense stain clearly showed that they had already died. Within a few days of a nerve being cut, Cajal found, sprouts could be seen shooting out from the distal stump in every direction. He described these sprouts as "exuberant," though they would face an "eventual collision with insurmountable obstacles that bend their course." At first, the sprouts would grow in various directions, before converging and forming bundles. Some sprouts never managed to push their way forward, stunted by a "sedentary reaction, without any power of vanquishing the surrounding obstacles, or energy to form new paths"—they seemed almost to lack the will. Others split up, as though trying to avoid obstacles, wandering in different directions in search of their target. Nerve regeneration, Cajal explained, is an absurd competition, "an obstacle race in the dark," taking place at the rate of one micron per day.

Eventually, some of the wandering fibers would reach the scar, and some of the sprouts would "throw themselves" into the breach like zealous soldiers charging onto the battlefield in an "invasion." The most intrepid fibers are not always the largest or strongest, Cajal observed, but rather the finest and most adaptable. Other fibers, when confronted with an obstacle, reversed course, their efforts "ephemeral and frustrated," and started growing in the other direction. They would begin to degenerate, retreating toward the central stump before eventually disappearing. Fibers do not always take the simplest pathways,

Cajal concluded. Often, they reach their targets "by devious routes," "straying" early in their journey before the "rectification of [their] lost course." For some hopeless fibers, "we can say nothing of their fate."

Even as all parts of a nerve cell have the capacity to grow sprouts, not all do, and Cajal observed that approximately one in six eventually managed to regenerate. In a neuron that does not regenerate, "something appears to be lacking . . . [something] capable of rousing it from its apathy and quiescence." The life of a neuron "does not obey an immanent and fatal tendency, maintained by heredity, as certain authors have defended, but it depends entirely on the physical and chemical circumstances present in the environment." Nerve fibers are coaxed forward, Cajal theorized, by some "growth promoting stimuli," a "physical influence" of some kind, as though they were following a scent. The exact mechanism remained unknown, but Cajal called it "chemotaxis" or "neurotropism." Irregeneratibility, Cajal concluded, results not from a flaw inherent to the cell itself—a "lack [of] growth potential"—but rather from an impoverished chemical environment surrounding the growing sprouts. He observed the same phenomenon in Spain: that there was nothing innately lacking in the national character, and that with better educational influences, Spanish culture, stunted since the Golden Age, could recover from its wounds.

EARLY IN THE MORNING ON FRIDAY, OCTOBER 26, 1906, THERE WAS A knock at Cajal's front door. Silveria rose from bed and received a telegram. She woke Cajal and relayed the message to him, which consisted of five German words: *Carolinsche Institut vertichen Sie Nobel-priess*— "Karolinska Institute awards you Nobel Prize." Certain that his students were playing a trick on him, Cajal went back to sleep. The next morning his picture appeared on the front page of *El Imparcial* under the headline "The Nobel Prize: Ramón y Cajal," and the announcement was reprinted in every Madrid daily (though the goring of the famed bullfighter Machaquito was given more space).

Cajal tried to downplay the news. He thought, once again, about refusing the honor. "I do not trust this flattering success," he said. He treated the award as "undeserved, abnormal, very dangerous to physical and mental health." Fame, after all, had done nothing but complicate his life.

Cajal received hundreds of letters and committed himself to answering every one, careful to avoid even the smallest grammatical errors. Such obligations required "a heart of steel, the skin of an elephant and the stomach of a vulture," Cajal said. The prize money—the equivalent of 115,000 pesetas, when one could buy a house on the outskirts of Madrid for 50,000—was cause for joy, however. *Heraldo de Aragon* reported that on Monday, the first full weekday following the announcement, Cajal was talking openly about the prize money, telling friends that he had already visited a bank director to discuss changing kroners to pesetas.

Only five years after its inauguration, the Nobel Prize was a sensation. When Cajal entered the classroom that morning, his students gave him "a delirious ovation," that, according to one newspaper, was "a show of appreciation so sincere, so enthusiastic, [and] so respectful." Stunned by emotion, Cajal could hardly summon words of thanks. When he left the classroom, he was followed by a horde of students who chased him down Atocha Street toward the institute, applauding and cheering. The young men tried hoisting Cajal in the air, but he wrestled himself free and hailed a carriage. Nearby women, drawn out onto their balconies by the commotion, shouted, "Long live Cajal!" The students ran after his carriage until it rumbled over the crest of Atocha Street and out of sight. They stopped, panting, their hands on their knees. "Yes, long live Cajal," one student said, "but why can't he live closer!" When he arrived home, students chanted his name until he appeared on the balcony, prompting a fresh ovation.

Many Spaniards, assuming that they had been deceived as well, accepted the news only after foreign sources confirmed it. For some Spaniards, the fact that, unbeknownst to them, a genius was living in their midst—or "a wise man among [Spain's] children," as one prominent journalist later put it—became a source of both pride and shame. Cajal's Nobel Prize victory, a prominent journalist exulted, proved that Spain "is not a country so deprived of eminent men, as many believe." The story of Cajal as a singular hero, already exaggerated in his own speeches and memoirs, became, in the public eye, even more mythic. *El Imparcial* pointed out that, "like a pure and good Spaniard," he wrote poetry in his youth before discovering "the intimate and supreme poetry that Nature had enclosed in the infinitely small, in the cell . . . in the subtle tissues that form the cerebral cortex." A former officer who served with Cajal in Cuba wrote him a congratulatory letter recalling how the health inspector there had once tried to place him under house arrest because his pants were torn and stained with ink from writing and drawing. Victoriano Cajal, Cajal's older cousin and now one of the wealthiest businessmen in Jaca, raced around the streets excitedly. He did not understand a single thing about Cajal's discoveries, but he bragged about his little cousin Santiagüé's triumph to anyone who would listen. All who had known the Nobel Prize winner as a young delinquent responded with the same expression: utter shock.

CAJAL HAD NOT EXPECTED TO WIN THE NOBEL PRIZE, OR SO HE claimed. What he did not mention is that he actually lobbied for it. As soon as the Nobel Prize was inaugurated, in 1901, he persuaded the University of Madrid medical school to print a detailed summary of his scientific activities, titled "Report of Titles, Honours, and Scientific Works of Dr. S. Ramón y Cajal," featuring a portrait of the author, which was sent to the world's preeminent researchers, including Golgi. Cajal's self-promotion bothered even Kölliker, among his staunchest defenders, who pointed out that many of the 117 citations listed in the pamphlet were repeat publications. "*Multa non prova il multum*" ("quantity is no evidence of quality"), he wrote in a letter to Golgi.

Each year, the Nobel committee solicited nominations from prom-

inent academics around the world. Golgi had received his first nomination in 1901 from none other than Kölliker. "Not even in my dreams
have I considered the possibility of being nominated for such a high
prize . . . ," Golgi wrote to Kölliker. "I consider that eventuality always
as a [fantasy]." In 1906, Kölliker proposed both Golgi and Cajal. Retzius had nominated Cajal every year since 1902, while sometimes adding Golgi, before clarifying in 1905 that Cajal should be considered
first and Golgi second. In 1906, Retzius advocated for Cajal and Golgi
both but emphasized that, if the award could not be shared, Cajal
alone should receive it. Four active committee members—three Swedes
and a German—supported Golgi's candidacy. In addition to Kölliker
and Retzius, Cajal's nomination was supported by one German and
one Swede.

Back in 1904, a Spanish colleague had written to Cajal to tell him
that he had been nominated for that's year's prize. "[My] nomination
has previously been made by other Spanish and foreign universities
with little success," Cajal replied. Besides, the Nobel Prize was typically reserved for "applied sciences," such as endocrinology and bacteriology; the previous year's winner was Koch, in recognition of his
lifesaving vaccines. Cajal was critical of the primacy of applied science.
"Let us cultivate science for its own sake," he wrote, "without considering its applications"; in the section of his *Advice* that he called "Be-
ginner's Traps," there is a chapter titled "Preoccupation with Applied
Sciences." Experimentation should come first, he stated—application
later. He figured that this philosophy disqualified him from receiving
the prize.

The Nobel committee commissioned an evaluation of the work of
Golgi and Cajal, detailing their contributions to science over the previous five years and weighing their relative merits. "If the achievements
by Golgi, on the one hand, and Cajal, on the other, in the research on
the nervous system are considered," Emil Holmgren, a professor of
histology at the Karolinska Institute, stated, "one cannot, in justice,
evade the final conclusion that Cajal is far superior to Golgi." In addi-

tion to Cajal's early breakthroughs, Holmgren cited his more recent work on neurofibrils and regeneration. "[He] has not served science by singular corrections of observations by others, or by adding here and there an important observation to our stock of knowledge," Holmgren concluded, "but it is he who has built almost the whole framework of our structure of thinking, in which the less fortunately endowed forces have had to, and will still have to put their contributions." No one outside Golgi's sphere of influence subscribed to his views anymore, and Holmgren argued that Golgi's persistent adherence to the reticular theory, in addition to his erroneous views on the role of dendrites and his misclassification of cell types, should be held against him.

The Nobel committee's deliberations lasted for several days, and the dispute was bitter. Golgi had not contributed original research on the nervous system since the early 1890s, but a new rule allowed for twenty years of a candidate's work to be considered, though more recent discoveries would be valued more. Golgi would be the first "retiree" to receive Nobel funds, one committee member pointed out. Either way, another member remarked, it would be meaningful to award the prize, for the first time, to a scientist of "the Latin race." In the end, the majority of voters supported awarding a joint prize, while two voted—anonymously—for Cajal alone. "I thought that he had deserved receiving a *full*, and *undivided* Nobel Prize," Retzius said, "and asked about this by the Nobel Council of the staff of professors at the Caroline Institute, I expressed this opinion of mine *decidedly*." Cajal could not have achieved his success without Golgi's method, but the understanding of the nervous system would not have advanced without Cajal. Golgi had found "the key to open the building that encloses the secrets of the nervous system," Retzius said, "but only Ramón y Cajal had taught how to use it."

On October 26, 1906, at his home in Pavia, Golgi received a telegram in German that was slightly different from the one that had reached Cajal the same morning: *Congratulations—Nobel Prize—you and Cajal.*

All Nobel laureates were expected to attend the award ceremony, where they would deliver a speech explaining their discoveries to the public. After his exhausting journey to America, Cajal was dreading yet another long trip. "Mr. Ramón y Cajal has written that his health does not permit him to make the voyage at this time," the president of the Karolinska Institute wrote in a letter to Golgi. (But, he said in parentheses, "I think that he will come.") Cajal eventually succumbed to the pressures of etiquette. "It was neither possible nor fitting," he wrote, "for me to try to evade the custom which, besides, implies a due and polite testimony of gratitude to the Trustees of the Nobel Foundation and to the generosity of the Scandinavian people." His departure time—Saturday, December 1, at 5:00 P.M.—was printed in the newspaper, and readers were encouraged to join the farewell party. On the platform at the Estación del Norte, ministers and other government representatives stood alongside students and faculty from every department of the university, waving banners. Officers of the Security Corps were deployed to preserve order. The crowd of young men cheered wildly, throwing their hats in the air and shouting, "Long live Cajal!" as the train pulled away.

ON DECEMBER 6, CAJAL ARRIVED IN STOCKHOLM, A CITY THAT WAS frigid and, for eighteen hours each day, completely dark. Stockholm was so cold that the coachmen's fur coats and large berets were white with frost. Retzius was waiting for him at the station with a delegation from various Swedish academies in tow. Golgi was in ill health himself, and the impending speech unnerved him; he wanted to arrive in Stockholm incognito and hurry to the hotel so that he could revise and rehearse his text, which was in French.

Though Golgi had tried to keep his travel plans secret, someone had sent advance notice of his train schedule. After missing Golgi in 1889, Cajal was now determined to meet his rival. A welcome party awaited the Italian on the platform, including Retzius, Retzius's wife, the president of the Karolinska Institute, a number of professors, and, completely unexpectedly, Cajal. Golgi was flabbergasted. He briefly acknowledged the party, walked off the platform, and went directly to his hotel. His wife, Lina, had slipped and fallen on the icy ground while exiting the ferry from Germany to Sweden. The fall—"like those that one makes in the skating rink," Lina wrote in a letter—made it difficult for her to stand. Golgi might have rushed past the welcome party out of concern for his wife's condition. Cajal mentioned nothing about the apparent snub, just as he had excused Golgi's earlier absence from Pavia.

"There was no better occasion to put a cap on the past and acknowledge the merits of both men," another scientist who was present—an Italian—said. "And it would have been truly agreeable if a gesture of reconciliation had originated from the older of the two, because Cajal, with his presence, had in a certain way encouraged him to do so. But the ice was not broken either that evening or in the subsequent unavoidable encounters during their stay in Scandinavia."

That two scientists with absolutely irreconcilable views would share the world's most prestigious prize was an absurdity that was lost on no one. "What a cruel irony of fate," Cajal wrote, "to pair, like Siamese twins united by the shoulders, scientific adversaries of such contrasting character!" Still, both managed to be polite. Golgi said that he was honored to share the prize with "the celebrated scholar Cajal." In his autobiography, Cajal referred to Golgi as an "illustrious" and "deserving" winner, but in enumerating the merits of his rival's work, he cited only the invention of the black reaction and none of Golgi's findings or theories; he also accused Golgi of working not to serve the truth but rather to advance his own ideas. Kölliker, a friend of both researchers, was outraged by how Cajal treated Golgi, the elder statesman. "Have you read in what insolent manner Ramón has expressed his opinion on your observations and conclusions and those of your students in his recent book *Textura del sistema nervioso?*" Kölliker asked Golgi in a letter. "It seems to me that Ramón believes himself the foremost histologist and the only one allowed to have an opinion on the structure of the nervous system." Kölliker agreed with Cajal's science but took exception to his arrogance.

THE GRAND HOTEL, SITUATED IN A BEAUTIFUL SQUARE ON THE waterfront overlooking Stockholm's old town and the Royal Palace, treated the Nobel laureates like kings. Their suites were adorned with plush carpets, giant mirrors, a walk-in closet, and velvet couches and divans. The bathrooms offered unlimited hot and cold water, the heating system was all steam, and electricity powered ornate chandeliers and bedside lamps. Each room was equipped with a telephone with which guests could summon a manservant or chambermaid at any hour. Cajal, indifferent to such luxuries, did not bother to describe them anywhere. On the other hand, Lina Golgi, a woman who, not unlike Cajal, hailed from a small rural village, found herself in a state of perpetual wonderment, giddily sharing details in letters to her mother. An editor from the Swedish morning newspaper *Stockholms Dagblad* asked Cajal for his impressions of the city. "I have noticed that it is colder here than in Spanish regions," Cajal replied, annoyed as ever at being pestered by reporters, refusing to engage beyond the function of a simple thermometer.

The Nobel Prize ceremony was scheduled for December 10, the anniversary of the death of Alfred Nobel. At 7:00 P.M., the laureates processed across the street from the hotel to the Royal Academy of Music where, at 7:30, the ceremony began in the Grand Salon, richly decorated by the royal architect for the occasion. Behind the stage, beneath an enormous laurel wreath festooned with blue-and-gold ribbons, was a towering obelisk featuring a white bust of Nobel. On the stage stood the flags of the laureates' home countries and four more obelisks inscribed PHYSICS, CHEMISTRY, LITERATURE, MEDICINE. Three armchairs were positioned in front of the stage to serve as the Swedish royals' temporary thrones, behind which a semicircle of chairs for laureates, presenters, and attendants had been arranged, and behind that, still, ample room was left for the country's most distinguished intellectuals, politicians, and military officers, accompanied by "many elegant ladies." After the royals had entered the hall and taken their seats, the orchestra, stationed on a great bandstand bedecked with pine boughs, struck up a pompous overture.

Cajal, the disheveled laboratory man, looked out of place in a tuxedo. The president of the Karolinska Institute addressed the overfull hall in French. "This year's Nobel Prize for Physiology or Medicine is presented for work accomplished in the field of anatomy," he pronounced. "It has been awarded to Professors Camillo Golgi of Pavia and Santiago Ramón y Cajal of Madrid in recognition of their work on the anatomy of the nervous system." The institute's president spoke about the brain as a great mystery, highlighted the complexity of its anatomy, and then returned to the laureates, praising their roles in establishing a new field. "Professor Golgi," he said, turning to Golgi and speaking in Italian:

The Staff of Professors of the Karolinska Institute, deeming you to be the pioneer of modern research in the nervous system, wishes therefore, in the annual award of the Nobel Prize for Medicine, to pay tribute to your outstanding ability and in such fashion to assist in perpetuating a name which by your discoveries you have written indelibly into the history of anatomy.

"Señor Don Santiago Ramón y Cajal," the president said next, turning to Cajal and switching to Spanish:

By reason of your numerous discoveries and learned investigations, you have given the study of the nervous system the form that it has taken at the present day, and by means of the rich material which your work has given to the study of neuroanat-

omy, you have laid down a firm foundation for the further development of this branch of science.

The president led Cajal forward to receive his diploma from the king of Sweden. "The ceremony of awarding the prizes," Cajal recalled, "was a pompous event and one of highest idealism." The royal orchestra played them all out with a march. Back at the Grand Hotel, a banquet had already been laid out in the Hall of Mirrors, a gilded, chandeliered room modeled after the one at the palace at Versailles. Three days of preparations had been required for the event—two hundred white-gloved waiters arranging flowers, setting plates, and polishing cutlery and glasses. A seven-course gourmet French meal was served, including turtle soup, fillet of perch pike, fattened young hen, cold quail in jelly, watercress ("princess") salad, and an ice-cream dessert shaped like a cannonball, along with six kinds of wine and champagne. Toasts were offered by ambassadors from various countries. Golgi, according to Retzius, "behaved extremely nicely and with dignity to Cajal," but he thought that Golgi's toast "displayed . . . Olympic pride and pretentious mien."

The laureates did not return to their rooms until around 2:00 A.M. At noon the next day, December 11, Golgi delivered his Nobel Prize speech, titled "The Neuron Doctrine—Theory and Facts." Shockingly, he began by alleging that the neuron doctrine "is generally recognized as going out of favour." The majority of scientists, Golgi claimed, accept the ideas of the neuron doctrine as "articles of faith . . . capricious . . . artificially distorted and falsified." Whoever believed in the "so-called" independence of nerve cells, he sneered, had not observed the evidence closely enough. "We suppose that they are independent," he chided, "only because we cannot prove that more intimate connections exist."

After delineating the neuron doctrine's principles, Golgi took direct aim at the law of dynamic polarization, the crown jewel of Cajal's theories. "This theory should not be considered as an essential part of the neuron theory," Golgi argued. "In fact it only expresses one interpretation of nerve function and does not exclude the possibility of others." He praised the work of his Spanish counterpart as "bold and clear . . . always thought-provoking, frequently plausible, and often prophetic," while at the same time denying the value of it. "While I admire the brilliancy of my illustrious Spanish colleague," Golgi said, "I cannot agree with him on some points of an anatomical nature which are, for the theory, of fundamental importance." Despite the mountain of counterevidence, Golgi continued to insist on the existence of a "diffuse nerve network," comprising "a true nerve organ," which, in his mind, was a "concrete anatomical fact." With unmistakable contempt, he boasted that he never once used the word *neuron*; the neuron was a historical artifact, he said, which "cannot claim any citizenship in science." The previous twenty years of findings had failed to make an impression on him.

It was customary for Nobel laureates to use their lectures to address their own achievements or discoveries, and so Cajal never imagined that Golgi would use the opportunity to launch a polemic. Confronting his rival for the first time, possibly anticipating a fight, Golgi had decided to fire an opening salvo. He also knew that he would be speaking in front of other neuron theorists, whom his speech would insult. "The noble and most discreet Retzius was in consternation," Cajal recalled. "Holmgren, Henschen, and all the Swedish neurologists and histologists looked at the speaker with stupefaction." Golgi's words were all the more surprising because, as Retzius attested, he was a "noble, friendly, and agreeable person who gained everyone's sympathies and esteem." The speech even stunned those audience members who did not identify with either side in the dispute. Despite Cajal's discoveries, Golgi had neither updated nor reexamined his own material; instead, he clung to his own findings, which were two or three decades old, ignoring not only Cajal's contributions but also those of Kölliker and Retzius, both of whom counted themselves as

Golgi's friends. "I was trembling with impatience," Cajal recalled, "that the most elementary respect for the conventions prevented me from offering a suitable and clear correction of so many odious errors and so many deliberate omissions."

The following day at noon, Cajal rose to the same stage and delivered his own speech, titled "The Structure and Connections of Neurons." From the start, he sought to differentiate his intentions from Golgi's: "In accordance with the tradition followed by the illustrious orators before me," he explained in French, "I am going to talk to you about the principal results of my scientific work." His case in favor of the neuron doctrine was forceful and unambiguous. In "twenty-five years of continual work on nearly all the organs of the nervous system and on a large number of zoological species," he declared, "I have never [encountered] a single observed fact contrary to these assertions." Thoroughly and confidently, he addressed Golgi's criticisms of his research one by one. Cajal took care to cite, among his supporters, Aldo Perroncito, a former student of Golgi's who had broken ranks with his master.

Cajal ended his speech with a jaw-shattering blow. "We mourn this scientist who, in the last years of a life so well filled, suffered the injustice of seeing a phalanx of young experimenters treat his most elegant and original discoveries as errors." Golgi—very much alive— was seated in the audience.

DESPITE THE HOSTILITY BETWEEN THEM, AND THE IRRECONCILABILity of their views, Golgi and Cajal were more alike than either man could appreciate. Golgi was born in the tiny village of Corteno, high in the central Alps, as rural and isolated as Petilla. Golgi's father was also a physician, and like Cajal, Golgi joined his father's profession. Golgi had to support his own histological research while simultaneously working at a hospital, where there was no laboratory, and so he created a makeshift one in his kitchen using a microscope and chemical reagents borrowed from the hospital supply closet. When he was not on shift, he experimented with brain tissue all through the night, by candlelight.

Both researchers came from countries outside the scientific mainstream, spoke unfamiliar languages, and felt that their work was looked down upon as a result. Though Cajal was less reserved than Golgi, both men preferred to be left alone. "My main wish," Golgi wrote upon receiving news of the prize, "would be to stay hidden." Cajal had taken his aloofness for pretension, but his attitude might have been a protective response to a social situation that threatened to overwhelm him. Constantly plagued by accusations of rudeness, Cajal, of all people, might have empathized.

But Cajal and Golgi approached the study of the nervous system from fundamentally different directions. Golgi, concerned with physiology, was asking himself how the nervous system could function as a whole, then searching for the anatomical correlates. The structure of the nervous system had to facilitate complex and extensive activities, somehow giving rise to mental phenomena, including consciousness, which he could not imagine the individual actions of nerve cells producing. Cajal, on the other hand, believed that physiological biases were hindering anatomists from seeing the truth. His stated aim was to "free histology from every physiological compromise," to adopt "the only opinion that is in harmony with the fact"; the word that he used in Spanish was *desligar*, meaning "to free" or "to separate," "to untie." Cajal was disposed toward the concept of cellular individuality, and his own interpretations were not entirely free of preconceived notions. In his early black reaction studies, he tended to interpret everything that he could—including samples that were ambiguous or imperfectly stained—in a way that would confirm his findings, even when conclusive evidence would not arrive for decades. He felt the need to dispel any hint of anastomoses in order to protect the neuron theory, and a few of his explanations from that period are surprisingly illogical.

Each scientist believed that what he saw was the objective truth and that the other was hopelessly deluded. Why did Golgi continue to insist on the existence of a reticulum? Cajal attributed this, once again, to suggestion; he called reticular theory "despotic" and "seductive," a kind of "mystical" thinking "projecting a priori theories on to facts that impede seeing reality." In Cajal's view, the irony of neuroscience is that the brain is at once "the masterpiece of creation," the highest achievement of evolution to date, and also a fragile and imperfect vehicle, vulnerable to "exquisite suggestibility" and profound error.

WHEN THE NOBEL SPECTACLE WAS OVER, CAJAL RETURNED TO SPAIN by rail. Retzius worried about him on the long journey, carrying a promissory note for his share of the prize money; distant and quiet, lost in his thoughts about the brain, Cajal seemed especially susceptible to a robbery. He was supposed to stop in Berlin for a ceremony to finally collect his Helmholtz Medal, a year after it had been awarded to him, and Retzius asked Waldeyer, a mutual friend, to wait for Cajal at the station. Cajal never arrived. Exhausted from all the festivities, he rode all the way back to Madrid. He never left the Iberian Peninsula again.

Meanwhile, when Golgi arrived in Pavia, a band was playing near the train station. He assumed that the music was meant for him, but the celebration was to honor a champion Italian bicyclist, returning home from a big race on the same train.

CHAPTER 25

"To Defend the Truth"

Winning the Nobel Prize brought Cajal even greater international fame. "Almost all the societies and academies related to his work did everything that they could to pay homage to him," Tello recalled, and in the years that followed, dozens of academic institutions named him an honorary member, including ones in London, Paris, Rome, Berlin, Dublin, Brussels, Edinburgh, Saint Petersburg, Buenos Aires, and Philadelphia. It was said that a letter could be sent to Cajal in Madrid by addressing it with only his name. Two doctors from Japan made a pilgrimage to meet Cajal and begged him for some souvenir, gladly settling for a few old *sacatintas*, or ink stamps, from the master's desk. One Spanish colleague kept a picture of Cajal with the Nobel Prize medal on his desk. "I will guard it always and I believe that even my children will keep this autograph after that," he said.

Businessmen even invoked Cajal's name to market their products. A company selling bottled mineral water listed Cajal—"the greatest bacteriologist in the world . . . who obtained the Nobel Prize in competition with the most eminent savants of all nations"—as its director and ran newspaper advertisements with Cajal's photograph in a seal. (Apparently, Cajal had once spent a few days at the springs and remarked that it had benefited his health.) Other manufacturers of mineral water as well as those of medicinal wines, disinfectants, cigarette paper, and toothpaste boasted that their products were "analyzed, or consulted, by Dr. Cajal." Years later, to correct the record, Cajal wrote a letter published in the newspaper *ABC* titled "How My Modest Name Is Exploited by Certain Unscrupulous Industries." He insisted that he had never performed any chemical analyses of commercial products. "Let the aforementioned industries, then, rectify their conduct toward me," he implored. "Erase my name from their claims and labels and reproduce that of the author of the analysis . . . Announce that, attributing to me a work for which I am not competent brings me into a ridiculous, unjust, and unlawful position." Cajal had to perform chemical analyses during the *oposiciones* for his doctorate, but he applied those skills only in special personal cases, such as studying the water quality of his summer home and of the home of his cousin Victoriano near Jaca, who repaid the favor with fruits and vegetables from his garden.

One time, Pedro was vacationing in the Spanish province of Galicia, in the northwest corner of Spain, when he decided to cross the border into Portugal. The customs officer on duty examined his identification card, then showed it to some colleagues. After talking among themselves for a while, they suddenly snapped to attention. Pedro's traveling papers would be issued at no charge. *Un sabio no paga*, the head officer said—"A savant does not pay." The customs officers recognized the surname, and Pedro still looked remarkably like his brother, just as he had when they were young. The officers telegraphed their Portuguese counterparts to inform them of the distinguished passenger's arrival. At the first stop, such a large crowd had gathered on the platform that policemen were called to keep the peace. Fireworks burst in the sky. A local band played. People chanted Cajal's name. Pedro had to tell the organizer that he was not, in fact, *that* Cajal. The welcome party presented him with a bouquet of flowers anyway. "Countrymen!" the organizer shouted to the crowd. "Long live the brother of the genius Spaniard Santiago Ramón y Cajal!"

Nearly every city and town in Spain named a street or plaza after Cajal. Students in Zaragoza, Valencia, and Barcelona, where he had lived and worked at one point or another, organized tributes and held rallies, as did those in Valladolid, Granada, Córdoba, and Cádiz, where he never had. A petition, supported by both workers' societies and the mayor of Madrid, was circulated to rename Atocha Street "Ramón y Cajal Street." (For bureaucratic reasons, the effort was unsuccessful.) Inmates at provincial and national charitable houses, such the Casa de Caritat in Barcelona, which published Cajal's first *Revista* papers, wanted to publish a magazine in his honor. A Valencian man wrote to Cajal asking to dedicate a *jota* to him. "People met him in the street, greeted him and asked for autographs," said Cajal's granddaughter María Ángeles Ramón y Cajal Junquera. Cajal had to change his family vacation plans because local residents would disturb him with tributes everywhere he went and pressure him to build a summer house there. Banquets were thrown and toasts were offered. One afternoon, when a crowd cheered for Cajal outside the institute, he stepped onto the balcony and requested that they let him work in peace.

"What can we do, Don Santiago?" a popular newspaper columnist wrote. "These sincere expressions of public admiration help to revive the spirits that have allowed themselves to become sadly overshadowed by the elegiac arguments of pessimism." The Spanish press began to cover Cajal like a celebrity. "One never has to rationalize a publication of the portrait of Doctor Ramón y Cajal," read the weekly magazine *Blanco y Negro*, referring to the glut of articles. During photo shoots, Cajal would end up instructing the photographer on how to light the room, asking incessant questions about exposure time, and carefully examining the photographer's camera. Newspaper articles implored the government not to squander the national budget for once, so that Cajal could receive enough money for his research. The annual funding for his laboratory, one commentator pointed out, was the same as what a star matador earned for a single bullfight.

One month after Cajal's Nobel triumph, in January 1907, King Alfonso authorized the foundation of a new institution dedicated to promoting science and culture in Spain, called Junta para Ampliación

de Estudios e Investigaciones Científicas, or Committee for the En-
hancement of Scientific Investigations and Studies. The Junta was
conceived as a grant-giving organization, encouraging students to
study abroad before starting their careers at home. Twenty-one members were named to its board, and the government appointed Cajal as
its first president. He was determined to provide young Spaniards with
the educational opportunities that he had never had.

The independence of the Junta was challenged almost immediately,
when, weeks after the decree, the Conservative party regained power
and demanded that the Ministry of Education be allowed to propose
three grant candidates and to choose the winner. Liberal representatives of the Cortes fought back, accusing Conservatives of trying to
control people's minds. The number of grant applications decreased to
almost zero, and the Junta's operations were effectively paralyzed. Opponents of the Junta were afraid of diluting Spanish culture through
contact with other countries, and Catholics and Conservatives specif-
ically saw the Junta as a vehicle for the secularization and liberalization
of Spain.

Cajal found himself thrust into politics once again. In the wake of
his Nobel Prize, a writer for *El Imparcial* proposed that he be named a
senator, and, in 1908, the University of Madrid elected him as its representative. Cajal had no interest in lawmaking, as revealed by the
Cortes archives, which document his lack of participation. Early on he
served on a commission to improve the national highway system near
Zaragoza, seeking to rectify the perennial headache of terrible Aragonese roads. Later he served on a commission to review the University
of Madrid's faculty of philosophy and law curriculum.

Cajal spoke on the floor of the Cortes only twice. In 1911, he addressed a budgetary session for the Ministry of Education and Fine
Arts, which threatened to cut the Junta's funding after Conservatives
attacked the scientific and cultural establishment for some perceived
left-wing bias. "I ask permission to speak," Cajal declared from his
seat, following protocol. "It is the first time that I am speaking here,

and secondly I am not an orator, as the Chamber will easily understand by the slovenly words that I will pronounce." Defending the Junta was one of the rare causes for which Cajal was willing to conquer his fear of public speaking. He sensed the wariness of some senators toward his organization, "and I want this distrust to disappear," he said. "Frankly, all of this produced a profound pain in me." The Junta should have the confidence not only of the government, Cajal contended, but also every political party, social entity, and every citizen of the country. "The majority of those who compose it are not Liberals but rather liberal-minded," Cajal assured the Cortes. Most who received grants belonged to parties on the right, and almost all were technically Catholic. "[The Junta] is not a political organism," Cajal says, "but a national organism that has no other objective than to stimulate the culture of our nation." He even denied his own son Jorge a grant to study in Italy, instead paying out of his own pocket, to avoid all appearances of corruption.

Most conspicuous about Cajal's time in the Cortes are the issues that he never debated. Despite losing its final colonies in 1898, Spain still clung to its overseas possessions in northern Morocco. In 1909, skirmishes broke out between Rifian natives of North Africa and Spanish workers at an iron mine. The Spanish prime minister called for military reinforcements. The streets of Barcelona saw riots and a general strike, as conscripts refused to board ships to North Africa. The government declared martial law. Anarchists and Radical Republicans burned down churches and convents. Civilians were imprisoned for life or executed. Catalonians refer to this violent interlude at the end of July as the "Tragic Week." Cajal, however, did not participate in any debates about the Tragic Week in the Cortes. His only activity in 1909 was to join six senators in proposing a fifty-thousand-peseta credit so that the government could acquire a statue of Diego Velázquez to place in front of the National Museum of Modern Art. Velázquez, Cajal's childhood idol, remained his favorite artist throughout his life.

In 1910, Cajal, in his capacity as head of the Institute of Serotherapy, Vaccination, and Bacteriology, to which he had been appointed in 1900, sent a group of doctors and scientists from the institute on a four-month expedition to Spain's African territories to study an outbreak of sleeping sickness. The commission included his son Jorge, who had become a pathologist. Cajal, who knew the danger of tropical disease from his wartime experience in Cuba, wrote the foreword to their report, his rhetoric as grandiose as ever. "The principal mission of science is to make habitable the whole planet and convert, by the gospel of hygiene, the most inclement territories into placid and comfortable residences."

Cajal's report did not challenge the official colonial policy. Rather, it sought to demonstrate to the Conservative government the value of science as an instrument of Spanish rule. Cajal argued that unhealthy colonies, plagued with malaria and dysentery, were costing the government lives and money, sabotaging industry, and driving capital away. "It is wise and prudent politics, upon occupying a country, that the doctor and naturalist form the vanguard of the administrator and soldier," he wrote. "Only science makes mastery bearable, and even desirable." By curing disease, Spain could subdue the local population and convince it to cooperate.

Cajal lacked empathy for Spain's colonial subjects. Absent from his account of his own experience in Cuba is any mention of the plight of the Cubans. Spain's campaign was undeniably brutal; soldiers were commanded to kill every suspected rebel, including women, children, and the elderly, no matter how defenseless. Any boy over the age of eighteen who was found away from his home was shot on sight. Sons were shot together with their fathers, and the women survivors, sometimes mutilated or raped, pleaded to be killed too, saying that they would rather die than live under the rule of Spaniards. The Spanish military referred to this form of combat as "war in accord with the Christian spirit of the age." Colonialism depended on a sense of racial superiority, and Cajal perpetuated the language of "civilization" and "savages."

After the Tragic Week, King Alfonso XIII was forced to call for a general election in 1910. Cajal voted for the Liberal candidate, José Canalejas, even though, as the owner of the *Heraldo de Madrid*, Canalejas was among those whom Cajal blamed for inciting the Disaster. After three years of Conservative rule, Cajal was desperate for change. Canalejas won the election, and in 1910 he named Cajal *senador vitalicio*—senator for life. The only official vote that Cajal ever cast on the Cortes floor was in support of Canalejas's controversial "Padlock law," which required all new ecclesiastical organizations to be approved by the government. Cajal was adamant about protecting secular education.

Under the Canalejas regime, the Junta began to thrive, creating the Residencia de Estudiantes—the Students' Residence—modeled after the colleges of Oxford and Cambridge. When Cajal first traveled to Britain for the Croonian Lecture, he admired its university system; when he met Lord Kelvin, Cajal was struck by how worldly he was, a product of a good education. The Residencia was similarly open to creative thinking and interdisciplinary dialogue. Its purpose was to supplement university education, and its facility included an excellent library, generous with its browsing and lending privileges. With oversight from Cajal, the Residencia became the cultural center of twentieth-century Spain.

WHEN HE FIRST ARRIVED IN MADRID, CAJAL PROMISED THAT HE would never end up wasting his money "like an American dentist." He preferred to rent his apartments, and in each city where he lived he moved often. According to his granddaughter, Cajal was "a lover of change"; that was the nomadic life that he knew as a young child. Money was always a concern, but winning the Nobel put him at ease. With Cajal's share of the prize money, he ordered the construction of a new house on Alfonso XII Street, bordering Retiro Park. The Institute for Biological Research was only a ten-minute walk, past the Royal Botanical Garden.

Cajal's house was a *palacete*, a mansion three stories tall with an arched doorway inspired by Islamic design, with wrought-iron doors, above which hung the intertwined letters *RC* etched in marble. The ceilings of the house were built especially high, in order to keep the rooms cool so as not to trigger Cajal's migraines. Though the house was enormous, there was hardly enough space for all Cajal's books and equipment. "It was the home of a person who only cares about working and studying," Cajal's secretary recalled. He poured all his love and almost all his savings into the project, ordering a terrace built off the study, encased in glass, so that he could look at the stars through his telescope. On the main floor, there was a drawing room with an ornate carpet, decorative silver, and portraits of Cajal and Silveria printed on gilded mirrors. The Cajal family now had plenty of money to hire servants. For the first time in her life, Silveria found herself with leisure time at her disposal, which she devoted to needle-work, embroidery, crocheting, and gardening. She focused on educat-ing herself with the aim of becoming more cultured. As the wife of a famous man, Silveria now had to play the role of an urban socialite, making conversation at dinner parties and responding to endless invitations.

Cajal had three sons—Santiago, Jorge, and Luis—whose full sur-names were Ramón y Cajal Fañanás, a combination of those of their parents. For secondary school, they attended the Escola Pia and the Instituto de San Isidro, the oldest school in Madrid, which counted the legendary Golden Age authors Lope de Vega, Pedro Calderón de la Barca, and Francisco de Quevedo among its former students. Cajal afforded his children the freedom to choose their own paths, Luis said. Nevertheless, both he and his brother Jorge became doctors. The youngest child of all, Luis, was docile and well behaved, less like his father than like his uncle Pedro. Jorge, headstrong and independent like his father and grandfather, rebelled against his mother's strict rules and harsh discipline; at seventeen he took a volunteer job just so that he could get out of the house. Jorge eventually abandoned biological research to start a pharmaceutical company, devastating his father.

Cajal's eldest, Santiago, was the spitting image of his father, with broad shoulders, a prominent nose, and piercing dark eyes. When Santiago was a boy, he suffered from typhoid fever, which delayed his intellectual development and brought on premature heart disease. In family portraits, Santiago's glance rarely meets the camera; his face bears a perpetual look of pain. Doctors gave up hope and predicted Santiago was destined to die young. Though he managed to graduate from college, his poor health prevented him from pursuing a career, and he fell into a crippling depression. Cajal empathized with his son, having suffered through a similar experience at the same age. He held on to the conviction that his son would prove the doctors wrong.

Cajal was constantly preoccupied with Santiago's condition. He tried to establish a career for Santiago as an editor and publisher, paying for him to open a bookshop and donating his own back catalog. To fund the venture, Cajal started writing a popular book on photography. He had never stopped experimenting with the medium, taking still lifes of bowls of fruit on patterned tablecloths and portraits of his daughters wearing colorful dresses. Beginning in 1900, he even rented a small studio on the top floor of 10 Prado Street, across from the Athenaeum. He published articles on innovative new photography techniques, and the Spanish Photographical Society named him its honorary president. "Only photography can satisfy the thirst for beauty," Cajal wrote. "Among the prosaicisms and miseries of the struggle of professional life, it inserts a bit of poetry and unexpected emotion . . . To photography I owe ineffable satisfactions and consolations." He intended for his son to help him with the book, as he had collaborated with his own father on their unpublished anatomical atlas.

Cajal's son Santiago died in 1911, at the age of twenty-nine. The following year, Cajal published his book *The Photography of Colors*, one

of the earliest manuals published on the subject in Spain, which he dedicated to Santiago. Cajal never discussed his son's death, just as he did not discuss the loss of Kety. "Let us not talk of sorrows," he wrote. "Why call up pains of which the only mitigant lies in forgetting!"

BY THE START OF THE TWENTIETH CENTURY, RESEARCHERS HAD begun to explore the potential relationship between glial cells—the nonneuronal cells of the central nervous system—and neurological illness, a possibility that Cajal himself first suggested in 1896. ("Neuroglia are found disturbed in mental illnesses!" he wrote excitedly. "Do they not relate in their physiological activity to the brutal shock of emotions?") Examining nervous tissue under the microscope in 1856, Virchow had spotted strange forms, clearly not nerve cells, which looked to him like "a sort of putty." The cellular material, described in his *Cellular Pathology*, proved ubiquitous throughout the nervous system. Virchow named them *nervenkitt*, or "neuroglia," from the Greek word *glia*, meaning "slimy" or "sticky" in appearance. Neuroglia became the name for all elements of the nervous system other than neurons, and they were typically dismissed as a kind of inanimate "glue." Golgi, who first observed glial cells in 1871, believed that they also provided nutrition for nerve cells. It was only later that histologists found evidence of neuroglial deformities in the brains of alcoholics, epileptics, paralytics, and the mentally ill. Cajal also thought that glia played a role in the functional activity of neurons, albeit a passive one. He believed that, by force of human will, glial fibers could contract, causing capillaries to expand and compress neurons and allowing for the total absorption that he experienced during his scientific work.

While studying abroad, working with the famed psychiatrists Alois Alzheimer and Emil Kraepelin, a young Basque doctor named Nicolás Achúcarro became fascinated by the subject. After directing a federal mental asylum in Washington, D.C., he returned to Spain and started working in Cajal's laboratory.

Achúcarro dedicated himself to inventing a pan-glial staining method capable of revealing all cells in the nervous system other than neurons. Far from being homogenous, as researchers had previously assumed, neuroglia proved to be a morphologically diverse group of cells. Cajal, seduced by the delicacy and variety of their spiderlike forms, invented two new histological techniques: formol-uranium and gold sublimate, revealing neuroglia more clearly than ever before.

Between 1911 and 1913, Cajal worked side by side with Achúcarro, and the laboratories at the Institute for Biological Research focused almost exclusively on neuroglia. A cultured man who read four languages, the younger researcher was brimming with idealism and passion, and his charming smile and good humor endeared him to many in the laboratory. Cajal liked to banter with him, trading sarcastic remarks. After finishing the morning's work at around 2:00 P.M., Cajal sometimes sat outside and chatted about biology, literature, or art with a small group of laboratory members, who recalled these fifteen- to thirty-minute "microtertulias" as open and cheerful. Every participant had a role in the conversation, and Achúcarro's was to provide a healthy dose of cynicism and wit. He was an expert at engaging Cajal, who, if not prodded, would gladly spend the day in silence. Achúcarro would ask Cajal pointed questions about subjects that he knew Cajal could not resist debating. But Achúcarro was fiercely loyal to the master. Some intellectuals in Madrid denied that a histologist could be considered a genius—only an artist or writer could. If Achúcarro ever heard such slander, the subsequent dispute was liable to come to blows.

In his research on neuroglia, aided by his new methods, Cajal described "a new class of cells that appeared to lack processes," by which he meant nerve fibers. They were neither neurons nor astrocytes—a known type of star-shaped glial cell—but rather a mysterious "third element," a term that he used to categorize all remaining cell types that were still too small to see. He tried to clarify the nature of the "third element" further, constantly tinkering with his gold sublimate method until he exhausted the process, even resorting to photomicroscopy—photographing his microscopic slides—which he only rarely did. No clearer images of glial cells could be produced. "In reality," Cajal eventually concluded, "no one knows anything about the nature or physiological significance of said corpuscles."

IN 1913, CAJAL PUBLISHED HIS BOOK *DEGENERATION AND REGENERATION in the Nervous System*, the second monumental work of his career. The two-volume tome, representing a definitive summary of twenty papers on the subject written over eight years, consisted of over eight hundred pages and three hundred illustrations. Writing it left him "profoundly exhausted." Publication was funded by a group of forty-seven doctors in Buenos Aires, who appreciated Cajal's work because it "counteracted at one stroke the prejudice, excessively accentuated among individuals at home and abroad, of our inability to cultivate the experimental sciences." Cajal sent copies to his Argentine subscribers, each printed on special paper and accompanied by a personal dedication, and he planned to ship the remainder to colleagues throughout Europe.

In this work, Cajal concluded that regeneration is clearly possible in peripheral neurons. As for the central nervous system, his findings were controversial; in the spinal cord, cerebellum, and cerebral cortex, he observed "aborted restorative processes incapable of bringing about a complete and definitive repair of interrupted paths." He maintained that the central nervous system has the *capacity* to regenerate, but that its environment, in relation to the peripheral nervous system, is "impoverished." In theory, he said, if there were different chemical influences in the brain and spinal cord, then regeneration might occur. "In adult centers [brain areas]," he wrote, "the nerves are something fixed and immutable: everything may die, nothing may be regenerated. It is for the science of the future to change, if possible, this harsh decree."

CHAPTER 26

"The Unfathomable Mystery of Life"

On June 28, 1914, a young Bosnian Serb anarchist murdered Archduke Franz Ferdinand, presumptive heir to the Austro-Hungarian throne, in Sarajevo. One month later, Austro-Hungary declared war on Serbia. Russia, wary of foreign influence in the Balkans, declared war on Austro-Hungary in response. Germany, allied with Austro-Hungary, declared war on Russia and, two days later, on France, which had come to the defense of Russia, its ally. In the span of four days, five major powers plunged into battle, which spread across three continents as conflicts exploded among colonial proxies. Twenty-two declarations of war had been signed by the end of the year. Two million young men mobilized to fight what became known as the Great War.

Cajal's *tertulia* at the Suizo was "overwhelmed with horror and abomination, erasing the last relics of our youthful optimism." Cajal struggled to focus on his work, constantly distracted by news of the war, chronicled daily on the front pages of the newspapers. He had always tried to avoid current events, which weakened his concentration, but it was impossible to turn away from the increasingly alarmist headlines. THE EUROPEAN WAR APPEARS INEVITABLE, read *El Imparcial*. GERMANY AGAINST ALL, read *El Heraldo de Madrid*. ITALY MOBILIZES, read *El Imparcial*. THE RESPONSE OF THE UNITED STATES, read *El Liberal*. FRESH MEAT, read *El Heraldo de Madrid*. "The horrendous

European war of 1914 was for my scientific activity a very rude blow," Cajal recalled. "It altered my health, already somewhat disturbed, and it cooled, for the first time, my enthusiasm for investigation." Cajal published only one new article in 1914, by far his lowest output ever.

Science was supposed to be universal, transcending national identities, but as a result of the Great War, the grand project seemed dead. Mail could not be reliably sent across the English Channel, telegraph lines were cut, trenches were dug from the North Sea to Switzerland, and borders were almost constantly closed. Scientists continued making discoveries, but they could not share their work. Cajal was supposed to attend a conference in Zurich, which was canceled, depriving him of a triumph. Scientists working abroad were suddenly perceived as enemy aliens, and some were arrested as spies. Universities expelled international students, and scientific academies severed ties with foreign members. Junta grantees studying in Germany were forced to return home, since only neutral countries were safe. Researchers from neutral nations were recruited to prove that the enemy belonged to a physically and morally degenerate race. No Nobel Prize in Physiology or Medicine was awarded between 1915 and 1918. The 1800s had seen the birth of the modern nation-state, and by the end of the century, those independent entities, with advances in travel and communication, had developed a complex network of international science. Now the system itself had degenerated.

Spain remained neutral in the conflict, at least officially. King Alfonso had connections to both sides: his mother was an Austrian archduchess, and his wife was a granddaughter of the British queen. "Without being at war," one foreign diplomat observed, "the war was in Spain's own backyard." The "Germanophiles"—for the most part, privileged members of the clergy, aristocracy, Spanish court, haute bourgeoisie, and military, as well as many Conservatives—preferred a strict, hierarchical social order and wanted the Central Alliance of Austro-Hungary and Germany to impose its will. The "Francophiles"— members of the professional middle class, petite bourgeoisie, and intellectuals, as well as many Liberals—favored intervention on behalf of the Triple Entente—the Russian, French, and British alliance.

On its war path toward Paris, the German army marched through Belgium, committing what became known as the "Sack of Louvain": demolishing the six-hundred-year-old city, killing about two hundred civilians, including children and the elderly, expelling the whole population, and burning their homes to the ground. Among the ruins was the university library, with more than three hundred thousand books, including irreplaceable medieval manuscripts and works by contemporary scientists, such as the oeuvre of Cajal's friend Van Gehuchten, who lived and taught in the city. German officers, quartered in his house, stabled a horse and pig in the drawing room and threw Van Gehuchten's whole library into the road. Eventually, his home, his laboratory, and his manuscripts were all set on fire. Inconsolably depressed, Van Gehuchten fled to Britain, where, a few months later, during a routine surgery, his heart stopped.

Cajal was devastated by the news of Van Gehuchten's tragic death. He was the first researcher from Cajal's generation to die, a reminder of his own mortality. Despondent, he turned to "his oldest and most persistent laboratory love": the retina. For a decade, Cajal's pupil Domingo Sánchez y Sánchez had been studying the nervous systems of invertebrates such as insects and leeches using Cajal's silver nitrate method. Sánchez's work became vital ammunition for the neuronists' campaign against the reticularists. Only eight years younger than Cajal, and having arrived at the institute the year it was founded, Sánchez, at the age of forty-two, was one of Cajal's first pupils, and the two collaborated on a comparative study of the retina. Sánchez contributed the majority of the research, while Cajal confirmed and expanded those results with his own work. Their 1915 paper, titled "Contribution to the Understanding of the Nerve Centers of Insects" and released in the *Trabajos*, was one of the few papers that Cajal published in the second year of the Great War.

In the nineteenth century, scientists conceived of evolution as a linear, hierarchical process. Cajal expected the retina to become more and more complex as species evolved. But with his exhaustive studies of ants and bees, Sánchez demonstrated that, though seen as occupying a lower rung on the evolutionary ladder, the nervous system of an insect was as complex as that of any mammal. Cajal came to believe that the hierarchy imposed on species was not only deceptive but degrading, a sign of "aristocratic disdain," casting the humblest creatures of the earth as somehow beneath humans. In his eyes, every creature was worthy of anatomical study. "Life never succeeded in constructing a machine so subtly devised and so perfectly adapted to an end," Cajal marveled about the insect retina.

The illustrations Cajal and Sánchez printed in their 1915 article are among Cajal's most exquisite. Almost all his past drawings were done in black ink. But the visual center of a housefly contains such rich and intricate connections that, even in a schematic image, Cajal had to differentiate between their fibers using blue, red, and green ink as well. The legends for his drawings were typically coded with alphabetical letters; here, Cajal ran out of characters, and his labels, which typically read A, B, and C, extended to hybrids like Acg' and Cmr'. "The complexity of the insect retina is stupendous, indeed disconcerting, and with no precedent in other animals," Cajal wrote in the paper. "Indeed, the retina of vertebrates seems by contrast gross and deplorably simple."

Cajal's insights into the insect retina threw his deepest-held principles into doubt. How could a "lower" species seem more complex than a "higher" one? Cajal even admitted that encountering the insect retina was the only time he questioned Darwin's theory, albeit only momentarily. "The more I study the organization of the eye in vertebrates and invertebrates," he wrote in a later edition of *Advice*, "the less I understand the causes of their marvelous and exquisitely adapted organization." The insect retina gave him a "terrifying sensation of the unfathomable mystery of life."

• • •

THAT SAME YEAR, JORGE ORTEGA Y GASSET FOUNDED A WEEKLY newspaper called *España* for prominent intellectuals to voice their fears and discontents about the war. The publication quickly established itself as one of the most important in the country. Thirteen public figures were asked to respond to the same prompt: What political, emotional, and ideological trends will dominate Europe in the future? "I have to admit," Cajal wrote in his response, "I have a very low opinion of human beings," who, as the *"last hunter animal,"* retain the "foul instincts" of beasts. "Our nerve cells continue to react in the same way as in the Neolithic Age," he lamented. Neurons are plastic, but only to a point—the brain is constrained by a countervailing force, which Cajal called "evolutionary resistance." Because of this "excruciating biological fact," he claimed that war will never be eradicated. All that civilization can hope to do is prolong the intervals of peace, but the "destructive phase" will always return, with each war becoming more horrifying. For Cajal, so sensitive to the inner life of organisms, the tragedy of lost life was multiplied exponentially. He mourned each premature death as the killing of a million invisible beings. "In about twenty or thirty years, when the orphans of the present war will be men, the same stupendous massacre will be repeated," he predicted with chilling accuracy. Suddenly, Cajal realized that the brain was not perfecting itself. "Our descendants will be as putrid as we are," he concluded.

Unable to bear the horrors of World War I, Cajal retreated to the garden of Cuatro Caminos in the spring of 1918, where he spent hour after hour with a straw hat on his head crouching on the ground with a magnifying glass in his hand, transfixed by the movements of ants. His latest mania was inspired by one of his favorite childhood authors, Jean-Henri Fabre, known as "The Prince of Insects," who, despite teaching six days a week at a provincial school, devoted his life to biological research, becoming the foremost expert on indigenous species in southern France. Darwin recognized Fabre as "an incomparable observer." Like many young readers, Cajal was enamored of Fabre's style, anthropomorphizing the insects that he studied with clear affection for the "humble," "lowly," and "hardworking" creatures. Fabre demonstrated that insects displayed not only instincts but intelligence, which Cajal interpreted as "the ability to triumph over unforeseen accidents." Disgusted by human ignorance, Cajal sought the wisdom of ants.

Fabre focused on the homing instinct, the innate ability of ants to move toward breeding spots or their nests, even across great distances and in unfamiliar terrain. Cajal conducted experiments with ants, delighting in creating new obstacles for them, each more imaginative and challenging than the last. He would carry them far from their nests and place scents along their route, kitchen spices like essence of bergamot and oregano, or everyday objects like bread, honey, sugar, meat, coins, and even water. He tried distracting them with noise, and lined their routes with paper or cotton to separate their feet from the ground. He placed mirrors in front of their nest and blocked its entrance with a rock. He covered one of their eyes, or both of their eyes, and even removed their antennae. He put an ant on a piece of wood, raised it in the air, and spun it around so as to make it dizzy. Then, with his magnifying glass, he observed his subjects as they tried to find their way home, timing them with his watch while sketching their routes on scrap paper with meandering, almost zigzagging lines. When he returned to Madrid, he continued his studies on a small hill in Retiro Park, across from his house.

Included in Cajal's archives are hundreds of pages about ants, with notes written on scraps of newspaper, the backs of envelopes, banquet invitations, and, eventually, full quarto pages. He even speculated about their character traits—"ants have bad memories for facts," "ants help the wounded," and "if ants could talk they would say my mother but not my father." Students recounted the "childlike joy" with which he would describe his experiments with ants, reveling in the fiendish tricks that he devised to stymie them. As Cajal found out, stopping ants from reaching home was almost impossible. Determined as young axons, they would relentlessly find their way forward.

Cajal released his findings in 1921, and he could have published a treatise about ants, but he decided that his work was not sufficiently original. Instead, he turned to an old pastime: writing fiction. Cajal imagined a letter that a scout from a parasitic species of ant called "slave-makers" would write to his queen after exploring the human realm in search of new labor for the colony. Human beings, the ant reports, "live almost as we do"; they are nothing more than "exceptional ants." The only difference is that they kill for pleasure and love nothing more than destruction. "Nothing transcendental has grown out of the human vermin," concludes the ant.

CHAPTER 27

"I Drown and I Awaken"

One afternoon near the end of the war, as Cajal was leaving his *tertulia* at the Suizo, he suddenly felt as though his head was on fire. He walked home in silence, and the next afternoon he took a cab instead of walking, but the pain persisted. Though a physician himself, Cajal had been skeptical of the medical profession since his stay in the sanatorium forty years earlier. He also knew better than anyone that the field of neurology was still immature. After a session in his photography studio, an attic where temperatures sometimes surpassed ninety-five degrees, he suffered such a debilitating migraine that he had no choice but to see a doctor. Cajal chose to consult his colleague Achúcarro, who at the time was ill himself, with cancer. After examining his master, Achúcarro produced his diagnosis: "My friend, the cerebral arteriosclerosis of old age has begun . . . Do not be alarmed."

Cerebral arteriosclerosis is a condition in which the walls of arteries become thickened or hardened, reducing blood flow to the brain, which doctors considered irreversible. Cajal could not help being alarmed. Confronted with another illness, his morbid thoughts returned to haunt him, and, already depressed by the war, all that he could think and talk about was death. He consulted textbook after textbook, obsessively checking his symptoms, but the only prognosis, as far as he could discern, was doom. "Some future!" Cajal recalled saying to himself. Achúcarro prescribed potassium iodide, the standard treatment, and recommended a strict regimen of isolation and silence, instructing Cajal to avoid warm temperatures and heated debates. According to one *contertulio*, Cajal had never missed a day at the café for thirty years, but now he no longer participated in *tertulias* or debates at the Athenaeum. He complained that the slightest conversation brought on a searing migraine. "I must now regulate my reading and writing, and speaking even more," Cajal lamented in a letter to Unamuno. The restrictions had the effect of "freezing my thoughts."

Tests showed nothing physically wrong with Cajal, yet he was sure that, at any moment, he would die of a massive stroke. "[He] only *believes* he suffers from cerebral arteriosclerosis," said a correspondent from the *American Journal of Medicine*, who was visiting the institute, because he is "prone to pessimism." Cajal's illness—which he called "this menace"—gave him an unassailable excuse to turn down invitations and appointments, the kinds of social obligations that he would have rather avoided anyway. Whenever he left the house, he carried a yellow parasol for protection from the sun, which embarrassed him. Before long, he was forced to stop his long walks and buy an automobile, which he called "the hated bourgeois artifact." Cajal blamed his condition on "thirty years of excessive and chaotic work," and he wondered if the states of "total absorption" that facilitated his discoveries might have led to "permanent brain congestion." The cruel irony was not lost on him—that devoting his life to the brain might cause it to give out at the premature age of sixty-six.

Before his illness, Luis recalled, Cajal liked to be the center of attention, holding forth at the head of the dinner table in the garden at Cuatro Caminos, where he would eat with family and friends, his dog—Lord Byron—at his feet. After dessert, he and his guests would relax and chat, a Spanish tradition known as the *sobremesa*. His favorite subjects to discuss were philosophy, botany, literature, and astronomy. After his diagnosis, however, he recounted the following dream: *I am sitting around the table at the* sobremesa *among friends. Astronomy comes up and I go on about the planets. I speak about astral collisions, a black star,*

perhaps a planet, collides with another. It starts to boil. It vaporizes. Good-
bye civilization, science, culture, religion. Nothing is more tragic.

No matter how little Cajal slept, he woke up each morning between six and eight o'clock and ate a light, simple breakfast in bed,
where he sat reading until ten or ten-thirty, before going to the medical school to teach his class. "Of my eight to ten hours of mental work
from a previous time," Cajal reported, "I should today content myself
with two or three (including the professorship). And this only in the
morning." The rest of his day was largely given over to medical visits;
his increasingly infrequent trips to the institute were made "to encourage my students and do some routine work," and only after the staff
had assured him that the heating in his laboratory would be off, so as
not to exacerbate his migraines. As soon as he walked in the door,
Tomás, the doorman, dutifully removed Cajal's coat but left on his
scarf and cloak, and students immediately flocked to him with com-
plaints and questions. His response was typically the same: yes, the
work is difficult, but just persevere. "I cannot get him to rest," Silveria
told two journalists who visited the house, "and afterward, he does not
want to accept that the years have not passed in vain. He wants to
work as when he was young." She took it upon herself to shield him
from disturbances. "The whole world wants an article, an autograph, a
sheet of paper, a portrait, an interview," she complained. "I avoid it all,
because it hurts him." Cajal's deepest fear was that his work would be
forgotten, and with that possibility on his mind, he could find no
peace.

After the war had ended, Cajal opened German books and journals for the first time and saw that his work was either not cited or
mentioned only in passing, and he convinced himself that, before long,
those books would replace his own in laboratories around the world.
For Cajal, according to one researcher visiting the institute, this constituted a "personal tragedy." Though Cajal printed the institute's journal in French, the majority of his work existed only in Spanish, and

older researchers, who had learned the language just to read him, were beginning to die.

In 1919, Anna Retzius wrote to Cajal with the news that Gustaf, his loyal friend, was on his deathbed. "When your dear letter arrived," Anna informed Cajal, "he was still able to hear me when I read it to him and he leafed through your book (the biography) looking at the illustrations and portraits with great interest, and he has begged me to give you his thanks. He loves you and admires you with all his soul and his heart, sir."

IN THE FOYER OF CAJAL'S HOME, BEHIND THE FRONT STAIRCASE, WAS a small, unassuming door that looked as though it might lead to a janitor's supply closet. Behind that door was a flight of stairs leading down to a large, unfinished basement, split into two levels and joined by a crude, broad step, since Alfonso XII Street was on a slight incline. Cajal converted the lower level into a small laboratory with a work-table, microscopes, microtomes, beakers, reagents, an old telescope; shelves with folders of annotated manuscripts, notebooks, and draw-ings; and boxes of histological slides in a special armoire. The upper level was ringed by unvarnished pine bookcases containing his per-sonal library, including thousands of volumes of fiction, poetry, philos-ophy, history, and popular science. He spent his days shuffling between the two halves of the basement, alternating between scientific observa-tion and writing in notebooks, between the microscope and the pen. Though large windows let in plenty of light from street level, the damp, narrow basement was always five or six degrees colder than the rest of the house. There, he said, "I could bury myself."

On April 23, 1918, Nicolás Achúcarro, Cajal's beloved pupil, passed away. After examining himself in the mirror and consulting a medical textbook, he had diagnosed himself with Hodgkin's disease. Realizing that his life was over, with absolutely no remaining hope, he had returned to his childhood home and established a laboratory in the attic, where Achúcarro's quixotic research career, like Cajal's, had begun. Cajal was deeply moved by Achúcarro's love for and devotion to science. In his final letter, Achúcarro's only complaint was that his death would put an end to his discoveries, a fate that seemed to Cajal like an "unimaginable torture." "[Achúcarro] would have been the most dignified successor of the great master," remarked Tello. Achúcarro was thirty years old, around the same age as Cajal's son Santiago was when he died.

Shortly after Achúcarro's death, Cajal began to record his dreams. Dreaming had interested him since the beginning of the century, when he first speculated about the role of neurons in complex psychological phenomena such as thought, will, and creativity. Cajal did not care about the content of the dreams themselves but instead focused on the mechanisms behind their visual, almost hallucinatory, qualities. His 1908 paper "On the Theory of Dreaming" included a study of a man who went blind as an adult yet still experienced visual dreams. Cajal concluded that the retina is not involved in dreaming and that the brain can generate images without sense-data from the eyes. The paper ended with "To be continued . . . ," yet Cajal did not continue his work on the subject for ten years.

Why did Cajal return to dreaming? He was motivated in part by his hatred of Sigmund Freud. The first Spanish translation of Freud—excerpts from what would become his 1896 *Studies on Hysteria*—had appeared in 1893. In 1908, when Cajal first considered the physiology of dreaming, Freud's theories were not yet popular in Spain. Three years later, after studying in Germany, Ortega y Gasset published an article on psychoanalysis that, though expressing skepticism, popularized Freud in Spain. Freud's theories about psychic life whipped Catholic Spain into a frenzy. *Tertulias* across Madrid debated scandalous new ideas about childhood sexuality; the id, the ego, and the superego; and the neurotic psyche. Many of Cajal's disciples were open to Freud and adopted some of his views, which they acknowledged as "revolutionary."

Cajal and Freud were born four years apart. Like Cajal, Freud studied medicine in college and trained in dissection and anatomy, writing his thesis on the gonads of male eels. Both Freud and Cajal chose biological research over the clinic. Freud excelled at the workbench, mastering histological techniques much like Cajal did. Freud even invented a new method for staining neurofibrils—gold chloride—which Cajal cited in his *Textura*. With his comparative studies of frogs, lampreys, and crayfish, Freud tried to identify structural distinctions between the invertebrate and vertebrate nervous systems. In seven years of research, Freud published fourteen articles on the nervous system, including original drawings. A decade before Cajal published his first paper, Freud's work appeared in the annals of the Imperial Academy of the Sciences, while Cajal was convalescing in Panticosa from what might have been the kind of nervous disease that Freud would later specialize in treating.

Initially, Freud's mission was the same as Cajal's: to finally reveal the true identity of nerve cells by visualizing the ends of their fibers. Freud was happy in the laboratory, but, as both a Jew and a junior assistant, his chances of promotion were minimal. To make his situation more precarious, anatomical research was not a lucrative career, and Freud was in need of money to get married. In 1885, the University of Vienna awarded Freud a grant to study in France, where he planned to investigate pathological lesions in infant brains. His real fascination, however, was with hypnosis. Freud worked at the Salpêtrière, where he was profoundly affected by Charcot's demonstrations, and when Freud returned to Vienna he opened a private psychiatric practice. As Cajal was beginning his black reaction studies, Freud was leaving neuroanatomy behind. Had the 1889 Berlin Conference convened five years earlier, Freud might have looked through Cajal's microscope at his slides and seen the images that had eluded him earlier in his career.

"At heart, Freud is trying to make psychophysiology lead into biology," Ortega y Gasset explained in one of his articles. At heart, Cajal was trying to do the exact opposite. The intellectuals of Madrid treated

Freud's theories as though they were biological realities, while Cajal considered them "collective lies." According to Cajal's "religion of the cell," Freud qualified as a heretic, a "surly and somewhat egotistical Viennese author" whose "pseudoscientific" ideas were "impregnated with mysticism."

In 1917, as the Great War raged on, Ortega y Gasset had convinced the Spanish publisher Biblioteca Nueva to buy the translation rights to Freud's complete works—the first full translation in any language. All correspondence for the project, including the initial proofs, had to travel by diplomatic courier. When the *Complete Works* finally came out, in 1922, students at the Residencia, the next generation of Spanish intellectuals, stayed up all night reading the book, running into one another's rooms to analyze their dreams. At the moment when Cajal felt as though the world had forgotten him, Freud was eclipsing his fame.

Cajal believed that Freud's psychoanalytic theory was dependent on the idea of childhood sexuality. Convinced that his dreams would not betray such conflicts, with pen and paper at his bedside, Cajal recorded hundreds of dreams, scribbling them down the moment he woke up or as soon as he remembered them the next morning. His first dream was dated April 1918 at 5:00 A.M., the same month as the death of Achúcarro. "I am delivering a speech before a popular auditorium. Athenaeum," Cajal noted. "I talk . . . about the neuron theory. Passion, eloquence while speaking about how the form was discovered and recounting the miracles of the Golgi and Ehrlich methods. I forget nothing. But I grow tired of talking and a moment comes when I ask myself am I not dreaming?"

In a dream recorded on May 18, 1918, Cajal jumped over the wall of an orchard, as he often did when he was a child. He saw women hanging clothes to dry and dropped to a crouch, running along the base of the wall, looking for a way out. Suddenly, he came face-to-face with one of the women, who froze, horrified at the sight of him, because he was not a child in the dream but an old man. Cajal apologized and invented a lie about losing his glasses on the other side of the wall. In his diary, Cajal made some general notes about the dream but did not see in it any emotional significance. "We were in winter," the elderly Cajal wrote. "There was no fruit."

Again and again, Cajal refused to interpret his own dreams, only pointing out their factual inconsistencies, insisting that they had no deeper meaning. *This cannot be explained by Freud*, he wrote in one journal entry. *Sea expedition at 4 in the morning*, another one of Cajal's dreams goes. *Near the port I fall into the water with a child. I drown and I awaken.*

CHAPTER 28

"Those Poisoned Wounds"

The death of Achúcarro devastated Cajal and left the histopathology laboratory at the institute without a director, jeopardizing its neuroglial research program and thus the place of the Spanish school on the world stage. To replace him as director, Cajal elevated Achúcarro's student, Pío del Río Hortega, who had been running day-to-day operations at the laboratory since Achúcarro had fallen ill.

Del Río was the hardest-working member of the laboratory. He was the first person to arrive at the institute each morning and the last person to leave, often late at night. His fastidious schedule irked Tomás, who kept the key to the building. A stout, muscular war veteran, proud of his mangled arm, which had been injured in battle, Tomás liked to talk about how much he loved wine, especially the wine from his hometown. While del Río was busy working, Tomás would have rather been home, drunk or asleep.

Cajal never hired a laboratory administrator, because he loved to interact with the staff himself. He liked chatting with the *sereno*, the neighborhood watchman, whom he always tipped extra each month, since Atocha Street had fewer buildings than average and so the *sereno* would not earn as much. Cajal also enjoyed bantering with his children's nanny, a fiery old Andalusian woman, whom he affectionately teased and who would tease him back. Though he had a reputation for rudeness toward people of his own social class or higher, he never mistreated working people. He asked them about their families, their problems, and what they thought about life in general, then listened patiently to their replies.

Cajal's biggest soft spot was for the *alimañero*—the exterminator—a character nicknamed "Ranero" (a reference to the swamps where frogs live). A connoisseur of the regional fauna, a savant of ponds, nests, and burrows, Ranero procured critters for the laboratory the old-fashioned way. Cajal would tell him which animals he needed for his experiments, and then Ranero would roam the outskirts of Madrid, wading into ponds to catch salamanders and newts and raiding farms to steal newborn rabbits and pregnant cats. He would show up at the laboratory door exhausted, limping from rheumatism, trailing a sack full of animals, his arms scarred with claw and fang marks, and Cajal would immediately stop whatever he was doing to greet him. Ranero was an alcoholic who hit the bottle first thing in the morning and kept on drinking throughout the day, and by the time he arrived at the institute, his face was flushed and his eyes were bloodshot and teary. The whole laboratory gathered to witness the dialogue between the Nobel Prize winner and the town drunk. Cajal paid Ranero whatever fee he set, no questions asked. When Cajal learned that Ranero's shack lacked a proper roof and floor, and that he and his wife shared their bedroom with their donkey, Cajal paid for renovations but never told Ranero, so as not to offend his pride. Ranero bragged about his toughness, boasting that whether in the city or the countryside, he could survive under any conditions. Cajal saw himself in this man, whom he treated like a brother. From time to time, Ranero would get into a brawl or be caught stealing, and none other than Dr. Cajal would appear at the police station, securing his release. Cajal's personal address book was filled with the names of Spanish politicians and intellectuals, as well as famous men of science from around the world. Right beside them is the contact information for Francisco Arías, "Ranero."

• • •

TOMÁS'S DUTIES INCLUDED RUNNING ERRANDS TO PURCHASE EQUIP-
ment and materials for the lab, and it was customary for researchers to
pay support staff extra commissions. Del Río was the only researcher
who not only never paid commissions but scrutinized Tomás's re-
ceipts, haggling over the cost of every item. Tomás insulted del Río
as he walked in and out the door.

In a letter to Cajal, del Río complained about Tomás's "attitude of
insolence" and politely asked for his own key. A few days later, Cajal
responded with the salutation "My dear friend and colleague," before
instructing del Río to move his laboratory out of the institute's building to a cold, small, and poorly lit room in the Residencia de Estudiantes. Cajal promised del Río that one day he would be able to return
to the institute and direct a fully independent department there. "You
can always count on my selfless protection," Cajal assured him. "I feel
as though I have been sentenced without being heard," replied del Río.
"In any case, you may be certain that my regard for you remains unchanged, and that despite what I have suffered these past days, I am
sorrier still for what you have endured."

To work alongside Cajal was del Río's lifelong dream. He grew up
in a small town in Valladolid province, in central Spain, where "[Cajal's]
precious name rang at all hours in my ears and dwelled in my spirit."
Cajal inspired many of the younger doctors in Spain to pursue science,
including del Río, who, when he was a child, convinced his father to
buy him a microscope as a gift. Del Río earned his medical license but,
like Cajal, had no desire to practice. A relative in Madrid claimed to
know Cajal from the Suizo, and del Río became one of those men who
sat at a nearby table, hoping to overhear a snippet of Cajal's conversation. "His face with energetic lines, his smooth and ovular forehead,
fringed with snow, his eyes sometimes distracted and other times inquisitive, his halting and rough voice, were my first impressions," del
Río recalled. When he shook Cajal's hand for the first time, he was
overwhelmed with emotion and started stammering.

Nineteenth-century biology laboratories were run like artists' workshops in the Renaissance, where apprentices worked under a master, assisting him with his projects while he supervised the development of their skills before freeing them to pursue their own paths. Unlike Achúcarro, who followed Cajal's program unquestioningly, del Río had aspirations of his own, and oftentimes he did not inform Cajal about his own research. Working in the same areas of the nervous system as Cajal did, del Río invented a staining technique based on Achúcarro's method called silver carbonate, which stained neuroglia in more complete detail than ever before, revealing finer structures than Cajal had been able to see. Del Río's new method allowed him to better visualize Cajal's mysterious third element. Rather than lump a diverse group of elements together, del Río managed to distinguish two distinct glial cell morphologies: oligodendrocytes—polyhedral bodies with only a few projecting fibers, which were long and smooth—and microglia, small, dark bodies with complex fibers, sprouting "a long, tortuous" bloom.

Del Río instantly recognized that this was his breakthrough, a discovery that would resolve a crucial problem in the field and establish his global reputation. But he was conflicted. For two decades, Cajal's concept of the third element, albeit vague, had reigned, and del Río's findings would overturn it. He feared the reaction of Cajal, whose temper had grown more volatile in old age. For a whole year, del Río kept his findings secret until he summoned the courage to walk across the laboratory and approach the master at his desk. Del Río read his paper aloud in an especially booming voice, since Cajal, nearing seventy, was going deaf. "I believe that your conclusions are too daring," he said to del Río. "They will unleash a storm of controversy."

Defying Cajal's advice, del Río presented his findings at a conference under the title "The Third Element of the Nervous System," immediately stating his intention to "criticize and correct" Cajal's theory. He submitted the paper to Cajal, along with a letter saying that "the fear of displeasing you has proved a heavy burden," while affirming his pride in the Spanish school. As editor of the *Revista*, Cajal had the power to accept or reject del Río's paper. In the following issue, del Río's paper appeared, next to a rebuttal from Cajal, who had produced his own slides using a variation of del Río's technique. He criticized del Río's conclusions and credited his findings to an earlier researcher, denying the novelty of his technique.

Months later, Cajal suddenly became even more hostile to del Río. He always tried to control his emotions when writing letters, often waiting a day to respond, but in an uncharacteristically long and passionate screed he accused the younger researcher of "bitter and severe affronts," including ingratitude and even sabotage.

One day, del Río walked up to his laboratory door and found a handwritten note from Cajal, firing him from the institute. Del Río left the building in tears. The next day, when del Río returned, Tomás, gleeful, refused to open the door. Ten days after the banishment, Cajal attempted to reconcile with del Río, having discovered that Tomás had invented and spread the rumors about him that had in part led to his exile.

Cajal had been inclined to trust Tomás. The two men had served together in Cuba forty-five years earlier, and Tomás was the only remaining figure from that period in Cajal's life. No one else understood what it took to survive in the jungle. Years later, a reporter from *El Sol* wrote about Tomás for a popular weekly column called Literary Views. He lived beside the institute in a shed filled from floor to ceiling with cages of laboratory animals including lizards, mice, and guinea pigs. The house stank of pigeon shit, and the smell of fried potatoes and train smoke permeated the hallway. Up a small spiral staircase, there was a dusty attic with a filing cabinet, some scattered dishes, a typewriter, and a photograph of Cajal. When the reporter asked Tomás about Cajal, his tired, heavy eyes were on the verge of tears. "He is very great, very great."

• • •

ONE WINTER, BEFORE THEIR FALLING OUT, DEL RÍO CAME DOWN WITH influenza, and Cajal grew so concerned about del Río's long absence that he asked Tomás to check on him. Cajal had no idea where del Río lived, and it turned out that he roomed in the same small boarding-house at 10 Prado Street where Cajal kept his secret photography stu-dio. Cajal sometimes photographed female models, most of whom were blond. Servants who entered the room to clean found a portfolio of artistic prints, in which some models appeared multiple times. It is unlikely that Cajal was having an affair—in private notebooks, he ap-peared to consider the possibility before concluding that romantic pas-sions could no longer be trusted, that he loved his wife, and that domestic harmony was what allowed his work to thrive. Yet Cajal must have felt that his photography studio was in some way transgressive, since he never spoke about its existence. In expelling del Río, Cajal might have been casting out the only person who knew his secret.

Above all else, Cajal envied del Río, and nothing angered him more than defeat in competition. Del Río's view of glial cells had tri-umphed over his own. Cajal was aware of this psychological tendency: "Mortified pride is defended always somewhat more fiercely when it is more unreasonable," he wrote. Del Río represented the next genera-tion of Spanish science, while Cajal's career was coming to a close. Del Río had just made his major breakthrough, whereas Cajal's glory days were long gone. "Like every old man," he said, "I also feel those poi-soned wounds of the heart and brain. They are the knocks of time at the door." Cajal never hesitated to criticize Golgi and the reticularists for clinging to their own ideas rather than accepting reality, but when Cajal was presented with "the new truth" about glial cells, he simply closed his eyes and ignored it, claiming that del Río had merely mod-ified his own stain—exactly what Golgi had said about Cajal's im-provements to the black reaction.

When Cajal realized what Tomás had done, he apologized to del Río and invited him back to the institute, and in the years that followed, the two became closer than ever. Years later, del Río wrote to Cajal wishing him a happy birthday, and Cajal replied by wishing him "a thousand congratulations" for his success. Del Río was one of the last people with whom Cajal socialized in his later years, when the two men would meet for coffee and discuss research and life.

Eventually, Cajal did cede priority to del Río for his discoveries— but in the definitive version of *Recollections of My Life*, published in 1923, Cajal cited the work of each of his disciples by name, except that of del Río.

CHAPTER 29

"No Solemn Gatherings"

In 1920, the Café Suizo—Cajal's home away from home, his intellectual and social refuge for thirty years—announced that it was closing forever. In twentieth-century Madrid, more old buildings were being demolished every day, and destruction of the Suizo would clear the way for a new outpost of the growing national bank chain BBVA. After drinking a ceremonial final coffee at the café, the poet Manuel Machado wrote a eulogy that was published in *El Liberal*, crystallizing the sadness that many felt with his description of the hanging wall mirrors as "giant, suspended tears."

To commemorate the Suizo, Cajal compiled a book of "thoughts, anecdotes, and secrets" inspired by his lifetime of *tertulias*, first titled *Café Chatter*, then renamed *Café Chats*, touching on a wide range of topics in a tone both humorous and reflective. The book contains 850 entries organized into ten chapters: on friendship, love, old age, death, genius, polemics, personal character, education, the arts, politics, and humor. The scientist and writer Gregorio Marañon complained that the title included the "unfortunate" word *café*, which he thought undermined the author's claim to wisdom. "Why unfortunate?" Unamuno replied to Marañon in a letter. "Deep down, Cajal was always a man of the café, and fortunately, that did not succeed in drowning out his histological research." Another colleague of Cajal's, upon seeing his formal portrait hanging at the Athenaeum, wondered why the artist did not paint him drinking coffee at the café.

Spanish booksellers ordered seven hundred copies of *Café Chats*, amounting to three editions in three years, and translations were published in two other countries. The book would have been of little interest if not for the identity of the author. Cajal was one of the most famous men in Spain. Though few understood his scientific work, his name had become synonymous with *wisdom*, much like Einstein's had with *genius*. One of the sentences in primary school calligraphy books was "El doctor Ramón y Cajal es un sabio eminente"—"Doctor Ramón y Cajal is an eminent savant."

After the Suizo closed, Cajal's *peña* moved to another café a few minutes away, but Cajal did not follow. He never joined a *tertulia* again. By that time, he had grown so deaf that even the loudest speech sounded muted and muffled. He would have had to rely on his neighbor to summarize the ongoing conversation in his ear. Friends and family had to shout to communicate with him, and before long, their voices would grow hoarse, so that they could manage only to whisper. Cajal assumed that they had lowered their voices to talk about him. After all those years, he concluded, people had finally realized the truth about him, which he had feared since he was a child: that he was unlikable.

AS HE GREW OLDER, CAJAL FOUND IT INCREASINGLY DIFFICULT TO connect with his seventeen- and eighteen-year-old students. "There is no lecture that they [do not] resist," he said of his students. "It is unbearable!" Little sunlight entered the classroom, because no one washed the windows, and in the springtime Cajal liked to keep them open to watch the acacias bloom. After he finished speaking, he would slouch down into his chair, while Tello taught the rest of the class and Cajal listened, his small gray eyes staring from corner to corner, as though examining invisible objects. He never looked at anyone directly, one student recalls. "I believe he sometimes forgot himself (or us)." Now and then, he would rub his eyelids, shake his sleeves, or, roused by a coughing fit, stand up suddenly and mutter something, aggrieved. Class would end when the porter's voice from the doorway boomed, "Don Santiago, it's time!" As the students filed out, Cajal would wash his hands at the broken basin in the corner of the room, as though he had just performed surgery.

Sometimes, in the middle of class, Cajal would leave his seat, walk to the blackboard, grab the chalk, and draw. When the janitor erased the board at the end of class with two brisk sweeps of his hand, recalled Marañon, even the most callous and indifferent students could not help but slightly mourn.

On May 1, 1922, Cajal turned seventy years old, the mandatory retirement age for public servants. When the bill was proposed by the Liberal party in 1900, Cajal had voiced his support, thinking that the law would help rid universities of archaic ideas, avoiding a "teaching routine, a medieval mentality." "I do not regret that," he maintained in an open letter to a newspaper about his stance toward the retirement age. "I consider it wise and reasonable." Cajal's supporters sent a letter to *El Heraldo* seeking an exception to the law in order to prolong Cajal's career. In the past, he had watched as professors practically begged for their retirements to be indefinitely delayed. Those men, Cajal observed, were fearful of a life without work. He refused to imitate their "painful odyssey."

In anticipation of his retirement, Cajal requested "no solemn gatherings nor panegyrical, rhetorical effusions." That day, *Nuevo Mundo* ran a tribute on its cover, accompanied by a full-page picture. "Ramón y Cajal," read the main headline of *El Imparcial*, "A Great and Stimulating Example." *ABC* published a retirement ode: "He was an exploring diver of the unknown, a great farmer that sowed tirelessly . . . all of Spain is a student that learned from Cajal." Another poem in *ABC* dubbed him a "modern Columbus." (The compliment did not flatter him. "Columbus was a man of will and faith but destitute of science," he wrote in a 1927 letter. "Columbus did not even know the size of the land.")

In the distant mountains of northern Spain, pharmacists, doctors, and medical assistants from all over Alto Aragon made a pilgrimage to Petilla de Aragon for a tribute. Cavalry rode in from the next town over, and the crowd waited three hours for them to travel ten or fifteen miles, since the mountain roads were still as treacherous as they had been a thousand years before. Cajal's disappointing visit thirty years earlier had spurred the villagers to repair the house where he was born, mounting an engraved stone plaque in the doorway, which read: "In this house on the first of May 1852, Dr. Santiago Ramón y Cajal, glory of Spanish Science, was born."

The University of Madrid held an "unequivocal tribute," with chairs from every department, prominent business owners, and government officials in attendance. One by one, students and professors rose to praise Cajal and read from his work. The applause was so loud it would have penetrated Cajal's increasing deafness, had he been there to hear it, but he had chosen not to attend. "All these sentimental outbursts confuse and stun me," he wrote in an open letter published in *El Sol*. He feared "not finding himself able to resist the emotion that the event could cause," according to one reporter who interviewed him. The Cortes held a ceremony for Cajal, which he also avoided, because "he found himself visibly moved . . . and his presence was unnecessary." Cajal was terrified of showing his emotions, which, as an old man, he found more difficult to contain than ever.

In honor of his retirement, the director of the Institute of Jaca, where he attended secondary school, wrote Cajal a letter of congratulations. Cajal apologized for his criticisms of the school in his autobiography. "My mischievous pranks as a student justify any disciplinary measure," he told the director.

That same year, his cousin Victoriano had visited Madrid while on business. Cajal's boyhood model of virility, now in his eighties, asked him for advice about his will, then died shortly thereafter. And so when Cajal found out that the government planned to build a new institute that would be named in his honor, he raised his cane in the air and threatened to hit the minister who delivered the news. There should be no tributes until after a person's death, Cajal shouted. Otherwise, he said, they only served as reminders of mortality.

In the summer after he turned seventy, Cajal and his friend Mariano Benlliure, who was working on a bust of Cajal at the time, got into an accident in Benlliure's car on the highway north of Madrid. Seeing that he himself was unscathed, Cajal started bandaging his friend's wounds. A nearby shepherd, who had witnessed the crash, rushed toward the two older gentlemen from the city. He shouted at Cajal to stop. "Let me do it, sir," the shepherd told the Nobel laureate in medicine. "I notice that you are not accustomed to handling bandages."

LEADING UP TO HIS RETIREMENT, IN 1922, LEGISLATION WAS PROPOSED on the Cortes floor to grant Cajal a lifelong annual salary of twenty-five thousand pesetas. The bill was overwhelmingly defeated. One senator who opposed the act argued that it would set a "dangerous precedent." "Santiago Ramón y Cajal Declared Dangerous Precedent," mocked the headline of Figure of the Week, a popular column in *Nuevo Mundo*, written by one of Spain's most influential journalists, who pointed out that the proposed pension would equal the cost of one police car. In 1931, *ABC* would compare the wages of Spanish celebrities, including athletes, bullfighters, politicians, actors, and singers. Cajal's income in retirement, absent the stipend, was the second lowest, at twelve pesetas and twenty-two centimos an hour. The famous bullfighter Cagancho brought in three thousand pesetas. Cajal's financial needs drove him to establish a neurology clinic in his home, a betrayal of his fundamental imperative. Experts from all over the world referred patients to him. A wealthy woman arrived from North

America and offered Cajal a giant sum to cure her hysteria. He gave her a common sedative and charged her three dollars and fifty cents. The Ramón y Cajal neurology clinic closed after just a few months.

In September 1922, the start of the first school year after his retirement, Cajal hired a secretary, a fifteen-year-old named Irene Lewy, from Chamberí, a working-class neighborhood on the outskirts of Madrid. Irene's father, a German-speaking Polish Jew, died when she was five, and to make ends meet, her mother rented out rooms in their house. Lewy was educated at the German School in Madrid, where she learned several languages, and Cajal trained her to translate foreign scientific texts. Her salary—sixteen pesetas and sixty-six centimos, arranged through a grant from the Junta—was more than Cajal's own. Irene also helped the lab technicians prepare histological slides and applied her artistic skills to help researchers compose their drawings. Cajal never relied on lab technicians, as a point of pride. They were nuisances to him, interrupting his routine, altering the perfect chaos of his worktable, causing him to waste time looking for bottles and slides. He never needed any help with his drawings.

"It impressed him greatly that I could type in German," Irene recalled. "In general, that I could type at all." Sometimes when she left her desk for a moment, she would come back to find Cajal tapping at the keys. Almost all of Cajal's hundreds of letters were written by hand. If his writing was destined for a typewriter, he reviewed the draft himself, using black ink for the first round of corrections and, when the page turned black, switching to red, creating manuscripts that del Río called "picturesque." Cajal's handwriting was naturally messy, though he could make his penmanship more presentable for official documents. Sitting at the desk, he bowed his head to the page, determined to perfect his grammar, as though a Jesuit friar lurked behind him.

In self-exile from his *tertulia*, Cajal went in search of a colder and lonelier café, where he could sit undisturbed for an hour or two after lunch and read the newspaper, hoping to prevent migraines. He found the Café Prado virtually empty during the day. In his old caramel-colored overcoat, threadbare and faded, Cajal sat in the far corner at the window facing the sun, twirling a spoon in his coffee, his bald head bowed, muttering to himself and jotting down notes. "All of him is bent under the weight of many thoughts," one visitor to the café observed. "This little old man, trembling and nervous, who appears dazed," wrote another regular at the café, "this is indeed the honor and glory and pride of Spain!" More often than not, an old waiter said, he would leave without his hat and cane, but he never forgot to leave a tip.

At a table on the other side of the café sat a boisterous *peña* of young artists, including Luis Buñuel, who was then twenty-two, and Salvador Dalí, who was eighteen. They declared themselves the Ultraists, the most radical of the avant-garde.

Buñuel's first ambition was to become a piano composer, but his father, the scion of an Aragonese business family, steered him in a more practical direction. When Buñuel arrived at the Residencia de Estudiantes in 1917, his principal area of focus was agronomy, but at the Residencia, as Buñuel recalled, "you could study any subject you wanted, stay as long as you liked, and change your area of specialty in mid-stream." Inspired by a childhood fascination with insects, Buñuel switched his focus to the natural sciences.

Behind the Residencia was the Museum of Natural History, where Cajal had established four of the Junta's laboratories. Buñuel joined the laboratory of Ignacio Bolívar, then in his seventies, a leading Spanish entomologist and one of the researchers who had encouraged Cajal to take up microscopy when he had visited Madrid decades earlier.

Two years after Buñuel, Federico García Lorca, a dark-haired, devilishly charming boy from a wealthy Andalusian family, arrived at the Residencia. "We liked each other immediately," Buñuel recalled. Lorca's first love was also the piano, and he was known to accompany himself as he sang in spontaneous public recitals. He would also perform his own death, in intense, terrifying detail, lying in bed and convulsing until he saw that his friends were afraid, at which point he would burst out laughing. Lorca and Buñuel would walk together

through the garden, shaded by poplar trees, and sit on the grass together, where Lorca read Buñuel his poetry.

Three years after Lorca, Salvador Dalí appeared, a boy from a small
city in Catalonia with long hair and sideburns, wearing knee-breeches
and stockings. Dalí's father was a strict disciplinarian; it was Dalí's
mother who nurtured his artistic talent. He had been thrown out of
secondary school for refusing to allow a panel of art teachers to examine him, since he insisted that he knew more than they did. When Dalí
arrived at the Residencia, there were no cubists in Madrid; he took his
inspiration from a French art catalog. Two of his early cubist paintings
appear to depict Lorca giving a poetry reading. In the annual Residencia production of *Don Juan Tenorio*, a nineteenth-century Romantic
interpretation of the Spanish legend, Dalí acted, Lorca either designed
or directed, and Buñuel, a notorious reveler and womanizer, played the
title role, which he seemed to have been born for.

Centuries after the Golden Age, Madrid was still a place where
literary or artistic disputes might end in violence. In 1922, the year of
Cajal's retirement, one writer slapped another in the middle of the
street, which led to a duel. The Ultraists organized a banquet to cele-
brate one of the writers because they considered the other to be a
pretentious fraud. They sent invitations to seemingly every intellectual
in Madrid, each of whom responded, except for Cajal. Since Buñuel
worked in one of Cajal's labs, the Ultraists nominated Buñuel, nicknamed "the Beast," to obtain the signature of the reclusive wise man
who sat alone on the other side of the café for their petition.

With his long black eyebrows; full, shapely lips; broad, misshapen
nose; and off-kilter, bulging eyes, Buñuel's face itself was vaguely cubist. Buñuel, who described himself as "a redneck from Aragon," grew
up in a rural town of Calanda, one province south of Cajal's. Both men
were macho, competitive, and self-aggrandizing. Buñuel once said that
if he could take only one book to a desert island, it would be *The Life
of Insects* by Fabre, also one of Cajal's favorites. Cajal likely did not
acknowledge his presence at first. "Don Santiago," said Buñuel, ad-

dressing him with the utmost respect. "I'm Buñuel. I don't know if you remember me. I prepared lots of flies' corneas for you. I worked with Bolívar." He then asked Cajal if he would sign the petition. Cajal looked up. "Buñuel, my friend, I won't sign." The "pretentious fraud" whom the Ultraists hated was a journalist who had written that, though he had interviewed millionaires, war heroes, and even royalty, he never felt that he was in the presence of real dignity before meeting Cajal. When Buñuel related the anecdote in his memoir sixty years later, near the end of his life, he not only doesn't mock Cajal's behavior; he excuses it: Cajal was only doing what anyone in his position would do.

Cajal, for his own part, despised modern art. He called the avant-garde "a multiform and contradictory hodgepodge of schools," baptized with "pompous names" like "cubism" and "expressionism," which were applied to works that looked to him like the paintings of lunatics or children. Picasso's work, according to Cajal, was nothing more than "deliberate nonsense." "In art, as in science," he wrote in a notebook containing observations of works at the Museo del Prado, "there are norms and canons that represent many centuries of progress." He described the distinctive, elongated heads in portraits by El Greco—whose reputation had recently been rehabilitated by the avant-garde—as "microencephalic," a medical term for brains that have failed to properly develop. Cajal believed that the ideal was nature itself, which artists should strive to see clearly, so that their paintings reflect the external world as accurately as possible. Surrealism violated the sanctity of objective, logical reality. He planned to write a book of art criticism called *Pathology of Art* but was preempted by another scholar, who had the same idea.

Though he rejected their style, Cajal ended up serving as a kind of inspiration for the most famous Spanish artists since the Golden Age. Scholars have pointed out that the drawings that Dalí and Lorca made while at the Residencia bore some resemblance to Cajal's own drawings. His legacy was omnipresent there.

In 1929, Buñuel and Dalí cowrote the iconic short film *Un chien*

Andalou. The film begins with Buñuel smoking a cigarette and sharpening a razor blade. Then he walks out onto the terrace and looks up, as thin clouds sweep across the night sky, in front of a perfectly full white moon. The next frame is a close-up of a woman's face, and a man's hand holds her eyelids open and begins to slice across the eye with a razor blade. One of the most indelible images in the history of film evokes the sectioning of a fly's cornea in Cajal's lab.

CHAPTER 30

"Marvelous Old Man"

To provide young Spaniards with the cultural opportunities that Cajal had missed, the Residencia invited leading figures from around the world to talk about their ideas, including Paul Valéry, Igor Stravinsky, John Keynes, Alexander Calder, Henri Bergson, Le Corbusier, and Marie Curie, whom Cajal greatly admired. As president of the educational Junta, Cajal signed the official invitation letter to Albert Einstein, who visited the Residencia in 1923 to deliver a series of lectures. Having won the Nobel Prize two years before, Einstein was renowned throughout Spain, though almost no one understood relativity theory; a chestnut seller on the streets of Madrid recognized him and cried out "Long live the inventor of the automobile!"

Everywhere he went, Einstein was mobbed. A reporter from *ABC* accosted him and his wife while they rode the train. Did he know Cajal? Einstein said he had been familiar with his work for twenty years. Einstein called his time in Spain "marvelous," and his wife, Elsa, concurred; it had been a long time since she had seen her husband in such good humor.

On Sunday, March 4, 1923, the Spanish Royal Academy of Sciences named Einstein a corresponding member. "From existing memory, no foreign savant has had such an enthusiastic and extraordinary welcome," one reporter wrote. Einstein delivered a speech, but Cajal was absent. In those days, he was invited to countless events and re-fused almost all of them, citing his ill health.

Einstein's schedule in Spain was hectic. On March 5, he delivered an address at the Spanish Mathematical Society, and after dining with a relative in Madrid, at around 8:30 P.M., he was driven to Cajal's house. Their visit had to be brief, because Einstein was due to explain relativity again in a second lecture that same night. No one knows what the two men said. They probably communicated in French, which both spoke poorly—not to mention that Cajal was largely deaf. Cajal never mentioned the meeting, but Einstein logged it in his journal: "Visit with Cajal, marvelous old man. Gravely ill."

SPAIN IN THE TWENTIETH CENTURY WAS PROVING AS VOLATILE AS IT was during the nineteenth. Fifteen different regimes had ruled during the previous six years, and, sick of failed governments, 50 percent of Spaniards did not vote. In July 1921, the Spanish army had been hu-miliated at the Battle of Annual, the decisive conflict of the Rif War, as the government made the same blunders as it had during the Disas-ter. Half of the twenty thousand Spaniards deployed died in the con-flict. Spanish society was fiercely polarized by the Rif War, forming two opposing camps: the "abandonists" and the "Africanists," led by King Alfonso, who inflamed the conflict. When news of the catastro-phe at Annual reached him, he was on vacation. "Chicken meat is cheap," the king allegedly said, before returning to his golf game.

Riots, strikes, terrorist attacks, and political assassinations broke out across the country. Antiwar protesters burned Spanish flags, and rumors spread of a Bolshevik revolution. The Spanish military, con-vinced that it remained the nation's last bulwark, became like "a loaded rifle without a target at which to aim," according to Ortega y Gasset. On September 13, 1923, Miguel Primo de Rivera, former captain-general of Valencia, Madrid, and Barcelona, staged a coup d'etat, de-claring his revolution with a *pronunciamiento*, in the style of Spanish rebels past. He promised to liberate Spain from its inept and corrupt political system, comparing his movement to Regenerationism.

Primo promised that his rule would be nothing more than a "brief parenthesis" that lasted only ninety days. Then he proceeded to suspend the constitution, disband the Cortes, eliminate political parties, and declare a state of martial law that lasted for eighteen months. He established the Office for Censorship and Information, with the goals of silencing "every single rebel and opposition manifestation" and of coordinating propaganda both inside Spain and abroad. Luis, unmarried and still living with his parents, recalled his father discussing the end of democracy at the dinner table. Alfonso had "jeopardized the throne," Cajal said, "because the people do not want a dictatorship, nor politicians, nor elites." The king supported Primo, once introducing him at a party as "my Mussolini." Though not a monarchist, Cajal considered the king to be an important symbol. Alfonso had corrupted his own power, as far as Cajal was concerned, breaking the most important political covenant—he had violated the people's will.

The first phase of the Primo regime was to eliminate all political opposition, primarily anarchists and communists, his most severe threats. The only people whom Primo despised more than politicians were intellectuals, a selfish minority, he thought, aloof from the masses and ignorant of their needs. He wanted to disband the Junta. When Cajal heard, he immediately sought an audience with the dictator. The elderly scientist in his threadbare overcoat and old suit, a veteran whose service ended in medical discharge, faced the country's most decorated general, twenty years younger, a celebrated hero of three wars, his uniform adorned with tassels, pendants, and medals. But Primo also revered Cajal. The motto of the dictator's regime was "Country, Religion, Monarchy," and no one was more patriotic than the great scientist, who worked tirelessly for the glory of Spain. Still, to that day, his old military comrades could be heard around their *tertulia* tables saying, "He is also one of us," and when Cajal heard anyone talk about military doctors, he would say, "I was also one of them."

While the Junta board awaited Primo's dictate, some members moved to resign en masse, but Cajal urged them to consider the consequences. The regime had not yet shuttered the Junta, he reasoned, which meant that the board still had time. "The politics of all or nothing is without a doubt more convenient and gallant but it can also be the most unskillful and counterproductive in the current circumstances," Cajal wrote in a letter to the Junta secretary. In a few years, he hoped, the regime might soften.

In the end, the Junta continued to function, but the regime installed its own official as vice president. As long as enough board members remained, Cajal argued, the "yeast" of the organization would survive. "Better a hair than the whole wolf," as Cajal said.

During the dictatorship, the salons of the Athenaeum, once home to wide-ranging, free-flowing debate, became singularly consumed by politics—the only question that mattered was how to oppose the regime. Sensing the threat, Primo sent a delegate to watch over the Athenaeum, then decided to suspend all events.

One day, the police entered the Café Prado to break up a secret *tertulia*. An officer approached the elderly man hunched over a small table in the far corner of the room, alone. "Are you an Athenaeum member?" he asked. "Yes sir," the old man replied. "I am Santiago Ramón y Cajal." The officer looked confused. "To be honest," the policeman told his colleague, "I do not know whether to detain him or not, because I cannot deny that his name seems to ring a bell." After discussing the matter among themselves, the police let Cajal go.

Cajal himself never fell out of favor with the regime. Once a researcher from Kiev wrote to Cajal, desperate for help. He had tried the master's staining techniques, he said, experimented with every variation, but the most delicate fibers of the nervous system still eluded him. Cajal invited the researcher from Kiev to the institute, and the Soviet government awarded him a two-month travel grant. Primo had intensified security at the border, refusing Soviet citizens entry, for fear of a Bolshevik revolution, and at the Spanish border, customs officers turned the foreign scientist away. The researcher reached into his bag and pulled out Cajal's letter of invitation and showed it to the captain, who saw Cajal's signature and telephoned the Ministry of Security, which forwarded the message to the minister of the Interior, who read the text to Primo in its entirety. Cajal followed up with a written appeal. The dictator personally approved the Soviet researcher's entry. "An invitation from D. Santiago Ramón y Cajal is the best passport that a foreign citizen can show," Primo said.

"I have good relations with the government," Cajal wrote to a friend during the Primo regime, "and they aim to please me."

CHAPTER 31

"Statues of the Living"

On February 14, 1925, Cajal received a letter from Margarita Nelken, a well-known young journalist. In 1922, when she was twenty-eight, she published *The Social Condition of Women in Spain*, a groundbreaking study in which she argued for a "new consciousness, a new morality, in the Spanish woman." She began her letter with "My illustrious and respected Don Santiago," before immediately apologizing for bothering him. She wanted to publish a collection of Cajal's writings about women, culled from *Advice for a Young Investigator*, *Café Chats*, and various newspaper articles. "Your incomparable authority constitutes a teaching of truly unequaled value," she wrote. Though he rarely acceded to journalists' requests, Cajal agreed to collaborate with Nelken, even suggesting that she take 75 percent of the royalties.

Nelken was not the first woman with whom Cajal had collaborated— which made him stand out in the Spain of that time. At the University of Valencia, in the 1880s, Cajal taught three of the first women medical students in Spain. They were required to come to school accompanied by a maidservant and wait in the dean's office until the male students settled into the lecture hall, so as not to distract them, before entering the room along with the professor and sitting in a special chair, apart from the rest of the class. "My child, you cannot imagine the impression [Cajal] made on us," recalled one of those students, Concepción Aleixandre, who went on to become a prominent gynecologist and women's rights activist. In 1908, another Valencian student, Manuela Solís, wrote a book called *Hygiene of Pregnancy and Early Childhood* and asked her Nobel Prize–winning former professor to write the prologue. Some newspapers praised Solís, while others dismissed her, claiming that she must have received favorable treatment because she was a woman. Cajal set the record straight: "A model of jealous and diligent students in Valencia, where we had the honor of teaching her Anatomy, she obtained only outstanding grades on her examinations, not due to gallantry but rather to hard and fast justice."

Cajal also welcomed women to the institute. In 1911, an Australian doctor named Laura Forster, who had already published articles on histopathology, spent a few months working with Cajal in Madrid, which she considered a great honor. Cajal suggested that Forster focus her research on degeneration in the spinal cords of birds, using his staining techniques and comparing her results to his own in mammals. In her subsequent paper, written in Spanish, Forster offered her "cordial thanks to Dr. Cajal for his amicable advice," and Cajal cited her work three times in his own.

In 1915, as president of the Junta, Cajal proposed that the Ministry of Instruction found a Residencia for women. And in a 1919 magazine article titled "The Intellectual Capacity of Women," he dispelled the myth that women are less intelligent because of their smaller average brain size. "Wait for society to concede to all young women of the middle class the same type of education and instruction as men," Cajal wrote. He always encouraged his daughters to continue their educations for as long as they wished. "In our home," his daughter Fe recalled, "there was always absolute freedom of thought." Two sisters—Manuela and Carmen Serra—were hired by Cajal to work in the laboratory as technicians or *preparadoras*, preparing tissue samples for microscopic slides and helping other researchers with their drawings. With the support of Cajal, Manuela performed her own research, and in 1921 she published a study on the glial cells of frogs in the *Revista*. She was listed as the sole author, a rare privilege that Cajal, as editor of the journal, must have encouraged or at least approved. In an official list of Spanish school researchers, Cajal included the names of two women.

In his brief prologue to Nelken's book, titled "Preliminary Warning," Cajal insisted that, "as a modest scientist," he had no authority to opine on issues of gender. "The female type, like the male type, is far from constituting all mental types," he wrote. "There are women, but not woman, although all possess certain common traits." Spanish Conservatives vilified Nelken either as "not a woman," because of her androgynous style, or else as a whore who had charmed her way to professional success. Like most Spanish men in the early twentieth century, Cajal opposed Nelken's brand of feminism, which he deemed "militant." "What feminist extremists call the emancipation of women," he wrote in a 1932 essay called "On Feminism," "is at heart nothing but the imposition of the formidable yoke of exhausting work without the consoling compensation of love and family." Freedom for women would decrease the number of marriages, he predicted, condemning women to "ugliness and old age." Though he disagreed with Nelken's views, according to his secretary Irene, Cajal treated Nelken with nothing but respect, always walking her to the door politely and signing his letters: "You know your respectful and sincere friend admires you."

"Like all husbands of the era, he was a little sexist," wrote Irene, who became a prominent feminist in her own right, in her memoir. At the same time, she celebrated his progressive views on women's education.

That year, when she was seventy-two years old, Silveria was diagnosed with myocardial atrophy, a condition that Cajal referred to as a "weakened heart." She and Cajal started traveling by car on their summer vacations, since getting on and off the train hurt Silveria's legs too much. In 1926, they vacationed in Sigüenza, a small, Romanesque city about ninety miles from Madrid, where they rented a large house that had cooler indoor temperatures. Conditions were perfect for Cajal but not for Silveria, who could not make it up the stairs. Cajal worried constantly about her, and her illness only added to his "fatigue and discouragement." With his mental state "lacking equilibrium," it was impossible for him to work.

LATER THAT YEAR, IRENE LEFT SPAIN TO MARRY A MAN TWICE HER age. Cajal needed a secretary more than ever. In a letter from around that time, he apologized to an acquaintance for sending him the same book twice, having forgotten that he already sent it. Though he rarely left home, Cajal returned to Sra. Lewy's house in the Chamberí and, despite his poor health, climbed four flights of stairs to the family's apartment. Irene's younger sister Enriqueta—whom everyone called "Kety"—also studied at the German school and knew multiple languages. Would Sra. Lewy let another daughter come to work for him? The institute was reputedly a hotbed for radicals—perhaps she felt that the atmosphere had corrupted Irene, who had just run off with a communist. Eventually, Sra. Lewy agreed to let Kety work for Cajal, but only beginning at the end of the year, once she finished secondary school. From time to time, Cajal tried to give Sra. Lewy a few dollars, in case Kety needed a new dress or a pair of shoes, or else he offered her items that strangers had sent his family, but Sra. Lewy always declined. "We live like queens," she said—but she still baked Cajal a German chocolate cake as a token of thanks.

Cajal subscribed to every important national and international journal related to science and medicine, and Kety's job was to translate and summarize French, English, and German texts, especially recently published articles, and to maintain Cajal's foreign correspondence. Crammed with books from floor to ceiling, the institute's library had been initially arranged by Tomás, who was hardly a bibliophile. Cajal taught Kety how to organize the stacks, with a special shelf for works by the Spanish school. For a period of time when he had no office at the institute, Cajal stationed his worktable in the middle of the library, next to Kety's desk. While girls her age went out on the weekends, she spent Saturday and Sunday afternoons talking to Cajal. "Very conservative but a stupendous man," Kety recalled. "He taught me to see life with clarity and simplicity. Also to be independent and express feelings." He treated Kety, who had the same name as his beloved daughter, like a *pequeña*, or "little one."

ON APRIL 24, 1926—AN EXCEPTIONALLY SUNNY MORNING— government officials, representatives from various learned societies, and a large contingent of Cajal's students and admirers gathered in Retiro Park to inaugurate a statue in his honor. With tensions between the Primo regime and intellectuals at their highest (that year, Primo would survive three attempted coups), the dictator decided to dedicate a monument to Spain's national scientific hero. On the morning of the ceremony, a notice appeared in the newspaper from the Office of Information and Censorship in response to rumors that a second inauguration of the monument had been planned by political dissidents. Unsanctioned demonstrations or any other acts of "infantile puerility," the regime warned, would be punished with hefty fines and jail time. Primo gave a brief address in which he pointed to the attendance of officials from both his regime and from the monarchy as proof that the country was in fact unified. But his presence was meant more to intimidate his opponents than to honor Cajal, who was not even present at the ceremony.

A band played the national anthem as schoolchildren laid flowers beside the monument. King Alfonso rose and pulled away the curtain, revealing a monstrosity of marble and bronze in the middle of a small pond, with the figure of Cajal reclining on top of a broad pedestal in the pose of a Roman god. Behind him was a shrine to Minerva, the goddess of wisdom, and on each side of her was a squat bas-relief pillar, shaped like a sarcophagus, streaming a single water jet. FOUNTAIN OF LIFE was inscribed atop one, with the figure of a family holding a newborn child, and FOUNTAIN OF DEATH was inscribed on the other, with the figure of a woman mourning a dead man.

Cajal's supporters were horrified. The hypocrisy of the government infuriated them, since a few years earlier many of the same officials had argued against his pension. Primo created a noble title for Cajal, dubbing him a marquis, but Cajal politely declined a place in the aristocracy. Primo tried to bestow a new medal on Cajal, the Plus Ultra, a reference to *Non terrae plus ultra*, the words that, according to Spanish legend, were inscribed on the Pillars of Hercules, which flank the Strait of Gibraltar, reminding travelers that there was "nothing more beyond" the Iberian Peninsula, which was thought to be the end of the world. The phase appeared on Spanish currency until the sixteenth century, when Charles V eliminated the *non*, leaving only *plus ultra*— "more beyond"—which became a motto of Spanish imperialism, urging Spaniards to explore "the New World."

Cajal understood that Primo was using him as a propaganda tool. The awarding of the Plus Ultra was "odd and deplorable," he wrote, and in letters to three different people he called the tribute "annoying and hyperbolic," "undeserved and exorbitant," and "insincere and undeserved." His rejection of the Plus Ultra medal embarrassed Primo, who refused to accept that Cajal was ill and sent his daughters to Cajal's house to deliver the medal. In the end Cajal took the medal, knowing that, whether he liked it or not, his actions would reflect not only on himself but also on the organizations that he represented. He was eager to put an end to all the commotion. Most of all, he hated tributes like the unveiling of the statue because, he said, "They rob me of what little useful time I have at my disposal to work." At seventy-four, he knew that this time was limited.

A week after the unveiling, Cajal was confined to bed for five days with the flu, an illness that he attributed to the stress of having to respond to hundreds of letters and notes of congratulations with letters of his own, "replete with cheap rhetoric." Strangers sought medical advice from Cajal, who had become a mythic figure in the countryside—

the *sabio* who could cure all. Crazy people seemed to gravitate toward him, but he took pride in responding to every letter, no matter how bizarre.

A German inventor sought Cajal's opinion on a special hat that he had developed to protect and ventilate the brain. "Do you believe it is possible to conserve fruit, for a long time, in hermetically sealed and practically empty receptacles?" asked a fruit grower from Valencia. A man fed his sick cat arsenic, to put it out of its misery, but the cat, miraculously, was healed. Could Dr. Cajal please explain this phenomenon? When a man from Petilla asked Cajal to intercede in a wrongfully adjudicated murder case, he responded that, though he would like to help, he was too old and sick and did not think he could make a difference. A schoolteacher sent him a butterfly as a gift, which Cajal repaid with a scientific summary of the creature. A Cuban man wrote to him in 1929 saying that his cousin appeared in a dream asking him to congratulate Cajal, after which Cajal would then make a new discovery. Cajal responded kindly, saying that he remembered the man from his days on the island during the war. "As for me," he said, "I am very old and sick and I do not expect to make any discoveries anymore."

Diagnosed with an infection in his bile ducts, Cajal remained in bed for two months; "the fever and pains have debilitated me considerably," he wrote on July 15. "It has only been 4 days since I returned to my laboratory, although with very little predisposition to work. However, I intend to resume my investigations when I recover my strength." In September, he tried to work again. "Even though I am pessimistic," he wrote on September 22, "I try to ensure that the hauntings of my brain do not stop my hands, and [that is how] for 50 years I have worked to create Spanish science." Cajal reported in a letter the next year that he had cirrhosis of the liver. "Every day I find myself more decayed," he observed. "And my wife does not improve. This creates a mental depression in me ill-suited for great projects and important undertakings. I don't know if I will leave this depressed state."

Cajal's monument is located on the Paseo de Venezuela, a wide, tree-lined path through the center of Retiro Park, beside the lake. "I disapprove, in principle, of statues of the living," he had said when he was first presented with the idea—as always, they reminded him of his own mortality. For thirty years, he had loved strolling the Paseo de Venezuela, but after the monument was raised, he never walked down that path again.

CHAPTER 32

"The Self Has No Mirror"

Ever since Cajal was a child, fantasizing in his dark bedroom, his mind had kept him awake at night. When he began his career, he worked long past midnight and woke up early in the morning, unable to rest while the microscope awaited him in the next room. During periods of intense research, he hardly slept for days, even weeks on end, and his insomnia returned with a vengeance whenever he was ill. Certain books in his library were so boring that he kept them on a special shelf, classified as "Sleeping Pills," but after his diagnosis of cerebral arterio-sclerosis, reading only intensified his headaches and prolonged his restlessness.

Cajal, who always prided himself on his abstemiousness, began taking prescription veronal, a barbiturate intended to be used as a sleeping aid, as early as 1907. The drug, which was supposed to calm him, instead transformed his mind into "a vortex of opposing currents." If inactive neurons were like unemployed workers, as he once wrote, they were now rioting. Early pharmaceutical trials showed that, under the influence of veronal, patients with "severe emotional repression" were known to suddenly overcome their inhibitions, and Cajal often experienced what he called "overexcited" states, including periods of graphomania, which had first occurred when he was eight years old. Swayed by barbiturates and sleeplessness, he would convince himself that his ideas were too precious to waste. A special desk had been built for him so that he could write in bed, and in the middle of the night, frenzied, he would reach for scraps of paper, newspaper pages, the backs of envelopes, and notebooks of all sizes and fill them with random thoughts, responses to things he had read, and fragments of dreams. After he finished, he would toss each sheet onto the floor. "We ask God that He grant us upon dying, as a supreme vision, the privilege of contemplating, in a synthetic vision, the flowers gathered by the path of life and the germs of ideas sown in our souls," Cajal wrote in *Café Chats*. The next morning, Silveria, old and frail, would bend down to collect the papers, like a bouquet of dead flowers, from the floor.

Sometimes, Cajal would take two doses of veronal, diluting the powder in sugar water to make it more palatable, with the second one coming at three or four in the morning. Once the mania dissipated, he would plummet into a depression and spend the next day crossing out what he had written the night before. After dinner, a time when he used to work, he would nap for three or four hours, sometimes sleeping for fourteen hours a day. In Cajal's basement, there was a stone mortar and pestle, which he used to grind chemicals to concoct the white crystalline powder for himself, which Kety weighed in cigarette paper on a laboratory scale. *I cannot sleep*, begins one of Cajal's dream journal entries. *I leave the boardinghouse in an unfamiliar city in search of a pharmacy. I am looking for a product that is not veronal and whose name I do not know. What I take to be a pharmacy turns out to be a bakery. More anxious I search along unfamiliar streets. And in a state of fatigue I awaken.*

Cajal admitted that he became addicted to veronal toward the end of his life, which he said created a "second nature" in him, the antithesis of his regular self. He recalled babbling nonsensically and overreacting to trivial things. Ashamed of his behavior, and certain that no one would want him around, he retreated further into silence and isolation, hiding in the basement, where papers layered the parquet floor and the shelves were bursting with books and manuscripts. Friends and colleagues nicknamed Cajal's basement "the Cave"; it was as though at the end of his life he had retreated to the place where his father had taught him his first lessons.

In his dream diary, Cajal recorded an early-morning vision, almost certainly written under the influence of veronal:

> *Believing that a representation is the self is like thinking that a photographic lens depicts itself. Maybe if there were a mirror in front. But in man the self has no mirror. The self is absolutely inaccessible. That which we take for a mirror, consciousness, only shows us the product of the [. . .] selection thought to be the object but what is thought to be the object is not what we think, but rather yet another image about which one thinks.*

In surprisingly unscientific terms, he defined the self as "an energy, an invisible pull like a god." And at that moment of epiphany, he awoke. "In sum, it is good to know ourselves," he observed in *Café Chats*, "but not so much that self-knowledge makes us pusillanimous and infecund."

IN THE SUMMER OF 1927, CAJAL RETURNED TO JACA FOR A SUMMER vacation, a trip intended as his "farewell tour to the Pyrenean valleys," the scene of both his childhood imprisonment and escape. Cajal, fanatical about automobile safety, always bought the latest American model and changed chauffeurs frequently, suspecting them of drunkenness. "My grandfather liked the good, the best," Cajal's granddaughter said. "But if he did not have it, he did not suffer." When Cajal's unmistakable yellow Buick appeared in the main square of Jaca, people whispered to each other, "It's him! It's him!" His chauffeur helped him down from the car and into the local social club, where Cajal sat reading the newspaper and writing in his notebook. Outside, his car idled, waiting for him to return, with Silveria in the back seat, transfixed by the bustle.

Cajal bore an air of great wisdom, his wide, bald head like a Pyrenean peak, wrinkled and fringed with snow-white hair. "We see him alone," one Jaca resident recalled, "always alone." In the afternoon, his chauffeur drove Cajal to his favorite spring, where he spent hours deep in thought. The grandchildren of Cajal's cousins, the youngest members of his family, recalled delivering fresh trout from the Aurín River—the local delicacy—to their famous relative.

Cajal visited Ayerbe to see if any of his old friends were left. Most had moved away, as he had, or else had died. But Pedrín, the shoemaker, was still alive. Embracing his former apprentice, whom he had likely not seen in fifty years, Pedrín was overcome with emotion. The boy who sewed boots and trimmed heels had gone on to become a world-famous brain scientist, while the old shoemaker had opened a shop on Ramiro el Monje, one of the main streets in Huesca. More and more clients brought their business to him, he told Cajal, and the shop had prospered. How had his young apprentice been? Pedrín asked. Cajal talked about his travels through Europe and to America. "But why did you leave?" Pedrín asked. "Poor Santiagüé. Here with this job you would have had a peaceful existence."

IN THE SUMMER OF 1930, CAJAL DID NOT TAKE A VACATION. SILVERIA'S condition had deteriorated, and he did not want to be apart from her. "When she is ill he does not talk to anybody. He is constantly preoccupied, silent . . . ," his chauffeur told a reporter. "He would not separate himself from her for anything." "My summer house will be my basement," Cajal wrote on June 21. He called it a "natural refuge for cockroaches."

On August 23, 1930, in the middle of the day, Silveria told the family that she needed to lie down. Cajal asked if anything was wrong.

"It is nothing Santiago," she said. Cajal, who knew the seriousness of her condition, arranged for a confessor. In her will, she professed her Roman Catholic faith and entrusted her soul to God. She had suffered a long and painful illness, but her death still came as a traumatic shock to the family.

Cajal always insisted that Silveria should receive equal credit for his achievements. "Half of Cajal is his wife," some said, and Cajal delighted in repeating the phrase. Jorge referred to his mother as his father's guardian. "She was a woman who loved him deeply and fully," Luis said. "In turn, Cajal favored her with a love of the same intensity." Cajal said that Silveria was a great beauty who "condemned herself joyfully to obscurity." "Only the unsurpassable self-abnegation of my wife made my scientific work possible," he wrote in his *Recollections*. Cajal envisioned their joint legacy as "two foreheads with a single halo."

After Cajal won the Nobel Prize, the wives of the faculty at the University of Barcelona sent Silveria a letter of congratulations. Her response is the only extant piece of writing in her own hand. "You exaggerate, benevolently, my role in the work of my husband," she claimed.

I did nothing but my duty, which every Spanish woman would have done, which consisted of shielding my partner from all the hardships and anxieties of a modest hearth, so that he could devote himself freely and comfortably to his favorite tasks. My only merit was having had blind faith in him and nurturing the hope that after the days of poverty and harsh economy would come others of comfort and well-being that would permit us to raise and educate our seven children in a Christian way.

CHAPTER 33

"Searching for Themselves in the Secret"

In a boardinghouse near Cajal's mansion, a first-year medical student, bored of studying, looked out from his balcony and through the glass terrace at number 64 saw a little old man with a white beard in an undershirt walking back and forth, as though in slow motion, between the center of the room and some cabinets in the corner, while holding something in his hands. After a while he stopped and looked up at the sky. Then, "with an air of great seriousness," he sat down, lowered his head, and stared at the floor for longer than the student could bear to watch. The next day, the old man was there in the same position, as though he had not moved. Day after day, the student became more and more curious about the identity of the old man, until finally he asked another boarder. "Don't you know him? That little old man, as you call him, is Don Santiago Ramón y Cajal."

Cajal's reclusiveness lent his character even more intrigue; people who came across him in the café or on the street all had anecdotes about him, which coalesced into a public image. *Cajal* the icon overtook the reality of the man himself in the collective imagination. What bothered him most, he said, were reports that he smoked cigarettes—he claimed to have never used tobacco in his life (in fact, his last housekeeper insisted that he smoked like a chimney). Cajal did not seek fame for fame's sake. He hoped that his prominence would inspire younger Spaniards to pursue science, yet few understood his work. "There is no one among our public figures more known nor more unknown than Cajal," proclaimed Ignacio Bolívar.

NOW HOME TO A NEW GENERATION OF *TERTULIAS*, THE PRADO HAD become too busy for Cajal, who searched for even more deserted cafés to haunt. On the Gran Via, Madrid's glittering new thoroughfare, was an awning with cursive, gold lettering above the basement of an old church that read CAFÉ ELIPA. When Cajal's signature yellow Buick appeared, his chauffeur eased him out of the car and helped him toward the front door. In his old age, Cajal had grown portlier, like a round-bottomed flask. Draped in a black cloak, shuffling along with his cane, he would stop at a kiosk in front of the café to leaf through magazines and newspapers, holding them at arm's length, wearing one of his three pairs of glasses: one for reading, one for seeing at a distance, and one for focusing on the covers of books through shop windows. A stray dog sat waiting in the doorway for the lump of sugar that Cajal always kept in his pocket for it.

Down two flights of stairs was a long and narrow room with low ceilings and glazed-blue-tile walls, less like a café than a catacomb. The waiters stood dutifully near his table, guarding his solitude, so that he could read, sip coffee, and jot his notes in peace. "I contemplate with resignation the dark tunnel beyond which nobody knows whether there awaits us an everlasting and life-giving garden or a tragic and interminable desert," he wrote, likely while at the Elipa. The only fantasy of immortality that seemed worthwhile was the survival of both body and soul, preserving the structure of each unique brain "with its miseries and limitations, together with the memory of our triumphs, loves, and failures."

Toward the end of his life, in answer to a letter inquiring about his religious beliefs, Cajal stated that he agreed with the sentiments expressed in "My Religion," a famous essay by Unamuno, who had written:

My religion is to look for truth in life and life in truth, despite knowing that I may never find them while I am alive. My religion is to struggle constantly and tirelessly with mystery; my religion is to wrestle with God from the break of day until the close of night, like they say that Jacob struggled with Him. I can never accept the concept of the Unknown—or the Unknowable, as some pedantic writers say—and I will also not accept any affirmation that says: "from here you can go no farther."

In the garden of the Residencia, Cajal liked to sit beneath the poplar trees on a stone bench surrounded by flowers in earshot of the athletic fields, where students shouted and played. He also liked to read a comic from Barcelona called *TBO*, a popular magazine aimed at children, never missing an issue, and even venturing out on Sundays, despite his frail state, to buy one for himself. Cajal spent so much time at the Residencia that he received mail there. Often, he would meet with the contemplative poet Juan Ramón Jiménez, who was delighted by Cajal's dry sense of humor and sardonic laments about the state of Spain.

In his poem "Santiago Ramón y Cajal (comes)," Jiménez captures the old *sabio*'s presence as he approached:

Absent, fine and realistic; always entangled in the subtle laces of the life of his microscope. I do not know a mind so much our own as his, strong, delicate, sensitive, brusque, pensive. The eyes look at no one—to nothing within limit—; they always wander lost, fallen, errant, as though searching for themselves in the secret, in order to finally see himself, face to face.

The great, deep questions increasingly consumed Cajal: "What is the nature of life? What is the nature of the universe? What is the nature of matter?" One rainy afternoon, taking shelter on the tram, Jiménez saw Cajal put on his glasses to start reading and then lean against the glass, "melancholic and careless," lost in thought.

• • •

SIX YEARS AFTER PRIMO HAD DECLARED his regime a "brief parenthesis," the dictator still refused to relinquish power, but now, under stress and ill with diabetes, he faced mounting pressure to resign. Late one winter night, he telegraphed all his captains-general, asking for a vote of confidence, which they declined to give. Abandoned by the military and attacked by the king, the economic elites, and the church, Primo de Rivera resigned on January 28, 1930. But it was the continuous opposition of the intellectuals, whom Primo persecuted but never managed to crush, that led to his eventual downfall.

Primo had destroyed the political landscape, decimating the old dynastic parties of El Turno Pacífico. Alfonso tried to reassume power, but almost thirty years after his coronation, Spaniards now viewed monarchical rule as a return to feudalism. On February 10, 1931, in an effort to rid Spain of the monarchy and reinvent the country, Ortega y Gasset, along with two other intellectuals, formed the Group at the Service of the Republic, publishing a manifesto in *El Sol* urging "all Spanish intellectuals" to devote themselves to a Republican victory. The group circulated pamphlets in the streets depicting Primo and Alfonso as dance partners. Cajal was approached with a petition condemning the king but declined to sign it. The king never treated him badly, he said; in fact, Alfonso had supported Cajal's scientific career from the beginning, and if not for the king and his authority, the institute and the Junta might not have existed. For that, Cajal was eternally grateful.

On April 12, 1931, the Republicans won the election in a landslide. Workers and students sang, danced, and hugged one another in the Puerta del Sol. When one police officer issued an order to rush the crowd, fellow officers escorted him back to the barracks. The Spanish people, stereotyped as violent, had risen up without bloodshed. The birth of the Second Republic was celebrated in Cajal's laboratory. Scientific organizations were venerated once again; the government issued a fifty-peseta note with the Retiro Park monument to Cajal—the one that he hated—on the back.

Despite his criticisms of the party, Cajal had never ceased to be a Republican in spirit, always maintaining that the Glorious Revolution of 1868 was the only true spiritual revolution in the history of Spain. Even as an elderly man, according to his grandson, Cajal still spoke longingly about Castelar and his government. Nearly sixty years after the First Republic's failure, the red, yellow, and purple flag once again flew from balconies, lampposts, and government buildings across Spain. It seemed as though the country might finally embody the nineteenth-century liberal ideals by which Cajal had guided his life. According to a former prime minister, however, the country was "a bottle of champagne, about to blow its cork."

Less than one month later, on May 20, another statue of Cajal was unveiled, this one in the courtyard of the medical school at the University of Madrid, near the entrance to his old lecture hall. The monument was not the work of the government; the students had paid for it themselves, selecting one of their classmates' work in an open design contest. The tall and thin figure, made of unadorned stone, stands perfectly upright, like a sentinel at attention—its nickname was "El Lápiz," or "the Pencil."

Three thousand people attended the ceremony. Tricolored flags flew, and a band played the Republican national anthem, the "Himno de Riego." A photograph published in a newspaper shows young men with dark hair in dark suits crowding around Cajal's statue. Two months earlier, there had been violent clashes at the gates of San Carlos between the FUE, a leftist student organization fighting to defend academic freedom, and the government in response to a strike by the students. Bullet holes could still be seen in the school walls. In the photograph, one young man stands in front of Cajal's statue with his fist in the air, delivering a rousing speech. "Soldiers, the homeland calls us to the fight," he cried. "Let us swear for her to vanquish or to die."

Cajal, then seventy-nine and ailing, was nowhere to be found, and no one was sure if he was coming. Tello looked resigned, since, as usual, it was he who would have to speak on the master's behalf. Kety ran down Atocha Street to the institute, where Cajal dictated a few pages of text to her, which she then gave to Tello to read aloud. Cajal's words sounded rushed and exhausted. "I do not know whether the statue looks like me," he could not help observing. But, he concluded, "it does not matter."

After the speeches, the crowd walked to the institute, filling the whole width of Atocha Street. Cajal was ensconced in the library; the only other person in the building was the janitor. Gathered underneath the balcony, where Cajal liked to watch the trains coming and going from Atocha Station, the crowd shouted and applauded, chanting his name. Slowly, he made his way to the window, bewildered. As soon as he appeared, the crowd fell silent. Cajal thought that younger generations of Spaniards had forgotten him. He was unable to summon any words. It was the only time that anyone had ever seen him weep.

CAJAL SURPASSED EIGHTY YEARS OF AGE AT A TIME WHEN LIFE expectancy for Spanish men was less than fifty. In a small black notebook, alongside thoughts inspired by his readings and sketches of the nervous system, he recorded his symptoms of physical deterioration hour by hour, down to the most minute detail, including the frequency and color of his bowel movements, in case the information might be useful to the medical field.

He decided to organize his reflections and reactions to old age in a book, titled *The World as Seen by an Eighty-Year-Old*, a work that is at times curmudgeonly, nostalgic, painfully bleak, and almost imperceptibly humorous in a way that his contemporaries identified as distinctly Aragonese. "When does old age begin?" Cajal wrote at the beginning of the book. "Time pushes along so subtly with the ceaseless flow of days that we scarcely notice that, far from our contemporaries, we find ourselves alone in the throes of survival." With the keenness of his scientific eye, he sorted the symptoms of his declining health into different categories—"sensory," "cerebral," "psychological," and "somatic" or "corporeal"—comparing the deterioration in human beings with that in other species. Death comes not all at once, he concluded, but successively, as it does to many insects.

A child of the highlands who enjoyed hiking as an adult, Cajal considered himself a mountaineer—only now the hill of Atocha Street, which he had summited daily for the past forty years, loomed as high as the Himalayas. After a few panting steps, he would rest on a bench, seated next to whomever was out on the street, whether they were nannies, soldiers, or *barquilleros*, who carried red tins with a wheel on top, engaging customers to play a game of roulette to win a wafer cookie. Though close to death, he retained his childlike sense of playfulness. His house was full of grandchildren, whom he adored. Having been raised in the capital and educated at the finest schools, they behaved in a manner that Cajal found too genteel, and so he always encouraged them to be more mischievous. On hot days, despite his deafness, he would sit in a wicker chair in the doorway of his house and chat amicably with his neighbors, as passersby stopped to inquire about his health.

IN 1932, THE NOBEL PRIZE WAS AWARDED TO EDGAR ADRIAN AND Charles Sherrington for their work on the functioning of neurons, a quarter century after Cajal won his own for describing their structure.

Four years earlier, Adrian, a Cambridge University physiologist, had succeeded in recording the transmission of a single nervous impulse, called an action potential, by placing electrodes on the optic nerve of a toad and amplifying the sounds—a series of static pops—through a loudspeaker. Adrian's research extended to the central nervous system a claim that physiologists had made about the peripheral nervous system since the 1870s: that the fibers of nerve cells function in a binary manner, either contracting fully or not contracting at all. Scientists call this the all-or-none principle. Adrian found that neurons can change the frequency and duration of their firing, thus sending messages to other neurons, which he compared to Morse code. Adrian was the first person to consider the possibility that nerve impulses contain information. He wished that he could pursue neurophysiology further, but he predicted that the discipline "will soon get into the realms of physics and chemistry and mathematics."

The independence of neurons left open the mystery of what took place in the spaces between them. Cajal had concerned himself with neurons themselves—not these spaces. He imagined that the gaps might be filled with some kind of substance, like a "granular cement," through which impulses could be conducted from fiber to fiber. Sherrington, on the other hand, hypothesized that these spaces contained morphological structures all their own, as highly specialized as neurons themselves. In an 1897 article, Sherrington had termed the structures "synapses," from the Greek word meaning "to clasp." The synapse and the neuron doctrine go hand in hand, both indispensable to our understanding of the nervous system. Yet Cajal almost never used the word *synapse*, even well after the term had entered the lexicon, and he rarely cited Sherrington's work even though he considered his British colleague a personal friend. The term appears only once in Cajal's two-thousand-page *Textura*, and his neglect of the word *synapse* calls to mind Golgi's own rejection of the word *neuron*. Both men, clinging to scientific primacy, resisted neologisms that were not their own.

In the early twentieth century, at Cambridge University, a researcher

named John Newport Langley—Charles Sherrington's mentor—
demonstrated that the effect of an electrical stimulus on nerve fibers
can be mimicked by a chemical stimulus. When muscles were injected
with isolated, purified adrenaline, they would contract with greater ra-
pidity, suggesting that adrenaline must already be present in the nervous system and used for regular signaling. Langley's student Thomas
Renton Elliott, in a speech at the Physiological Society in 1904, wondered if every time an impulse is transmitted through a nerve ending,
adrenaline might be "liberated"—the first identification of a "chemical
mediator," or "neurotransmitter." A quarter of a century earlier, when
du Bois-Reymond had theorized that the nervous system transmits
impulses chemically, no one took him seriously. The idea of any substance crossing the gaps between cells made no sense in light of the
reticular theory, since fibers were thought to be continuous.

In his later years, Sherrington criticized Cajal's law of dynamic
polarization—which defines the pathway of conduction from dendrites to axons—pointing out that, in some cases, cell bodies, and not
just axons, can transmit and receive impulses. Cajal's theory, Sherrington claimed, was "historically important," implying that, though
once epoch-making, it was now outmoded. When Cajal envisioned
the organization of the nervous system, he assumed that—in a reflex
arc, for example—impulses traveled from neuron to neuron along a
single, linear sequence of connections. His schematic drawings of the
cerebellum, cerebral cortex, and olfactory bulb show the direction of
impulses from neuron to neuron with fletched arrows, like those that
he once fashioned for war games as a child. Cajal was an orthodox
determinist: once an impulse reached the neuron, he believed, the fate
of that impulse was already decided—it would be propagated forward.
He accounted only for excitation and never for inhibition, a concept
ill-suited to his aggressive personality. Cajal's pupil Rafael Lorente de
Nó, who challenged Cajal's research whenever he could, demonstrated
in 1929 that, in the entorhinal cortex, located in the medial temporal
lobe, nervous impulses sometimes follow alternative pathways, so that

if a neuron is damaged or destroyed, the functioning of the nervous system is preserved. "The conception of the reflex arc as a unidirectional chain of neurons has neither anatomic nor functional basis," de Nó concluded. De Nó showed his results to Cajal but, out of respect, waited until the master had died to publish them.

CHAPTER 34

"My Strength Is Exhausted"

The Republican constitution of 1931, which defined Spain as "a republic of workers of all categories," enacted female suffrage, legalized civil marriage and divorce, and banned religious burials, celebrations, and symbols from public buildings. Crucifixes were banned from school, and in 1933 all monks and nuns were barred from teaching. Catholic churches around the country were burned. Violence was breaking out in the streets: four Civil Guardsmen were lynched by townspeople in Extremadura; in Basque Country, the Civil Guard shot eleven. Leon Trotsky, visiting Spain in 1916, had called the country "the Russia of the West." There were persistent fears of a Soviet takeover.

In a dramatic shift, the 1933 election brought to power the right-wing alliance of conservatives and Catholics known as the CEDA, whose leader, José María Gíl-Robles y Quiñones, was called "el Jefe," or "the Chief." Spain entered a period of violent uprisings known as the "Black Biennium." Cajal was sickened by the disintegration of Spain, but political crises were nothing new. In his Barcelona days, when Catalan separatism reached a peak, Cajal remained a firm centrist, maintaining that the country needed an organ of command and control based in the capital, just as a complex organism needed a brain to "enlace the parts with the whole." "If I could go back to being 25 years old, swelled with exasperated patriotism," he wrote in a letter about the threat of separatist rebellions, "I would answer without vacillating: reconquest *manu militari* [by force of arms], whatever the cost."

In October 1933, José Antonio Primo de Rivera, the deceased dictator's son, founded a nationalist party called the Spanish Falange, precursor to the infamous Falange of the Civil War, which sought to oust the liberal government. At their rallies, the Falangists wore blue shirts with the top button undone, a nod to Italian fascism and, in their minds, a symbol of virility. "Being smashed is the noblest aspiration of all ballot boxes," the younger Primo declared in his first speech. "There is no option left except fists and guns when someone offends the precepts of justice of the fatherland."

By this time, Hitler had consolidated power in Germany. According to Kety, who was Jewish, Cajal immediately understood that Hitler would be "one thousand times worse" than Mussolini. Cajal invited scientists fleeing fascism to work at the institute, including a French histologist who lived in his attic for several months. During the Weimar era, Germany and Russia had enjoyed an open intellectual exchange, but during Hitler's rule, Russian histologists could no longer publish their work in German journals. Cajal offered his Russian colleagues publication space in his institute's journal, originally dedicated to promoting the Spanish school alone. He could never forget the support of German researchers at the beginning of his career and wanted to repay the favor to foreign colleagues.

The following year, Mussolini's secret police accused the son of Giuseppe Levi, an Italian histologist, of anti-fascist activity, and arrested and imprisoned both father and son in a confrontation at their home. Cajal's pupil Fernando de Castro, whom Cajal had sent to study abroad in Levi's laboratory in Turin, informed Cajal of the predicament. At the time, Cajal was not pleased with Levi, whom Cajal claimed had committed the grave scientific error of not adequately citing his work. Still, he responded immediately, petitioning the Spanish ambassador to intervene, and listing every title and honor to his name in his letter, which he almost never did. Cajal even wrote a letter to Mussolini himself, on letterhead reading "Santiago Ramón y Cajal, Nobel Prize." Nowhere else in his archives does such stationery appear.

IN MAY 1934, AT THE HEIGHT OF THE BLACK BIENNIUM, CAJAL WAS working with Kety in "the cave" late at night, probably dictating a manuscript while she typed, when they heard loud noises coming from the street. His daughter Fe came downstairs, terrified, saying that a mob had gathered outside. Kety ran to call the institute for help, while Cajal slowly rose from his chair, climbed three flights of stairs, and opened the window to his balcony, which faced Alfonso XII Street. The mob burst into applause; they were students from the medical school who had heard that Cajal was ill and wanted to check in on him. Back in the basement, Cajal encountered de Castro, who had answered Kety's call and rushed over. Cajal told him what had happened, saying, "I can ask for nothing more in life."

Cajal's final cell was his bedroom, which he converted into a laboratory, adorned with certificates, diplomas, and testimonials from all over the world, and cluttered with stools, a shabby armchair, an antique lamp, and an old desk equipped with a microscope, test tubes, petri dishes, a microtome, an old-fashioned tissue culture incubator, and a large bottle full of distilled water with a rubber tube hanging from it. Cajal would sit on the old bed, writing in his notebook. Though he sometimes received visitors, he often interrupted them in the middle of a conversation to reread and correct whatever he was writing. As the country was imploding, he reminded them, "I serve Spain with my modest work."

One night, a former student, an ophthalmologist named Galo Leoz, paid Cajal a visit. Papers were strewn across the bed, where Cajal lay writhing; doctors had inserted a catheter just an hour before. He was unable to think, he told Leoz, because he believed that his brain would instantly become overheated, causing all his ideas to evaporate.

The conversation then turned to Cajal's plans for a new edition of his *Textura*, corrected and updated to include the work of the Spanish school—the culmination of his life's work and a final scientific testament. Suddenly, in his slippers and robe, Cajal lifted his skeletal legs in the air and sprang excitedly out of bed—"like Don Quixote ready to fight some wineskins," Leoz recalled—then struggled to climb the ladder in front of his bookshelves, grabbed a book, and flipped through page after page, each filled with notes and corrections, insisting that he had to edit it all again. He would finish the updated *Textura* no matter what, Cajal told Leoz. He had even drawn up the outline of a contract for the new *Textura* with Germany's Springer Press, and the obligation had kept him going, even when his vision was failing and he could hardly read. "In general, it can be affirmed that problems are never exhausted," Cajal told Leoz, "it is men who exhaust themselves in trying to solve them."

NEAR THE END OF THE SUMMER, CAJAL ASKED HIS DOCTOR IF HE could spend a few weeks in Cuatro Caminos, taking in the sunlight, flowers, and open air, the best medicine that he knew. *Country life is refuge*, one of Cajal's notes reads, *incomparable refuge*. His doctor approved the trip, and Cajal arranged for his lamp, desk, and books to be moved so that he could continue working on *The World as Seen by an Eighty-Year-Old*, as well as the last edition of the *Textura*. He had recently told a friend that he was almost finished. But in the end, Cajal was too ill to travel to Cuatro Caminos and was forced to remain in the city, "humiliated."

Cajal suffered from digestive problems, haunted by the ghosts of dysentery and malaria from Cuba. Veronal abuse had further exacerbated his stomach and kidney problems, and in September he started showing symptoms of colitis. He feared that he had a tumor, though he did not. After shedding so much weight, it seemed to him as though his body was regressing in the evolutionary sense, becoming increasingly weak, shriveled and helpless as that of a worm. His organs were abandoning him like a deserting army, he imagined. His vision was almost entirely gone and he was completely deaf; his brain was "reduced to a hermetic box floating on the human wave." All that he could hear were his own morbid thoughts and the arrhythmic beating of his heart. Near the end, Cajal wrote that he accepted his fate because his parents and ancestors had suffered it too.

On Sunday, October 14, 1934, Cajal asked for the papers and notes for his new edition of *Textura* to be brought up from the Cave to his bed. The mountain of folders could not be found. No one was allowed in the Cave without Cajal's explicit permission; politicians and intellectuals were sometimes admitted to ask for his opinion, colleagues visited in the evening to discuss the day's work, and former students were always welcome. A reporter once pretended to be one of these students in order to obtain an interview, and Cajal turned the basement light out, leaving the journalist in the dark.

Fe immediately assumed that her father had been robbed, while Cajal, Kety, and Dora Ballano, Cajal's housekeeper, searched every cabinet, shelf, and box, to no avail. "You can say that this constant and fruitless search worsened his condition and prostrated him in bed," Ballano recalled. She sometimes worked with a young assistant, Hilaria, who may have seen the messy stack of papers and thrown them in the trash, but the mystery remained unsolved. Cajal never left his bed again.

Cajal spent his final days surrounded by his doctors, children, and closest pupils, whom he referred to as his "spiritual grandchildren." He took all his pills despite the fact that none of them were working. When his throat became dry, one of the doctors suggested that he drink champagne. Cajal declined. "I want the garlic soup my wife makes," he said. He was preoccupied by the state of his laboratory animals. "How are my chickens?" he asked. "Are you taking good care of them?" At 4:00 P.M. on October 17, Cajal responded to a letter from de Nó, advising him on his experiments related to the staining of tissue in rabbits.

"Friend Tello," Cajal wrote later that evening. "No more food. My stomach tolerates nothing. I am aphonic, I cannot read nor write and my strength is exhausted."The script then became illegible and slanted off the page.

That night, Tello sent his son, a twenty-five-year-old doctor, to check on Cajal. Tello's son pushed past Cajal's general physician, Teófilo Hernando, who was standing, motionless, in the doorway of his room. Cajal was not breathing. His body was still warm. Tello's son pulled down the bedsheet and listened for Cajal's heart, which had stopped, then closed Cajal's eyes, just as he had been taught to do in medical school. Cajal's daughter Fe marked the date and time on her calendar: *Today, at ten-forty-five at night my father died.* Witnesses say that he spent most of his final days traveling with his oldest companion: his imagination. Those who saw his corpse said that his face showed no fear.

THREE HOURS BEFORE HE DIED, NO LONGER ABLE TO SPEAK, SEE, OR hear, Cajal wrote twelve lines of text, almost illegible, with words added, underlined, and crossed out. His parting words were printed in *El Sol* on the day of his funeral:

I leave you something greater than any wonder of the senses: a privileged brain, sovereign organ of behavior and action, which used wisely will immeasurably improve the analytical power of your senses. Thanks to it you can dive into the unknown and operate on the invisible, elucidating, as much as possible, the obscure questions of matter and energy (hidden from the common man); your inquisitive power shall be far from exhausted; in fact, it shall expand interminably so far that each evolutionary phase of Homo sapiens *will don the characteristics of a new humanity.*

Epilogue

Cajal's body was dressed in a black suit and placed in a simple mahogany coffin, which was set on the floor. His daughters adorned the outside of the coffin with flowers, while the walls and floor of the room were draped with black cloth. There was no crucifix. News of Cajal's death spread across Madrid, and a book was placed at the entrance for visitors to sign. First, in the morning, the doctors and public figures came, then the students and workers, and, by midday, people from all social classes walked into Cajal's house to pay their respects. The line extended out of the house and down many blocks—moving slowly. Women and girls laid flowers and stood by the body praying. A local sculptor cast a death mask. Amid the atmosphere of violence in Madrid, schoolteachers led their students down Alfonso XII Street—an army of little children marching with flowers. All told, enough wreaths were laid for Cajal to fill three trucks.

Though the funeral was to take place at four-thirty, by midmorning thousands of mourners had filled Alfonso XII and its side streets. A senior military official was there to oversee the proceedings, but Tello stopped him and presented Cajal's will, which called for a "purely civil" burial. A vanguard of young men held hands in front of the procession, clearing the path, as university students of all disciplines took turns carrying the coffin on their shoulders. The procession headed north. After a few hundred meters, the Civil Guard ordered the coffin to be transferred to an automobile, flanked by officers on motorcycles with machine guns. During the Black Biennium, any large gathering of people was treated as a potential riot.

At the Plaza de la Independencia, on the northwest corner of Retiro Park, the police dispersed the mourners. Cajal's first will, written jointly with Silveria in 1909, specified a Catholic burial. In his second will, he asked for a ceremony "with no type of pomp or circumstance" and for his remains to be buried in a common grave, together with those of his humblest countrymen. He would be at peace, he said, knowing his body would disintegrate "in this beloved land of Spain." When Tello, the executor of Cajal's estate, opened an envelope the day after Cajal died, he found a handwritten addition: *Bury me, if possible, next to my wife.* If it was not possible, because he would not have a religious grave, he asked to be buried in the Civil Cemetery, next to a man named Gumersindo de Azcárate, a fellow Republican freethinker from the Athenaeum.

The procession did not arrive at the Nuestra Señora de la Almudena cemetery, on the outskirts of Madrid, until nine-thirty, and by that time, despite the thousands in the crowd earlier in the day, there were fewer than a hundred people. "The burial has been the saddest I have witnessed in my life," wrote a reporter for *El Sol.* "Almost no one went." The turmoil in Spain prevented most people from attending.

In the end, Cajal was able to be buried next to Silveria. The two graves are covered with one slab, engraved with her name and, below it, his name and no epithet. Above the tomb is a giant cross. Under the Second Republic, Catholic cemeteries like Almudena were open to secular burials. But since the codicil to Cajal's will was written at the last minute, no new preparations had been made, and the mourners had to tear away moldings and ornaments to place the coffin in Silveria's tomb. Tello wept at the sight of such desecration and ended the ceremony early.

Cinemas around the world screened newsreels containing reports of a new king of Yugoslavia, a hurricane that devastated Osaka, and the burial of the Spanish savant Ramón y Cajal. A reporter for *La Libertad* interviewed people at a downtown bar about Cajal's death. "It is a spiritual pain on top of the other pains that Spain currently suffers," said an artillery commander. A woman sitting at the bar was actually reading Cajal's autobiography because "when she was a little girl her father had taught her to venerate him." A wealthy woman confessed that "I did not understand his science much, but it was enough for me to know that he was a national treasure." But scientists understood what had been lost. "It can be said that Cajal, alone, has given more than all of the other neurologists put together," wrote the Italian psychiatrist Ernesto Lugaro. One laboratory director was known to issue a warning to all young investigators about to embark on a research project. "Be careful," he told them. "Cajal researched it first."

EARLY ON THE MORNING AFTER CAJAL'S FUNERAL, A TRUCK ARRIVED at his home, and men from the institute began hauling his belongings away. On the nightstand beside his deathbed was a pile of books by Horace, Cervantes, Anatole France, and the Spanish essayists Angel Ossorio, Melchor Fernandez Almagro, and Ernesto Giménez Caballero (a leading proponent of Spanish fascism), along with a month-old issue of *El Socialista*. Once in a while, Cajal's granddaughter recalled, Cajal asked his chauffeur to buy a copy of the socialist newspaper, but she had no idea why he had kept this particular issue, which was heavily creased, suggesting that he had read it over and over again. Despite his air of neutrality, Cajal's sympathies always lay with the working class.

The last will and testament of Santiago Ramón y Cajal bequeathed his scientific possessions, such as his histological slides, laboratory furniture, microtome, microphotography camera, and two microscopes, to the institute that bears his name. Money never motivated Cajal, yet he died with almost six hundred thousand pesetas in assets. He owned several properties in the country, which he passed on to his daughters, Pabla, Pilar, and Fe, while to his sons, Jorge and Luis, he left his collection of personal manuscripts and books, along with a trove of medals, prizes, and diplomas, which they entrusted to the institute for safekeeping.

Among Cajal's manuscripts are a number of unfinished or unpublished books, including a treatise on hypnosis, spiritualism, and metaphysics, "which should have seen the light twenty years ago," he wrote, but which more recent findings had rendered obsolete. After Silveria died, Cajal arranged for a visit from a medium from Zaragoza who claimed that she could communicate with the dead. His grandchildren recalled their terror at seeing the witchlike stranger in the hallway of the third floor. Cajal watched her conduct a séance, and though his vision was failing, and he was completely deaf, he immediately recognized the "supernatural apparition" as a simple trick of light and mirrors and outed her, delightedly, on the spot.

The fallibility of the human mind was a perennial fascination of Cajal's. In another unpublished manuscript, *The Omnipotence of Suggestion*, he intended to expose the prevalence of false ideas in literature, religion, myth, politics, medicine, and especially—"who would have thought it?" Cajal wrote—the press. He knew that even the most eminent scientists—himself included—could become so attached to their own ideas that they failed to accept the truth. "Every man, if he is so determined, can become the sculptor of his own brain," Cajal observed. Though we may strive for perfect understanding, the human brain—the highest achievement of evolution—remains constitutionally flawed.

IN OCTOBER 1936, TWO YEARS AFTER CAJAL'S DEATH, GENERAL Francisco Franco's troops, with military support from Nazi Germany and Fascist Italy, besieged Madrid. The streets, once filled with Cajal's mourners, were now emptied, choked off by blockades and patrolled by cloaked vigilantes. The Republican government offered to move the

Cajal Institute to a safer location, in Valencia, but de Castro—Cajal's handpicked successor—refused. Bombs fell from the sky, partially destroying the building, while de Castro and Tello stayed behind, guarding their master's possessions against looters.

During the Spanish Civil War, of the forty-one researchers who could be counted as Cajal's heirs, virtually all lost their jobs, were exiled, or fled. "Death to the Intelligentsia" became a Falangist rallying cry; Franco equated scientists with enemies of the state. Del Río Hortega, a leftist, spent the rest of his life in Paris, Oxford, and Argentina; a lesser-known pupil, Facundo Villaverde, a rightist, was shot by Republicans. Kety, a proud Communist, fled across the Pyrenees on foot, living for forty years in Moscow and Beijing before returning to Spain once the Franco regime fell. Tello was removed both from his chair at the University of Madrid and his directorship of the Cajal Institute but continued to perform research in a private laboratory. Though de Castro was forced to work as a clinician to support his family, he never stopped doing biological research on his own, embodying his master's tenacity and determination.

Faced with death, Cajal found that only one thought could console him: that he would live on through the work of his pupils. He certainly has—if Cajal is the father of the neuron, then del Río Hortega should be considered the father of neuroglia, since he is credited with discovering two of the three types of these cells: microglia and (along with Wilder Penfield) oligodendroglia. Among the alternative impulse pathways that de Nó discovered were those that channeled activity back to the origin of the stimulus, later termed "feedback loops." Tello described regeneration in peripheral nerves, and his findings continue to aid researchers in the field of neuroembryology. De Castro's work on the sensory and motor ganglia helped illuminate the role of the autonomic nervous system, the largely unconscious system that regulates bodily functions. It would have pained Cajal, however, to know that the most prominent scientists of the Spanish school died on foreign soil. The only other Spanish-born scientist to win the Nobel Prize—Severo

Ochoa, in 1959—emigrated at the start of the Civil War, eventually landing in America. He had been a medical student at the University of Madrid but arrived the year after Cajal retired. "The greatest dream of my life was to be the student of Santiago Ramón y Cajal," Ochoa said. "I never met him but I consider myself his student."

In 1939, Franco purged the institute and replaced the Junta para Ampliación de Estudios e Investigaciones with the Consejo Superior de Investigaciones Científicas, or CSIC, a national science council whose mission was "to restore the classical and Christian unity of the sciences that was destroyed in the eighteenth century." Cajal and his pupils were known as agnostics, and Francoists dismissed their views as necessarily false because they did not have God's support. As the neuron doctrine spread throughout the world, Cajal's own country-men rejected it, and a revival of the obsolete reticular theory was cen-tered in Spain.

CAJAL'S LEGACY HAS BEEN APPROPRIATED FOR POLITICAL PURPOSES BY everyone from the Regenerationists to Primo, and accounts of Cajal have become more hagiographic since his death. Perhaps the most in-timate biography of Cajal was written in 1960 by Alonso Burón and Durán Muñoz, and though family members have contested aspects of the authors' portrayal, there are details and anecdotes in their book that are undoubtedly true and that do not appear anywhere else. Cajal was an ardent patriot but not an extreme nationalist. Ernesto Lugaro held up Cajal as an example of "sane nationalism." One former Liberal prime minister, reflecting on the collapse of Spain, wondered if Cajal might not have been the last symbolic figure capable of uniting the country.

During the siege of Madrid, after the home of Cajal's son Jorge was overrun by refugees, some of Cajal's manuscripts and papers were presumed lost, including a collection of loose sheets and scraps of newspapers and magazines that constituted his book on dreaming.

Cajal never published his descriptions of his dreams, but he did not destroy them either; instead, he entrusted the material to a former student of his, a psychiatrist named José Germain Cebrián, who both revered Cajal and promoted Freudian psychoanalysis. Germain typed up the diary, riddled with ellipses and cross-outs, and carried the manuscript with him secretly after fleeing the country during the Civil War. Before Germain died, he handed the manuscript over to another Spanish psychiatrist, José Rallo, who published it in 2015, once Cajal's work had entered the public domain. In the end, not even his dreams remained a secret.

Cajal's final book was to be called *Alone Before the Mystery*. The Spanish title, *Solos ante el misterio*, is in the plural form, implying that more than one person is alone—in other words, we are alone together. He never managed to begin this last work. "As long as the brain remains a mystery," he wrote elsewhere, "the universe, a reflection of its structure, will remain a mystery as well."

IN 1945, WHEN THE DAMAGE TO THE INSTITUTE FROM THE CIVIL WAR was finally repaired, the researchers there converted a room on the first floor, once intended to be Cajal's office, into a small museum for exhibiting his drawings. They did so without government funding or support. When the institute moved to its current site, on 37 Avenida Doctor Arce, in 1989, floor space was divided into laboratories, with no consideration given to the museum. For thirty years, Cajal's legacy—consisting of roughly thirty thousand artifacts, including two thousand photographs, two thousand handwritten manuscripts, twenty-five hundred letters, eleven notebooks, two thousand scientific drawings, and fourteen thousand histological slides—has been housed in a glorified storage closet, a nondescript sixty-square-foot utility room next to a working laboratory. The Cajal archives are difficult for researchers to access, and it is nearly impossible for the general public to see inside. Because Cajal left his possessions to the institute, which is controlled by CSIC, an official agency, it is the responsibility of the Spanish government to build the museum that Cajal deserves. Despite symbolic gestures and promises, the government has continually failed to do so.

THE SCIENTIFIC METHOD IS DESIGNED TO ASSESS OUR RECEIVED knowledge of the world and discard that which proves to be inaccurate. Yet a century after Cajal's Nobel Prize, an astonishing number of his findings have remained valid. Two structures that Cajal first identified—dendritic spines, now implicated in learning and memory; and the growth cone, a focal point of neurological development research—have become cutting-edge topics in neuroscience. His theory of neurotropism—that the growth and regeneration of axons are influenced by chemical signals—is now accepted and has been elaborated by further research. The foundational concept of plasticity, which Cajal first associated with the central nervous system, has been appropriated by other fields of study and even popular culture. Some have argued that Cajal's belief in the impossibility of nerve regeneration in the central nervous system hindered neuroscience for generations, but that view was based on a misreading of his conclusion. He encouraged future researchers to challenge his denial, and they have; neuroscientists have found evidence of regeneration in the central nervous system, revolutionizing the field.

"The job of the anatomist is to distinguish the apparent from the real," Cajal wrote. What makes his discernment even more remarkable is the fact that he was forced to rely on relatively primitive technology. He never observed the space between neurons—not directly. The black reaction stained fibers, and when Cajal saw the absence of color, he concluded that there must be a gap between cells. Definitive proof did not arrive until 1956, when, armed with an electron microscope, which provided two million times greater magnification, the U.S. neuroscientist Sanford Palay produced the first images of the synapse. "The absence of protoplasmic continuity across the contact surface between two members of the synaptic apparatus," Palay wrote in his landmark paper, "is impressive confirmation of the neuron doctrine enunciated and defended by Ramón y Cajal during the early part of this century."

How was Cajal so sure that the space between neurons existed? Why did he defend their independence so intensely? Separation between cells was a subject of major debate in nineteenth-century biology. Some researchers insisted that there were no boundaries between them, while some thought that membranes enclosed only some types of cells and, therefore, were not important. Cajal never doubted that membranes existed or that cells must be separated from each other, calling their individuality "a general property without exceptions." The French philosopher Henri Bergson attributed Cajal's discoveries to intuition more than intellect. "The findings of Cajal," Bergson concluded, "were no more than objective verifications of facts that his brain had foreseen as true realities."

We tend to think of science as strictly objective, especially in contrast to other intellectual disciplines such as art and literature. Researchers in Cajal's time tried to turn themselves into machines, aiming to represent nature perfectly by removing the self from the observation process. Cajal's idol, Rudolf Virchow, recognized that this ideal—to "desubjectivize" oneself, as he called it—was unattainable. "With each year," Virchow said in an 1887 speech, "I recognize yet again that in those places where I thought myself wholly objective I have still held on to a large element of subjective views." Cajal understood this in more intimate terms. "It is not enough to examine," he wrote. "One must contemplate. Let us pervade things observed with emotion and sympathy, let us make them ours as much by heart as by intelligence." Cajal's anatomical descriptions bear the unique stamp of his mind and his experience of the world in which he lived.

Cajal's metaphors, which reflected his experience of life, were particularly important to his understanding of the nervous system. After the telegraph, the next technological metaphor for the brain was the

telephone switchboard, and then the computer, which is now ubiquitous. Many neuroscientists today talk about the brain abstractly, in
terms of algorithms and statistics. But the brain is not a computer; it is
made of biological material that has evolved over millions of years.
What will be the next metaphor for the brain? Have the metaphors
become the ideas themselves? It would seem that data are unintelligi-
ble without metaphors or models through which to view them.

One metaphor that refuses to go away is the network. Only now,
instead of thinking of networks as contiguous webs of fibers, scientists
define networks in terms of functional connectivity. In systems neuroscience, as this new discipline is called, the basic unit of the brain is no
longer the neuron but rather collections of neurons, coordinating actions in multiple areas of the brain at once. Some researchers have
explained the workings of the brain as the result of a system of networks whose functions can be considered linked. The idea of interconnectedness, fundamental to the reticular theory, finds an echo in a
different form.

In addition, the workings of the brain may not rely as exclusively
on neurons as Cajal—and most neuroscientists since him—believed.
Recent studies have suggested that glial cells do far more than support
neurons; they may alter their activities. There are about ninety billion
neurons in the human brain and about ten times as many glial cells. In
the cellular world of the brain, there is still much to explore.

Since Cajal's death, neuroscience has made spectacular advances;
sixty-six Nobel Prizes have been awarded for work in neuroscience.
More powerful imaging techniques have revealed a staggering amount
of detail about the anatomy of neurons and synapses. Theoretical neuroscientists reconstruct the brain using supercomputers, aiming to
make their models consistent with experimental data, while other neuroscientists work to map all the connections of the nervous system, a
schema called a "connectome." Cajal was afraid that he would be forgotten by future generations of neuroscientists, but that is certainly not
so: his work is still cited hundreds of times each year in the scientific

literature, and no history of neuroscience is told without using his name. Part of the enduring appeal of Cajal's story comes from the relative isolation and poverty of his circumstances; working alone in his laboratory, with a cheap light microscope and some chemical reagents, he was able to change our fundamental conception of the brain. Based on the current practice of neuroscience, with international teams and multimillion-dollar projects, there may never be another figure like Cajal.

If Cajal were to open up an issue of a leading neuroscience journal, he would be astounded by the images that he would see. CT scans and fMRIs can produce images of living human brains in different cognitive or emotional states, and cognitive neuroscientists can now read people's minds in impressive, albeit crude, ways. A technique used in animals known as CLARITY can render the entire brain transparent so that it can be stained without damaging the sample. Different fluorescent proteins can be expressed within neurons, yielding fantastically colorful images. Histology is rarely practiced anymore as it was in the nineteenth century. Only a few labs around the world continue to use the black reaction, and rarely do neuroscientists draw what they observe by hand. But one part of Cajal's legacy that would have surprised him is the appropriation of his scientific illustrations as fine art, an exhibition of which recently traveled across North America. The aesthetic value of these images is not dependent on their scientific context, and no prior knowledge of neuroanatomy is required to appreciate their beauty. Moreover, his signature depiction of the Purkinje cell can still be seen in peer-reviewed neuroscience articles, anchoring schematic figures of cerebellar circuits.

Cajal claimed to have observed a million neurons with his own eyes; towers of blinking supercomputers can now process that amount of data in a nanosecond. We are masters of gathering data, but we know little more about the source of consciousness than we did while Cajal was alive. The standard neuroscience textbook keeps increasing in size each year, but our fundamental understanding of the mind does not. "We know the landscape," Cajal declared in the *Textura*, "but not how a vibrating movement of matter becomes a fact of consciousness." This statement remains true today.

A long line of investigators—including Cajal—have focused on the brain as an object of study, seeking to reduce it to its constituent parts, in the hope of ultimately isolating material correlates of the mind. With every advance in theory or technology, researchers have rushed to claim victory, announcing the finding of the latest Holy Grail. But perhaps the whole enterprise of reductionism will never suffice to answer the deepest questions. In the fifth century B.C.E., Democritus proposed the existence of "soul atoms," tiny and fast-moving particles, all alike, diffused throughout the whole of the body. His teacher, Leucippus, recognized the inescapable paradox inherent in this line of thinking, asking: What is the "soul of the soul?" Some theorists have moved in the other direction, asserting that the mind can include objects in the external environment, such as computers. One has proposed that consciousness is one and the same as the physical world surrounding us. Not all philosophers can agree about what the mind is, let alone how it operates.

Cajal was acutely aware of the epistemological conundrum of studying consciousness. He had no grand theory of the mind. As an anatomist of the brain, he understood its structural limitations more intimately than anyone. Yet this same flawed brain could also conceive of an idea so spectacular as complete self-understanding. Cajal had both grandiose ambitions and humble expectations. His goal was to found a new psychology, to make the mind and the brain meet. His "idealistic enthusiasm" eventually wore off, as he came to realize that his dream was impossible. Yet he persevered anyhow. "There is nothing other than self-awareness and self-improvement," he wrote. In honor of Cajal, may neuroscience always remain an act of self-reflection and self-discovery.

• • •

BY THE TIME THIS BOOK IS PUBLISHED, IT WILL HAVE BEEN TEN YEARS since I first encountered Cajal. I remember the moment I saw an image of one of his drawings—a single "psychic" neuron, vulnerable and alone, its spindly black form like a static-electric shock, which seemed both alien and familiar. It felt as though my neurons were finally recognizing themselves. I had no idea that the invisible world inside the brain could be so haunting and so beautiful. When Cajal sat down to draw the neuron, he believed that it might be the long-sought agent of our higher mental functions, what philosophers and scientists had been seeking for centuries. I felt an urgent desire to know the man responsible for this drawing, to understand what he was trying to communicate beyond the realm of science—some hard-won truth about humanity.

As soon as a person dies, his body starts to disintegrate, becoming less and less lifelike with every passing moment. The final physical traces of Cajal are rapidly vanishing. His home on Alfonso XII has been converted into luxury apartments. His tomb in the Almudena cemetery—not indicated on any map—was recently vandalized. His native village of Petilla has become nearly a ghost town, with a population of thirty-one, and a small museum in the home where he was born, open for roughly half the year, receives few visitors. His grandchildren have all died, as have the scientists who worked with him, and those who worked with those who studied with him will soon be gone.

Biology and biography contain the same Greek root: *bio*, meaning "life." Biology is the study of life, while biography is a life written. Cajal's method was to take dead tissue that he would dissect, preserve, and stain before studying it under a microscope and drawing its structures as though it were alive. With this book, I have aimed to follow Cajal's approach. From every source that I could find, I gathered every trace of him, every sliver of his life and scrap of his work, every piece of information about his science, his country, and his world. There may be distortions, like artifacts on microscope slides, that have been preserved from before my time, though I have tried to identify and root these out. To the available materials, I have applied certain literary and narrative treatments, not unlike staining techniques, which are meant to reveal deep aspects of Cajal's character in the same way that the black reaction exposed the most ethereal structures of cells.

Many of Cajal's illustrations are not reproductions of microscopic images but rather composite portraits, syntheses, and interpretations of his many observations, combining features that he deemed definitive and essential, highlighting and exaggerating certain elements to convey his distinctive vision of the neuron. According to de Nó, Cajal would observe neurons under the microscope, then go to the café, and, when he came back, produce an illustration from memory. Judged by the standards of realism alone, these illustrations might seem ill proportioned or inexact, and his rivals criticized him for his methods. But "a drawing is never an impersonal copy of everything present in the preparation," Cajal asserted in a letter from 1929. Otherwise, he explained, the image would be far too complicated, almost incomprehensible. "By virtue of an incontestable right, the scientific artist, for the purpose of clarity and simplicity, omits many useless details." He did not copy images—he *created* them.

So, too, was my portrait of Cajal formed from a synthesis of my ideas about him—a composite drawn from the observations of others, an interpretation that reflects the workings of my own mind. In that sense, my Cajal cannot be said to exist. Neither do many of Cajal's neurons, yet he called them "pieces of reality." May my biography of this complicated and monumental man, though subjective, serve as a representation of the truth of his life and work.

Notes

All translations, unless otherwise noted, are the author's own.

Prologue: "A Vehement Desire of My Soul"

4 *"To know the brain"*: Ramón y Cajal, *Recollections*, 305.

4 *The highest ideal*: del Río Hortega and Estable.

4 *"Only true artists"*: Quoted in DeFelipe, *Cajal's Butterflies of the Soul*, ii.

5 *"new truth"*: Ramón y Cajal, *Recollections of My Life*, 322.

5 *whistling of trains*: Lucientes.

5 *"Little Paris"*: Stoddard, 51.

5 *French jewelry [etc.]*: Finck, 10.

5 *brass-plated hats*: Huntington, 79–80.

5 *impossible to move*: Fetridge, 1273.

5 *Cajal liked living there*: Lewy Rodríguez, *Santiago Ramón y Cajal*, 81.

5 *directory of the school*: See the Institute of Jaca archives at the Ayuntamiento of Jaca.

6 *served as reminders*: See *Ramón y Cajal, 1852–1934: Expedientes Administrativos*, 99, 107, 110.

6 *"a vehement desire"*: Ramón y Cajal, *Recollections*, 13.

6 *"The brain is a world"*: Ramón y Cajal, *Charlas*, 156.

6 *maravedís*: Fermin Goñi, "Petilla, una deuda reyes," *El País*, March 17, 1979.

6 *hot springs*: Gitlitz and Davidson, 31.

7 *"inexplicable melancholy"*: Ramón y Cajal, *Recollections*, 9.

7 *"Señor"*: Ibid., 10.

7 layas: Raymond Carr.

1. "The Necessary Antecedent"

9 *cut off*: Belloc, 6.

9 *"great spiked collar"*: Huntington, 195.

9 *God's love*: Ford, *A Handbook for Travellers in Spain*, Part 2, 928.

9 The Song of Roland: Quoted in Matthew Carr, ch. 3.
9 *"peasant genius"*: Eccles and Gibson.
9 *"characteristic of inhospitality"*: Strabo, bk. 3, ch. 1.
10 caxal, *or* cajal: de Carlos Segovia, *Los Ramón y Cajal*, 17.
10 *"an exceedingly wretched place"*: Strabo, bk. 3, ch. 1.
11 *born in 1822*: F. Morales, 106.
11 *keep plots intact*: Ibid.
11 *mercantile and sacred*: Satué.
12 *shedding of blood*: Mumford, 31.
12 *open veins*: Ibid.
12 *"clears the mind"*: Geyer-Kordesch and MacDonald, 40.
12 *heat curling irons and wash shaving cloths*: Mew, 55.
12 *administer enemas*: Parkin, 223.
12 *Roughly 75 percent*: Viñao Frago, 578.
13 *"an ignorant pretender"*: Schultz, 129.
13 *three classes*: F. Morales, 110.
13 *"If you give an Aragonese"*: Mas, 14.
14 *two hundred thousand*: Walker and Porraz, 2.
14 *roamed the quay*: Trallero Anoro.
14 *allowed him to retain*: Moigno, 88.
15 *salary was less than*: Calvo Roy, 21.
15 *"sanitary services"*: See de Carlos Segovia, *Los Ramón y Cajal*, 29.
15 *"I cannot complain"*: Ramón y Cajal, *Recollections*, xxx.

2. "Perpetual Miracle"
16 *"a special branch"*: Ramón y Cajal, *Recollections*, v.
17 *"wayward, unlikeable creature"*: Ibid., 16.
18 *Moyano law*: McNair, 18.
18 *"Only my father"*: Pedro Ramón y Cajal, 20.
18 *"Truth on this side"*: Pascal, 844.
19 *public hygiene*: J. Sancho, 133–34.
19 *The family moved to Larres*: Calvo Roy, 25.
19 *"Our father was the best example"*: Fernández Aldama, 3.
20 *astonishing mental coordination*: Durán Espuny.
20 *"one of the unbridled tendencies"*: Ramón y Cajal, *Recollections*, 16.
21 jota: Campion, 157.
22 *concordat*: "Spanish Concordat of March 16, 1851."
22 *morning of July 18, 1860*: See De La Rue, 23.

3. "Plunging into Social Life"
24 Don, *an honorific*: Mainer Baqué, 3.
24 *with only one teacher*: Madoz, vol. 3, 198.

24 *primary-school teaching*: Boyd, 10.
25 *no friends*: Durán Muñoz and Alonso Burón, *Cajal: Vida y obra*, 29.
25 *"Things always interested"*: Ramón y Cajal, *Recollections*, 281.
25 *"From the heights"*: Ibid., 28.
25 *"necessity of plunging"*: Ibid.
26 *"blind desire"*: Fernández Aldama, 3.
26 *"excessive softness"*: Ibid., 43.
26 como chinches: Ford, *Gatherings*, 176.
26 *"absolute terror"*: Ramón y Cajal, *Recollections*, 44.
26 *"I am used to it"*: Fernández Aldama, 3.
27 *"irresistible mania"*: Ramón y Cajal, *Recollections*, 36.
27 *rocks in the fields*: Ehrlich, 108.
27 *"Castle of Science"*: Pedro Ramón y Cajal, 28.

4. "A Castle of Dreams"

29 institutos, *existed*: Mainer Baqué, 3.
29 *"You are where you did"*: Max Aub quoted in ibid.
29 *Parents often looked*: See Morales, 106.
29 romancistas: Atanasio Fuentes, 164.
30 *The journey took*: O'Shea, 525.
30 *"commanders of themselves"*: See Espoz y Mina, 182.
30 *Sunk in a furrow*: Armillas Vicente.
30 *The peak of Mount Oroel*: "Ruta a la Peña Oroel."
31 *few minutes' walk*: See Durán Espuny.
31 *"Who was incarnated"?*: Nieto Amada.
31 *"cerebral injection"*: Benito Peréz-Galdós quoted in Menéndez Pidál.
31 *"like those of a door knocker"*: Ramón y Cajal, *Recollections*, 55.
32 *"trembling with fear"*: Ibid., 57.
32 *"For if the world"*: Ibid., 38.
32 *"Before the grandeur"*: Ibid., 60.
32 *Juan's wife, Orosia*: Ibid.
33 *his cousin Victoriano*: Ibid.
33 *lowest possible grades*: de Carlos Segovia, *Los Ramón y Cajal*, 76.
34 *"Jesus did not"*: Durán Muñoz and Alonso Burón, *Cajal: Escritos inéditos*, 102.
34 *"Through every fanatical Catholic"*: Ramón y Cajal, *El mundo*, 52.
34 *"prefer to bury"*: Ramón y Cajal, *Recollections*, 16.

5. "The War of Duty and Desire"

35 *"a reaction against"*: Ramón y Cajal, *Recuerdos de mi vida*, iii.
35 *"the war of duty"*: Ramón y Cajal, *Recollections*, 41.
35 *Framed by reddish peaks*: Quadrado, 206.

35 *crude oval*: Armillas Vicente et al.
35 *its Gothic spires*: Ford, *A Handbook for Travellers in Spain*, Part 1, 936.
35 *"artistic instincts"*: Ramón y Cajal, *Recollections*, 25.
36 Memories and Beauties of Spain: Quadrado, 29.
37 *"absolute master"*: Ramón y Cajal, *Recollections*, 77.
37 *"Such a decision"*: Ibid., 99.
37 *"Spanish Lord Byron"*: *Encyclopedia Britannica*.
37 *"Every soldier"*: Quoted in de Espronceda, 352.
37 *José Cadalso*: See Wardropper.
39 *"the most disturbed"*: Ramón y Cajal, *Recollections*, 106.
40 *"Exactly when my soul"*: Ibid., 107.
40 *Barbershops were*: See Lewy Rodríguez, *Santiago Ramón y Cajal*, 32;
      Mas, 180–81.
40 *rural police*: Novales, 73.
41 *Officers in brown uniforms*: Ramón y Cajal, *Recollections*, 154.
41 *"inborn dislike"*: Ibid., 110.
41 *"the most reactionary nation"*: Armstrong, 785.
41 *"a tangle of revolutions"*: Latimer, 9.
41 *calling themselves Liberales*: Rosenblatt, 62.
41 *Liberal ideas*: Angel Smith.
41 *"capricious"*: Ramón y Cajal, *Recollections of My Life*, 159.
42 *"Cajal was a novelist"*: Ibid., 160.
43 antiesthetic: Ibid., 122.
43 *stains and cobwebs*: Ibid., 121.
43 *"transformed"*: Ibid.
43 *"Never did I live"*: Ibid.
43 *"Are you chastened?"*: Durán Muñoz and Alonso Burón, *Cajal: Vida y
      obra*, 38.

6. **"The Nasty and Prosaic Bag"**
45 *"to learn to realize"*: Alvira Banzo, 22.
45 *"What I really want"*: Abadías y Santolaria.
46 *"intoxication of the aesthetic"*: Ramón y Cajal, *Recollections*, 129.
46 *"more than once"*: Ibid., 130.
46 *"resurrectionists"*: Ibid.
47 *"incomprehensible pleasure"*: Ramón y Cajal, *Recollections of My Life*, 14.
47 *his father would look*: Bosch y Miralles.
47 *"fragments of solid reality"*: Ramón y Cajal, *Recollections*, 145.
48 *"If things are looked at"*: Ibid.
48 desamortización: Novales, 18.
48 *daguerreotype*: Márquez, 145.
48 *calotype*: Turner, 221.

49 *"delicious"*: Alschuler, 1183.
49 *"positively stupefied"*: Ramón y Cajal, *Recollections*, 140.
49 *"the theory of the latent image"*: Ibid.
50 *Early on the morning of*: See Strobel, 1–25.
50 *"the mortal enemy"*: Ibid., 23.
51 *host of gadgets*: See Turner and Weston Turner.
51 *"was the typical student"*: See Brioso.
51 *"stop in jail"*: Fernández Aldama, 3.
52 *"I must exchange"*: Ramón y Cajal, *Recollections*, 42.
52 *defending the thresholds*: O'Shea, 513.
52 *transformed into a hospital*: Ibid.
52 *Virgin Mary their captain general*: Ibid., 513–14.
52 *"The sight of it"*: Cervantes, *The Ingenious Gentleman*, 313.
53 *"sylvan glades"*: Ramón y Cajal, *Recollections*, 163.

7. "A Myth Concealed in Ignorance"

54 *no physiology laboratory*: López Piñero, *Santiago Ramón y Cajal*, 123.
54 *Florencio Ballarín*: See García-Mercadal y García-Loygorri, 375.
54 *Zaragoza Botanical Garden*: *Guía de Zaragoza*, 414–17.
55 *"first person whom I heard"*: Ramón y Cajal, *Recollections of My Life*, 164.
55 *Bruno Solano*: See Morales, 110–11; *Gran Enciclopedia Aragonesa*.
55 *"exquisite sensibility"*: Ramón y Cajal, *Recollections*, 167.
55 *"temple for the worship"*: Ibid.
56 *"the most difficult points"*: Ibid.
56 *constantly talked about medicine*: Durán Muñoz and Alonso Burón, *Cajal: Vida y obra*, 58.
57 *"You might be deterred"*: Leonardo da Vinci, 121.
58 *Don Justo would grab*: Bosch y Miralles.
59 *"manias"*: Ramón y Cajal, *Recollections*, 182.
60 *"wearisome"*: Ibid., 192.
60 *"Romance of Mari-Juana"*: See Cajal Institute archives.
60 *"The Pastor and the Serrana"*: Ibid.
60 *"Ode to the Student Commune"*: Durán Muñoz and Alonso Burón, *Cajal: Escritos inéditos*, 95.
60 *"Who does not feel"*: Ibid.
60 *"servile imitations"*: Ramón y Cajal, *Recollections*, 181.
61 *"MARIA"*: Durán Muñoz and Alonso Burón, *Cajal: Escritos inéditos*, 93.
61 *"massive and ungainly carcass"*: Ramón y Cajal, *Recollections*, 192.
61 *"no Adonis"*: Ibid., 184.
61 *"like a sideshow Hercules"*: Ibid.
62 *"The direct observation"*: Ramón y Cajal, *Advice for a Young Investigator*, 62.

62 *"fragment of solid reality"*: Ramón y Cajal, *Recollections*, 145.
62 *bedrock of scientific thought*: Fernández-Medina, 211.
63 *"We must not transfer"*: Wick and Grundtman, 5.
63 *"revolutionary"*: Ramón y Cajal, *Recollections*, 175.
63 *"monarchical principle"*: Otis, *Membranes*, 22.
63 *"As an investigator"*: Ibid., 18.
63 *Biologists began*: Akhtar Khan, 126.
63 *"Each cell forms"*: Franz Meyen quoted in Nicholson, 204.
63 *"distinct individuality"*: P.J.F. Turpin quoted in Nicholson, 206.
63 *"independent existence"*: Theodore Schwann quoted in Otis, *Membranes*, 14.
63 *"a life of its own"*: Timoner Sampol.
63 *"individually of its own account"*: Otis, *Membranes*, 14.
63 *"functional autonomy"*: Ramón y Cajal, *Elementos*, 109.
63 *"The body is a state"*: Majno and Joris, 14.
63 *"autonomous living being"*: Ramón y Cajal, *Recollections*, 175.
63 *"repugnance for all kinds"*: Ibid., 131.
64 *"a myth concealed"*: Ibid., 176.
64 *"There's a horse inside"*: Ramón y Cajal, *Charlas*, 247–48.
64 *"celestial anatomy"*: Ramón y Cajal, *Recollections*, 252.
64 *"academic reactionists"*: Ibid.
64 *"pure fantasy"*: Ibid.
65 *carmine*: Minot, 481.
65 *"enraptured and tremendously moved"*: Ramón y Cajal, *Advice*.
65 *"the sublime spectacle"*: Ibid.
65 *"I felt as though"*: Ibid.
65 *"Think of it!"*: Ramón y Cajal, *Recollections*, 182.

8. "Humbled by My Failure"

66 *decided to run away*: See Solsona.
67 *"I was left"*: Ramón y Cajal, *Recollections of My Life*, 204.
67 *"Zaragoza is a desert"*: Ibid., 205.
68 *On the floor of the Cortes*: Adee, 793.
68 *Al fin los hemos logrado!*: Quoted in Latimer, 347.
68 *"let us salute"*: Quoted in United States Department of State, *Foreign Relations of the United States*, 918.
68 *"rainbow of peace"*: United States Department of State, *Papers Relating to the Foreign Relations of the United States [and Spain]*, 919.
68 *quinta*: Castelar, 224.
68 *biggest mass conscription*: Hannay, 187.
68 *175,000 Spaniards*: Ibid.

69 *tobacco smoke*: Ramón y Cajal, *Charlas*, 113.
69 *Carlist stronghold*: Clare, 1717.
70 *"When a people arrives"*: Quoted in Zeigler, 1.
71 *"To be perfectly sincere"*: Ramón y Cajal, *Recollections*, 204.
71 *"milk sea"*: Verne, 172.
71 *"What a great alarm"*: Quoted in Triarhou, "Two Readings of Quixote," 157.
71 *"I lived as if in a dream"*: Ramón y Cajal, *Recollections*, 211.
72 *"foolishly quixotic"*: Ibid., 213.
73 *"the pleasantest period"*: Ibid., 220.
73 *"scruples of a nun"*: Ibid., 221.
73 *he had mocked*: Durán Muñoz and Alonso Burón, *Cajal: Vida y obra*, 83.

9. "Cells and More Cells"

75 *"blundering generals"*: Ramón y Cajal, *Recollections*, 226.
75 *One hundred and eighty thousand*: Guerra y Sánchez, 377.
75 *half of them*: Victimario Histórico Militar.
75 *"sinful"*: Ibid., 245.
76 *"I had to reshape my life"*: Ibid., 243.
76 *those who knew him*: Durán Muñoz and Alonso Burón, *Cajal: Vida y obra*, 69.
77 *"Nothing worth relating"*: Ramón y Cajal, *Recollections*, 248.
77 *"Considerations About the Organicist Doctrine"*: See Ramón Casasús.
77 *in Cajal's hand*: Martínez Tejero.
77 *"You are always great"*: See Ramón Casasús.
78 *"in love with his profession"*: del Río Hortega, 23.
78 *"His colleagues labeled"*: Ibid.
79 *"became practically synonymous"*: Everdell, 107.
79 *"Columbuses"*: Ramón y Cajal, *Manual de histología normal*, 145.
79 *Francisco Chenel*: See *El Imparcial*, July 8, 1876.
80 *nothing more than a novelty item*: Ruestow, 10.
80 *"for observing"*: Fahie, 210.
81 *"By the help of"*: Snyder, ch. 6.
81 *"It is the great prerogative"*: Hooke.
81 *"trifling"*: Hunter, 70.
81 *"a Sot"*: University of California Museum of Paleontology, "Robert Hooke (1635–1703)."
81 *"All the water"*: University of California Museum of Paleontology, "Antony van Leeuwenhoek (1632–1723)."
82 *"many very little things"*: Ibid.
82 *"like a new toy"*: Fernández Aldama, 4.

82 *"Come with us"*: Ramón y Cajal, *Recollections*, 295.
82 *"The Marvels of Histology"*: See Freire.
83 *"philosophic-scientific temerities"*: Ramón y Cajal, *Recollections*, 293.
83 *"overflowing with fantasy"*: Ibid.
83 *"comforting aroma"*: Ibid.
83 *"our organic edifice"*: Ibid., 295.
83 *"thousands of microscopic workers"*: Ibid.
83 *"in compact battalions"*: Freire.
83 *"It is certain"*: Ramón y Cajal, *Recollections*, 295.
83 *"the eternal and faithful"*: Ibid.
83 *"homeric struggles"*: Ibid.
83 *"the incessant war"*: Ibid.
84 *"Cells are what live"*: Ramón y Cajal, "La Teoría Celular."

10. "The Irremediable Uselessness of My Existence"

85 *"honeymoon with the microscope"*: Ramón y Cajal, *Recollections*, 252.
85 *quintessentially Spanish*: Pedro Salinas quoted in Kiddle, 255.
85 *grueling examinations*: Ibid.
85 *"cruel and always bitter"*: Ibid.
85 *"cancers"*: Ibid.
87 *"A physician rarely deludes"*: Ramón y Cajal, *Recollections*, 263.
87 *"the robber of youth"*: Frith.
88 *"the irremediable uselessness"*: Ramón y Cajal, *Recollections*, 265.
88 *between eighty-five and ninety-five*: Macpherson, 58.
88 *"I should like to die"*: Daniel, 1864.
88 *"particularly fond"*: Keats, 188.
88 *"poor melancholy angel"*: Daniel, 1864.
88 *"infinite grace"*: Walker.
89 *"delicately, morbidly angelic"*: Lougheed.
89 *"strangely luminous face"*: Barnes, 56.
89 *"The Staircase"*: Macpherson 118; Three Wayfarers, 53–56.
89 *women from the village*: Herrera y Ruiz, 6.
89 *The path was so narrow*: Three Wayfarers, 55.
89 *Guards were stationed*: Balneario de Panticosa (Huesca), 11.
89 *The sulfuric water*: Weld, 124.
90 *strict schedules*: See Cruz, 210–11.
90 *his angry outbursts*: Durán Muñoz and Alonso Burón, *Cajal: Vida y obra*, 103.
90 *"a blind and unshakeable"*: Laín Entralgo, "Estudios y apuntos," 40.
90 *"When least expected"*: Ramón y Cajal, *Recollections*, 266.
90 *"still the vibrations of sorrow"*: Ibid., 268.

11. "Not for the Living but for the Dead"

92 *"in almost flourishing health"*: Ramón y Cajal, *Recollections*, 269.

92 *"They have named you"*: de Carlos Segovia, *Los Ramón y Cajal*, 69.

92 *"The histologist is a physician"*: Edward Schafer quoted in Anctil, 99.

93 *For histologists*: Shepherd, 81.

93 *in cold weather*: Merchán et al., 140.

93 *if the tissue is hardened*: See Ramón y Cajal, "Coloración por el método."

93 *has to be washed*: Ibid.

94 *The dehydration process*: Heinbockel and Shields, 8.

94 *"clearing"*: See Lee, 9.

94 *paraffin*: Ibid.

94 *"ideal sanatorium"*: Ramón y Cajal, *Advice*, 87.

94 *Working alone*: Ibid., 93.

94 *the ideal mentors*: See Ramón y Cajal, *El mundo visto*, 189.

95 *practice alone*: Ramón y Cajal, "Contribución al estudio."

95 *"For the union"*: Ramón y Cajal, *Advice*, 60.

96 *"Needless to say"*: Ramón y Cajal, *Manual de histología normal*, 34.

96 *"In making [illustrations]"*: Turner, 215.

96 *"can't draw"*: Van Leeuwenhoek quoted in Turner, 20.

96 *"No matter how exact"*: Ramón y Cajal, *Advice*, 132.

96 *"Drawing develops"*: Alvira Banzo, 4.

96 *rifle was no better*: Ramón y Cajal, *Recollections*, 511.

96 *"I have endured"*: Ibid., 217.

96 *emergency call*: Durán Muñoz and Alonso Burón, *Cajal: Vida y obra*, 179.

97 *with blunderbusses*: de Carlos Segovia, 69.

97 *cesarean section*: Luis Ramón y Cajal, 73.

97 *"nothing more than"*: Lewy Rodríguez, *Santiago Ramón y Cajal*, 28.

97 *7,500 pesetas*: López Piñero, *Santiago Ramón y Cajal*, 158.

98 *Torrero*: Durán Muñoz and Alonso Burón, *Cajal: Escritos inéditos*, 241.

98 *preference for blondes*: Durán Muñoz and Alonso Burón, *Cajal: Vida y obra*, 338.

98 *"fair and slender"*: Ramón y Cajal, *Recollections of My Life*, 117.

98 *"Marriage is born"*: Durán Muñoz and Alonso Burón, *Cajal: Escritos inéditos*, 91.

98 *"I have a brain"*: Quoted in Laín Entralgo, *Cajal por sus cuatro costados*, 86.

99 *"a sensitive compliance"*: Ramón y Cajal, *Advice*, 104.

100 *"ordinary bourgeois newlyweds"*: Ramón y Cajal, *Vacation Stories*, 3.

101 *"undeniable satisfactions"*: Ramón y Cajal, *La fotografía de los colores*, 15.

101 *local bullfighting ring*: Durán Muñoz and Alonso Burón, *Cajal: Vida y obra*, 126.

101 *"goblins or necromancers"*: Ramón y Cajal, *Recollections*, 268.
101 *The neighbors wondered*: Perrín G. Thomas.
101 *"The sensible person"*: Durán Muñoz and Alonso Burón, *Cajal: Vida y obra*, 115–16.
102 *"probably the most ignorant"*: Núñez Florencio, 230.
102 *"inquisitorial, ignorant, fanatical"*: Juderías, 111.
102 *"derogatory foreign judgments"*: Ramón y Cajal, *El mundo visto*, 201n133.
102 *"One should publish"*: Ramón y Cajal, "Observaciones microscópicas," 55.
103 *"like pebbles"*: Ramón y Cajal, *Advice*, 64.
104 *"excursions"*: Ramón y Cajal, "Investigaciones experimentales," 10.
104 *"a defect of interpretation"*: See ibid., 4.
104 *"the innate tendency"*: Ibid., 2.
104 *"It is a general ailment"*: Ibid., 1–2.
105 *grease [etc.]*: Ramón y Cajal, *Manual de histología normal y técnica micrográfica*, 125–27.
105 *"Because we love them"*: Ramón y Cajal, *Advice*, 93.

12. "The Role of Don Quixote"

106 *"I was excessively surprised"*: Ramón y Cajal, *Recollections*, 252.
107 *"the simplicity"*: del Olmet and Torres Bernal, 373.
107 *"the garden of Spain"*: O'Shea, 263.
107 *The Moors*: Sime, 15.
107 *salt and sugar*: Fischer, 5.
108 *hoarfrost*: Ibid.
108 *"one continual delicious spring"*: Fischer, 331.
108 *golden light*: O'Shea, 472.
108 *smell of fresh oranges*: Davillier, 70.
108 *strong winds blew*: Ibid.
108 *"the Aragonese"*: Vera Sempere.
108 *shiny oranges*: Ford, *A Handbook for Travellers in Spain*, Part 1, 317.
108 *long pink couch*: Lathrop, 178.
108 *Agricultural Society of Valencia*: Vera Sempere, 400.
108 *tertulias*: See Crow, 237.
108 *"Children of the flesh"*: Ramón y Cajal, *Recollections*, 280.
109 *3,500 pesetas*: Durán Muñoz and Alonso Burón, *Cajal: Vida y obra*, 136.
109 *The streets of Valencia*: See Ridaura, *Vida cotidiana*.
109 *After finishing their chores*: See Beltrán de Lis, *Reglas de urbanidad para señoritas*.
109 *The Valencia faculty*: Durán Muñoz and Alonso Burón, *Cajal: Vida y obra*, 136.

109 *they quickly realized*: Ibid., 137
109 *"a millionaire in time"*: Letter to Mihály Lenhossék, 1925, in Fernández Santarén, *Santiago Ramón y Cajal: Epistolario*, 419.
109 *"It was necessary to allot"*: Ramón y Cajal, *Recollections*, 311.
110 *Chimney Salon*: Vera Sempere.
110 *two thousand photographs*: See see the Cajal Legacy website (cajal.csic .es/LegadoCajal/index.php/FondoFotografico) for a full inventory of photographs in the Cajal archives.
111 *"No one skipped"*: Concepción Aleixandre quoted in Calvo Roy, 87.
111 jota: *Dance and society*, 138.
111 *"It was as though"*: Concepción Aleixandre quoted in Calvo Roy, 87.
111 *"Why not be honest?"*: Bartual Pastor.
112 *"Soon, very soon"*: Ibid.
112 *"There, in his laboratory"*: Ibid.
112 *"[Cajal's] aspect"*: Ibid.
112 *"so surprising"*: Bartual Vicéns.
112 *"Dear Professor"*: Quoted in ibid.
114 *the man's card*: Ramón y Cajal, *Charlas*, 251–52.
114 *"My motto was always this"*: Letter to Luis Roberto Simões Raposo, November 5, 1925, *Epistolario*, 523.
114 *"the Dulcinea"*: Ramón y Cajal, *Discursos leídos . . . el día 7 de mayo 1922.*
114 *"I took the role"*: Laín Entralgo, *Cajal por sus cuatro costados*, 32.
114 *"large, solid and black"*: Ramón y Cajal, *Manual de histología*, 14.
114 *"cut from glass"*: Ibid.
114 *"diffuse, proceeding from"*: Ibid., 3.
115 *Enrique Ferrer y Viñerta*: Vera Sempere, 398.
115 *Amalio Gimeno*: Ibid.
115 *Peregrín Casanova*: Ibid.
116 *"What can be more curious"*: Darwin, 472.
116 *"drive to formation"*: "Johann Wolfgang von Goethe (1874–1832)."
116 *"having something before me"*: Ibid.
116 *"pure abstraction"*: Ramón y Cajal, *Expedientes administrativos.*

13. "The Religion of the Cell"

117 *victims expel torrents*: "The Cholera in Spain," June 6, 1885, 1170.
117 *"roof gutter"*: Kousoulis, 540.
117 *dark blue and leathery*: "The Cholera in Spain," June 6, 1885.
117 *four to twelve hours*: "The Cholera in Spain," July 11, 1885, 77.
117 *in 1832*: See Hubbard, 126.
117 *"bad air"*: Frerichs.
117 *He invented vaccines*: Science History Institute.

118 *"a little bent"*: Bailey, *Cholera*, 54.

118 *under quarantine*: United States Bureau of Foreign and Domestic Commerce, 77–79.

118 *water contamination*: "The Cholera in Spain," June 6, 1885, 1170–71.

118 *In Burjassot*: *British Medical Journal*, June 6, 1885.

118 *water closet*: *British Medical Journal*, August 29, 1885.

118 *Filth and garbage*: *British Medical Journal*, July 11, 1885, 77.

118 *In April 1885*: Hays, 303.

118 *Valencia's circulatory system*: Bornside, 519.

118 *The wealthy fled*: *British Medical Journal*, August 29, 1885.

118 *Public officials abandoned*: *British Medical Journal*, June 6, 1885, 1171.

118 *Bonfires were set*: *British Medical Journal*, July 18, 1885, 114.

118 *"Do not shake hands"*: *British Medical Journal*, July 4, 1885, 33.

118 *A rumor spread*: *British Medical Journal*, July 25, 1885, 164.

118 *"cholera wagon"*: *British Medical Journal*, June 27, 1885, 1308.

118 *"the suspicious illness"*: "La enfermedad sospechosa," see *British Medical Journal*, June 6, 1885, 1170.

118 *Mobs stoned doctors*: *British Medical Journal*, June 6, 1885, 1261.

119 *"faith-healers"*: *British Medical Journal*, July 18, 1885, 114.

119 *a few blocks away*: Cabot, "Cajal Frente a Ferrán," 77.

119 *almost next door*: Ibid.

119 *suspended in beef tea*: "Cholera," *British Medical Journal*, June 27, 1885, 1308.

119 *"the Ferrán question"*: See "Cuestión Ferrán" in *Las provincias: Diario de Valencia*, 38.

119 *"There is not a tittle"*: Cameron, 338.

119 ferranista: See López Piñero, *La Facultad de Medicina*.

119 *bared arms*: Bornside, 523.

120 *two and seven dollars*: "The Cholera in Spain," *British Medical Journal*, June 6, 1885.

120 *"I have frequent news"*: López Piñero, *Santiago Ramón y Cajal*, 185.

120 *"the invisible enemies"*: Otis, *Membranes*, 119.

120 *"all-out war"*: Ibid., 67.

120 *"We had to fight"*: Quoted in Juarros, 87.

121 *"I have to keep my secret"*: Quoted in Tardieu, 1063.

121 *"With regard to"*: Ramón y Cajal, *Recollections*, 287.

121 *"The Safest and Easiest Way"*: See Ramón y Cajal, "El más Seguro y sencillo de los métodos de coloración de los microbios."

121 *"even more beautiful"*: Quoted in López Piñero, *Santiago Ramón y Cajal*, 186.

121 *best in optical amplification*: Durán Muñoz and Alonso Burón, *Cajal: Vida y obra*, 151.

121 *"rickety door bolt"*: Ramón y Cajal, *Recollections*, 290.
122 *"I am going to leave"*: Letter to Jaime Ferrán, November 1885, *Epistolario*, 237.
122 *"impatiently champing"*: Ramón y Cajal, *Vacation Stories*, xxii.
122 *"no more than"*: Andrew Perez.
123 *"unbosoming"*: Ramón y Cajal, *Vacation Stories*, xxii.
123 *"dynamic compensation"*: Ibid.
123 *"as happily"*: Ibid., 1.
123 *"inoculated with"*: Ibid., 2.
124 *"not lacking in talent"*: Ibid.
124 *"unsuccessfully, but with honor"*: Ibid., 124.
124 *"the spirit of science"*: Ibid., 131.
124 *"There was a whole world"*: Ibid., 150.
124 *"a science that deals with"*: Monlau, 167.
124 *"The interior observation"*: Makari, 11.
124 *"the anomalies"*: Ibid., 13.
125 *"I succeeded in performing"*: Ramón y Cajal, *Recollections*, 313; my translation.
125 *"sanguine-lymphatic"*: Stefanidou et al., 355.
126 *"Your lids are closing"*: Bernheim, 2.
126 *"You will have consciousness"*: Stefanidou et al., 355.

14. "Moved by Faith"
127 *"perfectly irreconcilable"*: Lewy Rodríguez, *Asi era Cajal*, 44.
128 *presence of granules*: Shepherd, 10.
128 *"ganglionic corpuscles"*: Liddell, 103.
128 *Ehrenberg also assumed*: Shepherd, 11.
129 *"where sight may stare"*: Ibid., 137.
129 *microdissection*: Ibid., 34.
129 *Otto Karl Friedrich von Deiters*: Ibid., 33.
130 *"Every modest, sensible"*: Shepherd, 38.
130 *"one falls back"*: Ibid., 39.
130 *"the patience of a Benedictine"*: Ramón y Cajal, *Recollections of My Life*, 35.
130 *"tormented soul"*: Ramón y Cajal, *Histology*, 21. *Histology* is the French title of *Textura*.
130 *"technical bible"*: Quoted in DeFelipe, *Cajal's Neuronal Forest*, 9.
131 *"I have obtained"*: Mazzarello, *The Hidden Structure*, 63.
131 *"even to the blind"*: Ibid.
131 *the black reaction*: See Merchán et al., *Cajal and de Castro's Neurohistological Methods*, 71.
131 *"the marvelous beauty"*: Mazzarello, *Golgi*, 77.

131 *"the inextricable thicket"*: Ramón y Cajal, *Recollections*, 324.
132 *"The dream technique"*: Quoted in DeFelipe, *Cajal's Neuronal Forest*, 25.
132 *dried them with blotting paper*: Sharpey-Schäfer, 114.
132 *a brownish cloud*: Ramón y Cajal, *The Structure of the Retina*, xxi.
133 *"capricious and highly uncertain"*: Mazzarello, *Golgi*, 77.
133 *"technique and science"*: Pittaluga.
133 *"Ideas do not show"*: Ramón y Cajal, *Recollections*, 325.
133 *"suggests rather than"*: Ibid., 309; italics in original.
133 *letter to a fellow anatomist*: Letter to Van Gehuchten, 1913, *Epistolario*, 534.
134 *"no time to work"*: Letter to Óloriz, 1886, quoted in Cortezo.

15. "Free Endings"

135 *"the seat of courtesy"*: Cervantes quoted in Baedeker, *Spain and Portugal*, 198.
135 *twice as dense*: Ibid.
135 barracas: Ealham, 6.
135 *A housing report*: Gilmour.
135 *the Raval*: Cruz, 223.
136 *industrialists built*: Brugmann, 240.
136 *Lluna and Riera Alta*: Rocha Barral and Estévez i Torrent, 128.
136 *"It was difficult"*: Augusto Pi Suñer quoted in Durán Muñoz and Alonso Burón, *Cajal: Vida y obra*, 189.
136 *"without pretensions"*: Ramón y Cajal, *Manual de anatomía patológica general y de bacteriología patológica*, 1.
136 *"dedicated to students"*: Ibid.
136 *"I have four"*: A. Pérez Morales.
137 *with a basic education*: López Piñero, *La Facultad de Medicina*.
137 *"There is certainly"*: Shepherd, 84.
137 *"lose their individuality"*: Otis, *Networkings*, 60.
137 *"an act of rebellion"*: See Ramón y Cajal, *Histology*, 27.
137 *broad promenades*: Stoddard, 14.
138 *Paperboys*: Matto de Turner.
138 *flower girls*: Ibid.
138 *birds in stacked cages*: Ibid.
138 *Antoni Gaudí*: Sobrer, 162.
138 *"constant and rude"*: Millán Bermejo.
138 *"cutting ties"*: Ramón y Cajal, *El mundo visto*, 113.
138 *"Modern winds"*: Quoted in Ramon Resina, 26.
138 *"needs at all hours"*: Quoted in Vaill, 132.
138 *"severe abstention"*: Ramón y Cajal, *Advice*, 34.

138 *"the creative tension"*: Ibid., 38.
138 *"To bring scientific investigation"*: Ibid., 33.
138 *"One must achieve"*: Ibid.
139 *"When he worked"*: García del Real.
139 *"The laboratory man"*: Ramón y Cajal, *Recollections*, 319.
139 *Café Pelayo*: See Barcelonauta, "Café Pelayo."
139 *"the Aviary"*: See Barcelonauta, "La Pajarera."
139 *Patrons signaled*: O'Shea, 55.
139 *panorama showcased*: Barcelonauta, "La Pajarera."
139 *men expressed*: Ramón y Cajal, *Charlas*, 113.
140 *The most important role*: Ibid., 115.
140 *"To conserve"*: Ibid., 119.
140 *"Have I learned"*: Ibid., 124.
140 *"Withdrawn, isolated"*: Roca, 18.
140 *"so absentminded"*: Ibid., 38.
140 *left tail of his jacket*: Ibid., 18.
141 *At the restaurant*: Ibid., 17.
141 *Since ancient times*: McDonogh and Martinez-Rigol.
141 *Ildefons Cerdà*: Herzog, *Return to the Center*, 94; Pallares-Barbera et al., 122–26.
142 *"sunny, spacious"*: Ramon Resina, 189.
142 *"If we view"*: Ramón y Cajal, *Histology*, 41.
142 *"Since the full grown"*: Ramón y Cajal, *Recollections*, 324.
143 *shaped like teardrops*: Koob, 30.
143 *"complex and bizarre"*: Rapport, 115.
143 *"the metamorphosis"*: Ramón y Cajal, *Recollections*, 18.
144 *his hand-cut sections*: Roca, *Tribut*, 18.
144 *"most elegant"*: Ramón y Cajal, *Recollections*, 364.
144 *"pear-like"*: Quoted in Shepherd, 42.
144 *"basket"*: Ibid., 143.
144 *"new truth"*: Ramón y Cajal, *Recollections*, 381.
144 *Golgi's illustration*: Shepherd, 150.
145 *"We have never"*: Quoted in ibid., 148.
145 *forty-three papers*: See Sala Catalá.
146 *"By this time"*: Shepherd, ch. 12.
146 *"The Science Pavilion"*: Ibid.
146 *"Four pots"*: Nieto-Galan; my translation.
147 *Casa de Caritat*: Dunfort i Coll, 7.
147 *"entirely swallowed up"*: Shepherd, 140.
147 *took them to mass*: "Silveria Fañanás García."
147 *"She divined"*: Ramón y Cajal, *Recollections*, 326.
147 *"spiritual director"*: Ibid., 272.

147 *"not sullen"*: Ramón y Cajal Junquera, "D. Santiago, mi abuelo."
147 *"authoritarian"*: Ibid.
147 *"Preoccupied with my studies"*: del Olmet and de Torres Bernal, 308.
148 *"like a god"*: Ramón y Cajal Junquera.
148 *"deluded"*: Shepherd, 167.
148 *"pretender"*: Ibid.
148 *"a state of restlessness"*: Ibid.
148 *"was a little alarmed"*: Ramón y Cajal, *Recollections*, 352.
148 *"do not come into"*: Anctil, 102.
148 *"never actually coalesce"*: Ibid., 62.
148 *"as startling"*: Romanes, 519.
148 *"at first sight"*: Actin, 103.
148 *"So far as I know"*: Ibid., 108.
149 *"Connections of nerve cells"*: Shepherd, 107.
149 *"I believe"*: Ibid., 108.
149 *"By means of these"*: Ibid., 112.
149 *"It was as though"*: Ibid., 113.
150 *"But why do we always"*: Forel, 163.
150 *"Well, our two papers"*: Ibid.
150 *"The facts described by Cajal"*: Van Gehuchten, 1913, quoted in *Epistolario*, 534.
151 *"During the sacred fever"*: Ramón y Cajal, *Recollections*, 413.
151 *"I work when"*: Ibid.
151 *"I rest when"*: Quoted in Durán Muñoz and Alonso Burón, *Cajal: Escritos inéditos*, 22.
151 *"Ideas boiled up"*: Ramón y Cajal, *Recollections*, 325.
151 *"the* neura*"*: Herophilus, 241.
151 *"a river rising"*: Pearce, 287.
151 *"Ah, Mr. Beatty!"*: Prockop, 481.
152 *"impenetrable thicket"*: Ramón y Cajal, *Recollections*, 305.
152 *"Continually awake"*: Ibid., 381.
153 *when he finally exited*: Durán Muñoz and Alonso Burón, *Cajal: Vida y obra*, 194–95.
153 *"loved his children"*: Ibid., 191.
153 *"What ignorance"*: Luis Ramón y Cajal, 76.
153 *"pale and suffering face"*: Ramón y Cajal, *Recollections*, 380.

16. "Doubting Certain Facts"
154 *Praxagoras*: See Konstantine and Peter, 3.
154 neurons: Scarborough, 142.
154 *"gripping the notches"*: Homer, 57.
154 *puppet strings*: Lewis, 26.

154 *Luigi Galvani*: McComas, 101.
155 *Alessandro Volta*: Ibid.
155 *"The possibility"*: Isaac, 17.
155 *In each of*: Jones, "Cajal's Debt to Golgi."
155 *"clearly meant"*: Otis, *Networking*, 61.
155 *"infinite number"*: Ibid.
156 *"isolated transmission"*: Ibid.
156 *"an anastomosis"*: Ibid., 52.
156 *"unfathomable"*: Ramón y Cajal, *Recollections*, 336.
156 *"What is the direction"*: Ibid., 382.
156 *"applicable to all cases"*: Ibid., 542.
157 *"cerebral polarization"*: Ramón y Cajal, *Advice*, 32.
157 *within a circular region*: Shepherd, 77.
157 *"Nothing convinces"*: Ramón y Cajal, *Recollections*, 354.
157 *"[Cajal] went"*: Roca, 19.
157 *He sold*: Lewy Rodríguez, *Santiago Ramón y Cajal*, 121.
158 *"My discoveries"*: Letter to Golgi, August 27, 1889, *Epistolario*, 389.
158 *"Sir and dear colleague"*: Ibid.
158 *aggressive and pretentious*: Mazzarello, *The Hidden Structure*, 347.
158 *"I intend to visit"*: Ibid.
158 *"next October"*: Letter to Golgi, August 1889, *Epistolario*, 389.
158 *olive skin*: See Cannon, x.
159 *"such a poor environment"*: See de Hoyos Sainz.
159 *"handsome old man"*: Minot, 480.
159 *"knew more"*: Ibid.
159 *"biological novel"*: Ramón y Cajal, *Recollections*, 182.
159 *Forty years earlier*: Laín Entralgo, *Cajal por sus cuatro costados*, 26.
160 *"based upon deception"*: Shepherd, 38.
160 *"enchanted"*: Quoted in Rapport, 139.
160 *"I have discovered you"*: Ramón y Cajal, *Recollections*, 357.
160 *"extremely excellent"*: Quoted in Actil, 121.
160 *Varese*: Email with Paolo Mazzarello, April 18, 2019.
161 *"The laboratories of Europe"*: Shepherd, 179.
161 *"This time"*: Chryst, 47–48.
161 *"I have shown"*: Letter from His, August 14, 1890, *Epistolario*, 405.
161 *"perfectly free"*: Levine and Marcillo.
161 *"a typical case"*: Figueres-Oñate et al., 2.
162 *"Even if"*: Mazzarello, *Golgi*, 225.
162 *"Palm Sunday"*: Ramón y Cajal, *Recollections*, 373.
162 *"Truth Against Error"*: Ramón y Cajal, "La verdad contra el error."
163 *"Why should I not"*: Ramón y Cajal, *Recollections*, 365.
163 *"No one has seen"*: Ibid., 368.

163 *"a conical lump"*: de Castro et al., 484.
164 *"relentlessly forging"*: Quoted in Nieto, 1039.
164 *"a living battering-ram"*: Ramón y Cajal, *Recollections*, 369.
164 *"breach[ing] the membrane"*: Ibid.
164 *"assaulting the interior"*: Ibid.
164 *"If the route"*: Ramón y Cajal, *Studies on Vertebrate Neurogenesis*, 54.
164 *"leaning"*: Ibid.
164 *"They apparently"*: Ibid.
164 *"the region of the future"*: Ramón y Cajal, *Cajal's Degeneration and Regeneration of the Nervous System*, 381.
164 *"Must we accept"*: Ramón y Cajal, *Studies on Vertebrate Neurogenesis*, 113.
164 *"Thus the total arborization"*: Ramón y Cajal, *Texture of the Nervous System of Man and the Vertebrates*, 578.
165 *"without deviation"*: Quoted in de Castro et al., 484.
165 *"exquisite chemical sensitivity"*: Ibid.
165 *"chemotactic ameboidism"*: Cajal, *Studies on Vertebrate Neurogenesis*, 95.
165 *"disoriented and often tortuous"*: Ibid., 256.
165 *"after hesitation"*: Ibid.

17. "The Only Opinions That Matter to Me"

166 *"If by an implausible chance"*: Letter from Luis Simarro, June 11, 1890, *Epistolario*, 296–97.
167 *"How can you deal"*: Cortezo, 207.
167 *"Who in Spain"*: Lewy Rodríguez, *Santiago Ramón y Cajal*, 77.
167 *different font*: Anctil, 122.
167 *"more fully"*: Sherrington, "Santiago Ramón y Cajal 1852–1934," 433–44.
167 chromosome: Jones, "The Neuron Doctrine 1891."
168 *"prince of the academic world"*: Ramón y Cajal, *Recollections*, 404.
168 *"demagnetized"*: Ibid., 407.
168 *Santa Inés Street*: Juarros, *Ramón y Cajal*, 139.
168 *religious experience*: See *El Siglo Médico*, vol. 3, November 3, 1934.
168 *traditional Spanish cape*: Luis Ramón y Cajal, 73.
169 *old brown suit jacket*: Ibid.
169 *"The man's appearance"*: Julián de la Villa quoted in de Carlos Segovia and Alonso Peña, loc. 3454.
169 *wiped the classroom chalkboard*: Durán Muñoz and Alonso Burón, *Cajal: Vida y obra*, 237.
169 *Van Baumbergen [etc.]*: Lewy Rodríguez, *Santiago Ramón y Cajal*, 65.
169 *the spiral shape*: "La Catedra de Cajal."

169 *"those emigrant leukocytes"*: Lewy Rodríguez, *Santiago Ramón y Cajal*, 65.

169 *Horse Yard*: Giménez-Roldán, "Monument to Cajal in Madrid," 5.

169 *"The bodies"*: Ibid.

169 *"not an orator"*: "En Honor de Ramón y Cajal," *La Época*, 1.

170 *"The majority, recently"*: José Álvarez Sierra quoted in de Carlos Segovia and Alonso Peña, loc. 3466.

170 *"Aragonese aftertaste"*: Ibid.

170 *better to learn*: Fernán Pérez.

170 *"a collection of donkeys"*: Leoz.

170 *"pedagogical dictionary"*: de Carlos Segovia and Alonso, loc. 3253.

170 *"Both my father"*: Ramón y Cajal, *Recollections*, 119.

171 *"The Uses of Wine"*: "El uso del vino en las comidas," in Durán Muñoz and Alonso Burón, *Cajal: Escritos inéditos*, 19–20.

171 *"I do not abuse bread"*: Ramón y Cajal, *El mundo visto*, 70.

171 *German chocolate cake*: Naño.

171 *"an archipelago"*: Eugenio Montes quoted in Jiménez, "Cajal en los cafés de Madrid," 34.

171 *The mirrors*: Giménez.

171 *self-renewing caste*: See Herrero Pérez, 13.

172 *"Cajal, man"*: Fernández Aldama, 5.

172 *"Pure, but sanctified"*: Ramón y Cajal, *Recollections*, 408.

172 *"Certainly, no one"*: Ibid., 405.

172 *"the axis"*: "Café Suizo."

172 *news was discussed*: Luis Vélez de Guevara, paraphrased in "Café Suizo."

173 *"torrent"*: Ramón y Cajal, *Recollections*, 446.

173 *"unimagined splendors"*: Ibid.

173 *"in ecstasy"*: Ibid.

173 *"You do well"*: Ibid.

173 *"those tiny cells"*: Ibid.

18. "The Absolute Unsearchability of the Soul"

174 *"advanced our knowledge"*: Greef.

174 *"[Its] pyramidal cells"*: Ramón y Cajal, *Recollections*, 415.

174 *"I have learned"*: Letter from Kölliker, May 29, 1893, *Epistolario*, 553.

175 *hundreds of researchers*: Køppe, 9.

175 *"Every new work"*: Letter from Retzius, June 25, 1891, *Epistolario*, 488.

175 *"mental histology"*: López-Muñoz et al., 11.

175 *"It is well known"*: Ramón y Cajal, *Advice*, 99.

175 *"When Cajal opened"*: Paul Flechsig quoted in Lewy Rodríguez, "Las ideas sociales de Cajal."

175   *"Cajal has given"*: Arthur Van Gehuchten quoted in Aldecoa de González.
175   *Ideas about the brain*: Ehrlich, 4.
175   Australopithecus: See Finger, 3.
176   *called the pylorus*: O'Connor, 141.
176   *The Epic of Gilgamesh*: See Cobb, 15.
176   *we indicate* I: Lewis, 32.
176   *"Men ought to know"*: Hippocrates, 366.
177   *"unlock the secret"*: O'Connor, 140.
177   *"Where is the soul"*: Wuerth, 149.
177   *"know the nerves"*: Quoted in Bell, 146.
177   *"solely for the purpose"*: Gross.
177   *"Where there is"*: Finger, 32.
178   *"It is a rule"*: Ramón y Cajal, "Lecture III: The Sensori-Motor Cortex," 38.
178   *"exact science"*: Quoted in Banerjee, 231.
178   *"the utter darkness"*: Ramón y Cajal, *Cajal on the Cerebral Cortex*, 479.
178   *"with some reservations"*: Quoted in ibid., 80.
179   *"with a certain amount"*: Ramón y Cajal, *Recollections*, 446.
179   *"cannot be"*: Ramón y Cajal, *Cajal on the Cerebral Cortex*, 81.
179   *"no hypothesis"*: Ibid.
179   *"very special activity"*: Ibid.
179   *"Neither materialism"*: Ibid.
179   *"The Spaniard of Barcelona"*: Allen, 428.
180   *"I feared that I should not"*: Ramón y Cajal, *Recollections*, 417.
180   *"Bristling with oddities"*: Eccles and Gibson.
181   *"sadly imperfect"*: Sherrington, "A Memoir of Dr. Cajal," ix.
182   *"a pyramidal attachment"*: Ramón y Cajal, *Recollections*, 424.
182   *"an essentially physiological slant"*: Ibid., 418.
182   *"You will understand"*: Ramón y Cajal, *Cajal on the Cerebral Cortex*, 84.
183   *"men devoted"*: Ibid., 81.
183   *"cerebral gymnastics"*: Ibid., 87.
183   *"everything communicates"*: Fresquet Febrer.
183   *"the greatest anatomist"*: Sherrington, "A Memoir of Dr. Cajal," xii.
183   *"Cajal treated"*: Ibid., viii–vix.
183   *"dramatic history"*: Ramón y Cajal quoted in Lewy Rodríguez, *Así era Cajal*, 47.
183   *"groped to find"*: Ibid., xiv.
183   *"protoplasmic kisses"*: Ramón y Cajal, *Recollections*, 373.
183   *"If we would enter"*: Sherrington, "A Memoir of Dr. Cajal," xiv.
183   *"Listening to him"*: Ibid.
184   *Saint-Germain*: Kurzban.

185 *"The wonder of our time"*: Otis, *Networking*, 11.
185 *"It is a kinship"*: Ibid.
185 *Samuel Morse*: Ibid., 120–21.
185 *Werner von Siemens*: Ibid.
185 *"Is there in our parks"*: Ramón y Cajal, *Recollections*, 364.
185 *"in the manner of moss"*: Quoted in Palay and Chan-Palay, 142.
185 *"a short, delicate"*: Ibid.
185 *"knotty thickenings"*: Peters and Rockland, 81.
185 *"nest endings"*: Garcia-Lopez, 198.
185 *"climbing fibers"*: Ramón y Cajal, *Recollections*, 370.
185 *"a forest"*: Catani and Sandrone, 139.
185 *"orchard"*: Laín Entralgo, *Cajal por sus cuatro costados*, 32.
185 *"inextricable grove"*: Ibid., 44.
185 *"full grown forest"*: Ramón y Cajal, *Recollections*, 324.
185 *"young wood"*: Ibid.
186 *"terrifying jungle"*: Ibid., 431.
186 *"the tangled jungle"*: Ramón y Cajal, *Discursos leídos . . . 7 de mayo 1922.*
186 *"I regret that I did not first see"*: Ibid., 8.
186 *"A continuous pre-established"*: Quoted in DeFelipe, *Cajal's Neuronal Forest*, 42.
186 *"dynamism" [etc.]*: Ramón y Cajal quoted in DeFelipe, "Brain Plasticity and Mental Processes: Cajal Again," 811.
187 *"invisible and molecular"*: Ferreira et al., 5.
187 *was not translated*: Nubiola, 7.

19. "Grand Passion in Service"

188 *"long regarded"*: Shepherd, 278.
188 *"from celebrated histologists"*: Ibid.
188 *"the determination"*: Ramón y Cajal, *Recollections*, 440.
189 plumillas: Fundación de Ciencias y Salud, "Cajal y su legado."
189 *"the surface"*: Quoted and translated by Yuste.
189 *"onus of proof"*: Shepherd.
189 *"To discover the brain"*: Ramón y Cajal, *Recollections*, 305.
191 *Crock Shop*: "Cacharrería"; see Lewy Rodríguez, *Santiago Ramón y Cajal*, 95.
191 *"aristocracy of the intellectuals"*: Martinez del Campo.
191 *"the brain of Spain"*: Cortezo.
191 *The Athenaeum began to focus*: Villacorta Baños, 101.
192 *"agitate"*: Moret, 1.
192 *"peseta per minute"*: Villacorta Baños, 115.
192 *"Structure and Activity"*: Garcia Camarero.
192 *221 students*: Ibid.

192   *"organ of equilibrium"*: Cortezo.
192   *"extraordinary interest"*: Ibid.
192   *"amazing faculties"*: Lewy Rodríguez, "Cajal: Pensamiento social, espiritu democrático."
193   *"How can he be"*: Moreno.
193   *"to awaken"*: Ramón y Cajal, "Discurso del Sr. D. Santiago Ramón y Cajal: Fundamentos racionales y condiciones técnicas de la investigación biológica."
193   *"I got no such"*: Ramón y Cajal, *Advice for a Young Investigator*, xiii.
193   *"the living logic"*: Cannon, *Explorer of the Human Brain*, 227.
193   *"grand passion"*: *Ramón y Cajal, 1852-1934: Expedientes Administrativos de Grandes Españoles*, 303; my translation.
193   *"fervor extends"*: Ramón y Cajal, *Advice*, 45.
194   *"has an entirely"*: Ibid.
194   *"eager to enhance"*: Ibid.
194   *"Aside from its determinism"*: *El Día*, December 5, 1897.
194   *"the great Spanish savant"*: *El Globo*, December 5, 1897.
194   *"the sketch"*: Rodríguez Mourelo, 14.
194   *"to fight war"*: Offner, 12.
194   *"last peseta"*: Quoted in Smith, 9.
195   *Within their first two months*: John Lawrence Tone, 8.
195   *for every battlefield*: Zeigler.
195   *only way to lose*: Balfour, *The End of the Spanish Empire, 1898–1923*, 33.
195   *four hundred thousand*: Sartorius, 202.
195   *one hundred and fifty thousand*: Pitzer.
195   *Dante's* Inferno: de Quesada, 332.
195   *"war of extermination"*: *El Liberal*, October 26, 1898.
195   *"almost in sight"*: Louis A. Pérez, 27.
195   *"natural appendage"*: John Quincy Adams quoted in Gott, 58.
195   *Millions of U.S. dollars*: Library of Congress.
195   *"It was not civilized"*: McKinley.
195   *vying for the highest circulation*: Ibid., 85.
195   *yellow journalism*: PBS.
196   *savage rapist*: Sternlicht, 43.
196   *"the cause must be removed"*: Companys Monclús, 325.

20. "From Catastrophe to Catastrophe"

197   *"a scientific fight"*: Lewy Rodríguez, *Santiago Ramón y Cajal*, 85.
197   *"bewildering meshwork"*: Ramón y Cajal, *Comparative Study*, 347.
197   *"marvelous alchemy"*: Bohren and Clothiaux, 213.
199   *photograph of the two men*: Epistolario, 45.
199   *"My bugle"*: Balfour, *The End of the Spanish Empire, 1898–1923*, 44.

199 *"The Great Sin"*: "El Gran Pecado," *El Imparcial,* July 3, 1898.
199 *"common cloud"*: Rodríguez Quiroga.
199 *"an epilogue"*: "Dolor y Gloria," *Heraldo de Madrid,* July 3, 1898.
199 *"Everything is broken"*: Quoted in Balfour, "Riot, Regeneration and Reaction," 406.
199 *"The Nation Speaks"*: "Habla el país," *El Liberal,* October 26, 1898.
200 *"renounce our* thuggery": Ibid.
200 *"The misfortune of a country"*: Ramón y Cajal, *Recollections,* 465.
200 *"blunders and derails"*: See Sánchez Granjel.
200 *"Cajal comes"*: *El Correo Militar,* October 28, 1898, 1.
200 *"Shoemaker to your shoes"*: Ibid.
200 *"Despite this"*: Letter to Retzius, September 10, 1898, *Epistolario,* 497.
200 *In 1896 and 1897*: Rodríguez Quiroga.
200 *"a death trance"*: Ramón y Cajal, *Recollections of My Life,* 467.
201 *"considerably more"*: Durán Espuny.
201 *Two-thirds*: Balfour, *The End of the Spanish Empire, 1898–1923,* 53.
202 *Illiteracy remained*: Viñao Frago, 586.
202 *market squares*: Balfour, *The End of the Spanish Empire, 1898–1923,* 111.
203 *Some blamed*: Balfour, "Riot, Regeneration and Reaction."
203 *"[that] enigma"*: Ramón y Cajal, "Prólogo."
204 *"convoluted brain warp"*: Ramón Alonso, "Cajal, explorador."
204 *"sappers"*: Ibid.
204 *maternity hospital*: See letter to Salomon Eberhard Henschen, September 1929, *Epistolario,* 402.
204 *colitis*: Letter to Gustaf Retzius, January 28, 1900, *Epistolario,* 500.
204 *"profoundly surprised"*: Ramón y Cajal, *Recollections of My Life,* 484.
205 *propensity for migraines*: Rodríguez Lafora.
205 *"the stupendous city"*: Ramón y Cajal, *Recollections,* 486–87.
205 *"aerial steps"*: Ibid.
206 *"We live in America"*: Ibid., 490.
206 *"a Spaniard with"*: Palanca.
206 *"Rarely have I"*: Letter from Stephen Salisbury, October 24, 1899, *Epistolario,* 803–804.
206 *"known so intimately"*: Haines.
206 *"Spanish Scientist Will Speak Today"*: Haines.
206 *"blazed new paths"*: Ibid.
206 *"Comparative Study of the Sensory Areas"*: Story, 311.
206 *"The brain is only"*: Ibid., 382.

21. "The Mysterious Butterflies"
207 *an admitted hypochondriac*: See Ramón y Cajal, *Recollections,* 505.
207 *thwarted by chimneys*: Ibid., 505.

207 *"their needles"*: Ramón y Cajal, *Charlas*, 69.
207 *under the sky*: Fernández de Molina.
207 *"Whatever the preoccupation"*: Ramón y Cajal, *Recollections*, 142–43.
208 *twenty-minute ride*: See de Montalbán, 388.
208 cochifras: "Los Maestros: Ramón y Cajal," in *Por Esos Mundos*, November 1904.
208 *Every resident*: Ibid., 387–88.
208 *"being bored"*: Durán Muñoz and Alonso Burón, *Cajal: Vida y obra*, 329.
209 *beans, peas, and strawberries*: Ibid., 315.
209 *"Spain is truly"*: Ibid., 315.
209 *"Like the entomologist"*: Ramón y Cajal, *Recollections*, 363.
209 *"Congratulations"*: "Hablando con Cajal," *Heraldo de Madrid*, August 9, 1900.
209 *"in the nick"*: Letter to Dr. Decref, August 11, 1900, *Epistolario*, 974.
210 *"a very illustrious man"*: See *Heraldo de Aragon*, August 18, 1900.
210 *"It's Spanish"*: "Ramón y Cajal," *La correspondencia médica*, August 19, 1900.
210 *"of such cardinal importance"*: *Heraldo de Aragon*, August 18, 1900.
210 *"The winner"*: Rivas Santiago.
210 *national triumph*: *Blanco y Negro*, August 18, 1900, 10.
210 *"Once again the name"*: Ángel Del Río, "Congreso Internacional de Medicina: Premio á Cajal," *El Liberal*, August 10, 1900, 1.
211 *"Who is D. Santiago Ramón y Cajal?"*: Márquez, 218.
211 *sent a reporter*: A. Muñoz, "Hablando con Cajal," in *Heraldo de Madrid*, August 9, 1900.
211 *"Look"*: Ibid.
211 *chemical composition*: "A Ramón y Cajal," *El Imparcial*, October 4, 1900.
211 *"but that is not"*: Ibid.
211 *"When one thinks"*: *Por Esos Mundos*, November 1904, 388.
211 *"And now"*: *El Imparcial*, October 4, 1900.
212 *"It must be advised"*: Ibid.
212 *a banquet*: See "En honor de Ramón y Cajal," *La Correspondencia Militar*, October 4, 1900; de Cávia, "A Ramón y Cajal," *El Imparcial*, October 4, 1900; *Heraldo de Madrid*, October 5, 1900.
212 *"I say to you"*: See Durán Muñoz and Alonso Burón, *Cajal: Vida y obra*, 281.
212 *"the blessed history"*: Ibid.
212 *"leisure and rest times"*: A. de Montalbán, "Los maestros: Ramón y Cajal," *Por Esos Mundos*, November 1904, 39.
213 *"the story of a common life"*: "Prólogo a la segunda edición."
213 *"a teaching"*: Quoted in Jiménez, 19.

213 *"the rich experience"*: Azorín, 76.
213 *"fundamental to the ideology"*: Ibid., 75.
213 *"truly literary"*: Ibid., 77.
213 *"Cajal is the most important"*: Pío Baroja quoted in Calzada.
213 *cultured and urbane*: This description of Cajal's linguistic affect came from a conversation with Pablo García López.
213 *"lot of muck"*: Letter to Azorín, January 5, 1915, *Epistolario*, 628.
213 *"My arrogance"*: "Advertencia al lector," in *Recuerdos de mi vida*, viii.
213 *"more interesting"*: Letter to Azorín, January 5, 1915, *Epistolario*, 628.
214 *"on par with"*: Quoted in Sánchez Martin.
214 *two thousand pesetas*: Bartual Pastor.
214 *"Surely there is no savant"*: Muñoz, "Hablando con Cajal," 1.
214 *A proposal*: González de Pablo, "The New Lifeblood of the Spanish Nation," 29.
214 *"Spain should do"*: *El Imparcial*, August 11, 1900, 1.
214 *The major liberal newspapers*: González de Pablo, "The New Lifeblood of the Spanish Nation," 35.
214 *"Spain, not Cajal"*: A. Muñoz, "Hablando con Cajal."
214 *Councilmen from Zaragoza*: Ibid.
214 *eighty thousand pesetas*: González de Pablo, "The New Lifeblood of the Spanish Nation," 33.
215 *"Fortunately"*: Letter to Retzius, January 23, 1902, *Epistolario*, 502.
215 *"[He] makes a personal impression"*: Barker, 403.
215 *"When working"*: "Intimidados y curiosidades," *La Libertad*, June 6, 1932.
215 *Tello was convinced*: Durán Muñoz and Alonso Burón, *Cajal: Vida y obra*, 237.

22. "The Summit of My Inquisitive Activity"

216 *"We washed our"*: Granjel, 29.
216 *"did not move"*: Rodríguez Quiroga.
216 *"We regenerators"*: Ramón y Cajal, *Recollections of My Life*, 468.
217 *"always on the margins"*: Pedro Sainz Rodríguez quoted in Ramón Alonso, "Cajal, politico."
217 *"Is it urgent?"*: Artigas Sanza.
217 *Cajal brought*: Lewy Rodríguez, *Santiago Ramón y Cajal*, 71.
217 *"Old People"*: Durán Muñoz and Alonso Burón, *Cajal: Vida y obra*, 238.
217 *chops and steaks*: Ford, *A Handbook for Travellers in Spain*, 28.
217 *"If those mental tournaments"*: Natalio Rivas quoted in Duran Muñoz and Alonso Burón, *Cajal: Vida y obra*, 284.
217 *"to the excellency"*: Ramón y Cajal, "Prólogo" in Zapata.
218 *"You ask me"*: Ibid.

218 *Between 1900 and 1904*: Ramón y Cajal, *Recollections*, 536.
218 *crayfish*: Parker.
218 *"threaded"*: Frixione, 396.
219 *"We must stop"*: Ibid.
219 *"fanatics"*: Ramón y Cajal, *Recollections*, 536.
219 *"an anarchical"*: Ibid.
219 *"ancient banner"*: Ibid.
219 *"contagion"*: Ibid., 537.
219 *"virulent and widespread"*: Ibid.
219 *"astonishment"*: Frixione, 396.
219 *"great progress"*: Ibid.
219 *"He would never say"*: Rodríguez Lafora.
219 *"Now it is necessary"*: Frixione, 396.
220 *"I found myself"*: Frixione, 398.
220 *"It was interesting"*: Barker, 403.
220 *"understand him better"*: Letter from Bethe, October 12, 1904, *Epistolario*, 325.
221 *"What a disappointment!"*: Ramón y Cajal, *Recollections*, 519.
221 *"pale filaments"*: Frixione, 396.
221 *"as creepers"*: Quoted in ibid., 404.
221 *"I burned"*: Ramón y Cajal, *Recollections*, 520.
221 *"I was seized"*: Ibid.
222 *"hot, free nitrate"*: DeFelipe et al., 11.
222 *"splendidly impregnated"*: Ramón y Cajal, *Recollections*, 523.
222 *"I did not sleep"*: Ramón y Cajal, *Recollections of My Life: Part 1*, 523.
222 *"Nor did I sleep"*: Ibid.
222 *"brain throbbing"*: Ibid.
222 *"complicated, time-robbing"*: Barker, 403.
222 *"The simplification"*: Ibid.
223 *"Just like in"*: del Río Hortega, "Arte y artificio de la ciencia histológica," 196.
223 *"For the histologist"*: Ramón y Cajal, *Recollections*, 526.
223 *"absolutely objective"*: Campos-Bueno and Martín-Araguz, 17.
223 *"panreticularism"*: Frixione, 399.
224 *"Abandon all hope"*: Ibid.
224 *In the vote*: Campos-Bueno and Martín-Araguz, 12.
224 *"Essentially one thing"*: Quoted in Todes, 264.
224 *"explain the mechanism"*: Quoted in ibid., 1.
224 *"Now we must learn Russian"*: Ibid., 515.
224 *"the summit"*: Quoted in García Carraffa and García Carraffa, 199.
225 *bronchopneumonia*: in Durán Muñoz and Alonso Burón, *Cajal: Vida y obra*, 179.

23. "The Most Highly Organized Structure"

226 *three thousand pesetas*: Pasik and Pasik, "Prologue," in Ramón y Cajal, *Texture of the Nervous System of Man and the Vertebrates*.
226 *"the respectful praise"*: Ramón y Cajal, *Recollections*, 530.
226 *"trophy placed"*: Calvo Roy, 51.
227 *"the greatest Spanish"*: Kölliker quoted in Durán Muñoz and Alonso Burón, *Cajal: Vida y obra*, 228.
227 *"Genius lives on"*: Catani and Sandrone, 31.
227 *"an infinite company"*: Hooke.
228 *"thread-like"*: Shepherd, *Foundations*, 16.
228 *"an extremely fine"*: Ibid.
228 *"knitted tangle"*: Ibid., 52.
228 *"a blanket"*: Ibid., 54.
228 *"a kind of braid-work"*: Ibid., 53.
228 *"a sheet of muslin"*: Ibid., 62.
228 *"a dreadful felt-like maze"*: Ibid., 115.
228 *"an echo"*: Lewy Rodríguez, *Asi era Cajal*, 47.
229 *"the most highly"*: Quoted in Pieribone et al., 231.
229 *"still largely"*: Ramón y Cajal, *Histology of the Nervous System of Man and Vertebrates*, 7.
229 *"this ideal"*: Quoted in Marijuán.
229 *"It appears to us"*: Quoted in Azmitia, 401–402.
230 *"I never dared"*: Ramón y Cajal, *Vacation Stories*, xxi.
230 *He worried*: Otis, *Membranes*, 87.
230 *"literary trifles"*: Ibid.
230 *Royal Academy of Letters*: Lazar.
230 *"The truth is"*: Ramón y Cajal, *Recollections*, 556.
231 *"Minister?"*: Olmet and de Torres Bernal, 17.
231 *"Beware of orators!"*: Ramón y Cajal, *Charlas*, 18.
231 *"Thank you!"*: F. González-Rigabert, 6.
232 *"At a recent and solemn"*: Quoted in Lewy Rodríguez, "Ramón y Cajal. Madrid, 'tierra de amigos.'"
232 *"duplicity"*: Calzada.
232 *"Cajal serves"*: Lewy Rodríguez, *Santiago Ramón y Cajal*, 111.
232 *"unhealthy worship"*: Cajal, "Prólogo," in *Superorganic Evolution*.
232 *"The world is"*: Ibid.
232 *"The only legitimate"*: Ibid.
232 *"The earth for all"*: Ibid.
233 *"that aurea mediocritas"*: Ibid.
233 *"parasites"*: Ramón y Cajal Junquera, "D. Santiago, mi abuelo."
233 *"unconsciously"*: Lewy Rodríguez, "Las ideas sociales de Ramón y Cajal."
233 *"Pure aristocratic"*: Ibid.

233 *"notorious leftist tendencies"*: Otis, *Vacation Stories*, 17.
233 *"We all feel"*: Ramón y Cajal, "Prológo," in Maestre, *Introducción al estudio de la psicología positive*.
233 *"the subject of science"*: Ramón y Cajal, *Programa*.
234 *"all can taste"*: Ramón y Cajal, "Prólogo," in Lluria Despau, *Superorganic Evolution*, xiii.

24. "A Cruel Irony of Fate"

235 *"made fecund"*: Ovid.
236 *Aristotle*: See Odelberg, 1068.
236 *Fishermen had*: Lenhoff and Lenhoff, 60.
236 *To fight back*: Rowland, 436.
236 *"the great controversy"*: Ramón y Cajal, *Recollections*, 535.
237 *"prolonged agony"*: Ramón y Cajal, *Cajal's Degeneration and Regeneration of the Nervous System*, 100.
237 *"death struggle"*: Ibid., 14.
237 *"sprouts" in the process*: Ibid., 88.
237 *"completely destroyed"*: Ibid., 106.
237 *"exuberant"*: Ibid., 51.
237 *"eventual collision"*: Ibid.
237 *"sedentary reaction"*: Ibid., 615.
237 *"an obstacle race"*: Ibid., 8.
237 *"invasion"*: Ibid., 109.
237 *"ephemeral and frustrated"*: Ibid., 2.
238 *"by devious routes"*: Ibid., 7.
238 *"straying"*: Ibid., 80.
238 *"rectification"*: Ibid.
238 *"we can say"*: Ibid., 572.
238 *"something appears"*: Ibid., 615.
238 *"does not obey"*: Ibid., 55.
238 *"growth promoting stimuli"*: Quoted in Reinoso-Suarez, 32.
238 *"physical influence"*: Ramón y Cajal, *Cajal's Degeneration and Regeneration of the Nervous System*, 92.
238 *"lack [of] growth"*: Reinoso-Suarez, 31.
238 Carolinsche Institut: Loewy, 9.
238 *playing a trick*: Olmedo de Cerdà, 13.
238 *"The Nobel Prize"*: El Imparcial, October 27, 1906, 1.
238 *Machaquito*: See Calvo Roy, 173.
239 *"I do not trust"*: Bartual Pastor, 20.
239 *"undeserved, abnormal"*: Quoted in Ramón Alonso, "Cajal y el dinero."
239 *"a heart of steel"*: Quoted in Cannon, 216.
239 *changing kroners*: See Calvo Roy, 174.

239 *"a delirious ovation"*: See *El Liberal*, October 25, 1906.

239 *Stunned by emotion*: "Ramón y Cajal," *El Liberal*, October 28, 1906.

239 *"Long live Cajal!"*: Esteban.

240 *"a wise man"*: Tedeschi, "El Premio Nobel: Ramón y Cajal," *El Imparcial*, October 27, 1906, 1.

240 *"is not a country"*: "Ramón y Cajal en su laboratorio," *El País*, October 28, 1906, 1.

240 *"like a pure"*: "El Premio Nobel: Ramón y Cajal," *El Imparcial*, October 27, 1906, 1.

240 *"the intimate"*: Tedeschi, "El Premio Nobel: Ramón y Cajal," *El Imparcial*, October 27, 1906.

240 *Cajal had not expected*: Letter to Retzius, March 24, 1901, *Epistolario*, 501.

240 *"Report of Titles"*: Mazzarello, *The Hidden Structure*, 277.

240 "Multa non prova il multum": Ibid.

241 *"Not even in my dreams"*: Mazzarello, *Golgi*, 309.

241 *"[My] nomination"*: Letter to León Corral y Maestro, October 17, 1904, *Epistolario*, 121.

241 *"Let us cultivate"*: Ramón y Cajal, *Advice*, 19.

241 *"If the achievements"*: Gunnar Grant, "How the 1906 Nobel Prize in Physiology or Medicine Was Shared Between Golgi and Cajal," 493.

242 *"[He] has not served"*: Ibid., 494.

242 *"the Latin race"*: Kock.

242 *"I thought that"*: Quoted in Gunnar Grant, "How the 1906 Nobel Prize in Physiology or Medicine Was Shared Between Golgi and Cajal," 495.

242 *"the key to open"*: Quoted in Olmet and de Torres Bernal, 353.

243 *"Mr. Ramón y Cajal"*: Mazzarello, *The Hidden Structure*, 311.

243 *"It was neither"*: Ramón y Cajal, *Recollections*, 550.

244 *"And it would"*: Rapport, 161–62.

244 *"What a cruel"*: Quoted in Finger, 215.

244 *"the celebrated scholar"*: Mazzarello, *Golgi*, 347.

244 *"Have you read"*: Mazzarello, *The Hidden Structure*, 277.

245 *The Grand Hotel*: Henschen.

245 *Lina Golgi*: Mazzarello, *The Hidden Structure*, 314.

245 *obelisks*: Henschen.

246 *"many elegant ladies"*: Nicolelis, 41.

246 *"This year's"*: Ibid., 42.

247 *"was a pompous"*: Ramón y Cajal, *Recollections*, 550.

247 *a seven-course*: See "Nobel Banquet Menu 1906."

247 *"behaved extremely nicely"*: Mazzarello, "Camillo Golgi's Scientific Relationships with Scandinavian Scientists," 159.

247 *"displayed . . . Olympic pride"*: Mazzarello, *The Hidden Structure*, 314.

247 *"is generally recognized"*: Quoted in Valenstein, 3.
247 *"articles of faith"*: Daston and Galison, *Objectivity*, 116.
247 *"We suppose that"*: Quoted in Otis, *Networking*, 58.
248 *"This theory should"*: Golgi.
248 *"bold and clear"*: Berlucchi.
248 *"While I admire"*: Golgi, 192.
248 *"diffuse nerve network"*: Ibid.
248 *"a true nerve organ"*: Ibid.
248 *"concrete anatomical fact"*: Ibid., 193.
248 *"cannot claim"*: Letter from Golgi, undated, *Epistolario*, 395.
248 *"noble, friendly"*: Grant, "Gustaf Retzius and Camillo Golgi," 161.
249 *"I was trembling"*: de Carlos Segovia, 15.
249 *"The Structure and Connections"*: Ramón y Cajal, "The Structure and Connections of Neurons."
249 *"twenty-five years"*: Ibid., 221.
249 *"We mourn"*: Ibid., 253.
250 *"My main wish"*: Mazzarello, *The Hidden Structure*, 311.
250 *"free histology"*: Otis, *Networking*, 63.
250 *"the only opinion"*: Quoted in Bennett, 20.
250 desligar: Otis, *Networking*, 63.
250 *surprisingly illogical*: See Piccolino.
251 *"despotic"*: Otis, *Networking*, 63.
251 *"seductive"*: Ibid.
251 *"projecting a priori theories"*: Letter to de Nó, May 18, 1926, *Epistolario*, 91.
251 *"exquisite suggestibility"*: Ramón y Cajal, *El mundo visto*, 162.
251 *his share of the prize money*: Gunnar Grant, "How the 1906 Nobel Prize in Physiology or Medicine Was Shared Between Golgi and Cajal," 98.
251 *when Golgi arrived*: Mazzarello, *Golgi*, 257.

25. "To Defend the Truth"

252 *"Almost all"*: Tello, 120.
252 sacatintas: Durán Muñoz and Alonso Burón, *Cajal: Vida y obra*, 321.
252 *"I will guard"*: Letter from Luis de Zulueta y Escolano, April 26, 1919, *Epistolario*, 749.
252 *"the greatest bacteriologist"*: Triarhou, "Ramón y Cajal as an Analytical Chemist of Bottled Water?," 2.
252 *medicinal wines [etc.]*: Quoted in ibid., 7.
252 "analyzed, or consulted": Ibid., 2.
253 *"How My Modest Name"*: Quoted in ibid., 7.
253 *"Let the aforementioned"*: Ibid., 8.
253 *"Erase my name"*: Ibid.

253 *fruits and vegetables*: Durán Espuny.

253 *One time*: See Fernández Aldama, 4.

253 *Nearly every city*: Durán Muñoz and Alonso Burón, *Cajal: Vida y obra*, 303.

253 *Zaragoza [etc.]*: Fresquet Febrer, 23.

254 *A petition*: See *El Sol*, June 21, 1931.

254 *"People met him"*: María Ángeles Ramón y Cajal Junquera quoted in "La herencia de Cajal," *El Periódico de Aragon*, November 28, 1999.

254 *"What can we do"*: Mariano de Cavia, "Actualidad," *El Imparcial*, October 29, 1906, 1.

254 *"One never has"*: "El doctor Ramón y Cajal," *Blanco y Negro*, February 24, 1906, 9.

254 *Newspaper articles*: Fresquet Febrer, 23.

254 *The annual funding*: Ibid.

255 *"I ask permission"*: "Presupuesto de gastos de la Sección 7a, Ministerio de la Gobernación, para 1911," 1219.

256 *"and I want this distrust"*: Ibid.

256 *"Frankly, all of this"*: Ibid.

256 *"The majority of those"*: Ibid., 1220.

256 *out of his own pocket*: Cortezo, 206.

256 *His only activity*: *Diario de las Sesiones de Cortes*, October 15, 1909.

257 *four-month expedition*: López Piñero, *Santiago Ramón y Cajal*, 319.

257 *"The principal mission"*: Durán Muñoz and Alonso Burón, *Cajal: Escritos inéditos*, 148.

257 *"It is wise and prudent"*: Quoted in ibid., 151.

257 *"Only science"*: Ibid.

257 *"war in accord"*: Ponce de León, v.

258 *"a lover of change"*: María Ángeles Ramón y Cajal Junquera quoted in *Heraldo de Aragon*, November 28, 1999.

259 *The ceilings*: "Santiago Ramón y Cajal," *Alma Historia*, vol. 16, 1999.

259 *"It was the home"*: Lewy Rodríguez, "Santiago Ramón y Cajal y la juventud española."

259 *Silveria found herself*: "Silvería Fañanás y Garcia," *Heraldo de Aragon*.

259 *had to play*: Cabot, "Silveria."

259 *Escola Pia*: Ramón y Cajal Junquera.

259 *Cajal afforded*: Ibid.

259 *The youngest child*: Durán Muñoz and Alonso Burón, *Cajal: Vida y obra*, 193.

259 *Jorge, headstrong*: Ibid., 340.

260 *When Santiago was a boy*: Ramón y Cajal Junquera.

260 *"Only photography"*: Ramón y Cajal, *La fotografía de los colores*.

261 *"Let us not talk"*: Quoted in Rapport, 190.

261 *"Neuroglia are found"*: Cajal quoted in del Río Hortega, *El maestro y yo*, 39.

261 *"a sort of putty"*: Jabr.

262 nervenkitt: Verkhratsky and Parpura, 14.

262 *"slimy" or "sticky"*: *Glial Physiology and Pathophysiology*, ch. 1.4.

263 *Achúcarro would ask*: See ibid., 39.

263 *the subsequent dispute*: See Rodríguez Lafora, 32.

263 *"a new class"*: Rezaie and Hanisch, 12.

263 *"third element"*: García-Marín et al., 374.

263 *"profoundly exhausted"*: Laín Entralgo, *Cajal por sus cuatro costados*, 35.

263 *"counteracted at one stroke"*: Ramón y Cajal, *Cajal's Degeneration and Regeneration of the Nervous System*, 19.

264 *printed on special paper*: Ramón y Cajal, *Recollections*, 575.

264 *"aborted restorative processes"*: Ramón y Cajal, *Cajal's Degeneration and Regeneration of the Nervous System*, 12.

264 *"impoverished"*: Ibid.

264 *"In adult centers"*: Quoted in Colucci-D'Amato et al., 136.

26. "The Unfathomable Mystery of Life"

265 *"overwhelmed with horror"*: Quoted in López Piñero, *Santiago Ramón y Cajal*, 333.

265 "THE EUROPEAN WAR": *El Imparcial*, August 1, 1914, 1.

265 "GERMANY AGAINST ALL": *El Heraldo de Madrid*, August 14, 1914, 1.

265 "ITALY MOBILIZES": "Italia moviliza," *El Imparcial*, May 23, 1915, 1.

265 "THE RESPONSE": "La respuesta de los Estados Unidos," *El Liberal*, February 3, 1917, 1.

265 "FRESH MEAT": "Carne Fresca," *Heraldo de Madrid*, February 19, 1915.

265 *"The horrendous European war"*: Ramón y Cajal, *Recollections*, 583.

266 *Spain remained neutral*: Romero.

266 *"Without being at war"*: García Sanz.

267 *about two hundred*: "Blame for the Sack of Louvain."

267 *more than three hundred thousand*: "'Sack of Louvain—Awful Holocaust.'"

267 *Van Gehuchten*: See Bhattacharyya, 84.

267 *"his oldest and most persistent"*: Quoted in Shepherd, 155.

268 *"aristocratic disdain"*: Ramón y Cajal, *Recollections*, 579.

268 *"Life never succeeded"*: Ibid., 576.

268 *"The complexity of the insect"*: Ibid., 577.

268 *"The more I study"*: Ramón y Cajal, *Advice*, 27.

268 *"terrifying sensation of the unfathomable"*: Ramón y Cajal, *Recollections*, 576.

269 *"I have to admit"*: Quoted in Durán Muñoz and Alonso Burón, *Cajal: Vida y obra*, 355.
269 *"Our nerve cells"*: Ibid., 356.
269 *"evolutionary resistance"*: Ibid.
269 *"excruciating biological fact"*: Ibid.
269 *"destructive phase"*: Ibid.
269 *"In about twenty"*: Ibid., 357.
269 *"Our descendants"*: Ramón y Cajal, *In the Aftermath of Peace*.
269 *"an incomparable observer"*: Susan T. Grant, 397.
270 *"the ability to triumph"*: Quoted in Rapport, 190.
270 *"ants have bad memories"*: Cajal Legacy archives, box 3.2, folder 8.05A.
270 *"ants help the wounded"*: Ibid.
270 *"if ants could talk"*: Ibid.
270 *"childlike joy"*: del Río Hortega, *El maestro y yo*, 48.
271 *"live almost"*: Ramón y Cajal, "Letter from a Slave-Making Ant."
271 *"exceptional ants"*: Ibid.
271 *"Nothing transcendental"*: Ibid.

27. *"I Drown and I Awaken"*

272 *"My friend"*: Ramón y Cajal, *El mundo visto*, 30.
272 all that he: Durán Muñoz and Alonso Burón, *Cajal: Vida y obra*, 359.
272 *"Some future!"*: Giménez-Roldán, "Cajal's Unbearable Cephalalgias," 5.
273 *"I must now"*: Letter to Unamuno, March 6, 1917, *Epistolario*, 672.
273 *"freezing my thoughts"*: Giménez-Roldán, "Cajal's Unbearable Cephalalgias" 4.
273 *"[He] only believes"*: See "Madrid: Our Regular Correspondent," 425.
273 *"prone to pessimism"*: Ibid.
273 *"this menace"*: Letter to Carlos María Cortezo, January 9, 1926, *Epistolario*, 775.
273 a yellow parasol: See Calvo Roy, 177.
273 *"permanent brain congestion"*: Letter to Carlos María Cortezo, January 9, 1926, *Epistolario*, 776.
273 Before his illness: Santiago Ramón y Cajal Junquera, "D. Santiago, mi abuelo."
273 Lord Byron: Ibid.
273 I am sitting: Ehrlich, 119.
274 *"Of my eight"*: Letter to Unamuno, March 6, 1917, *Epistolario*, 672.
274 *"to encourage"*: Letter to Carlos María Cortezo, January 9, 1926, *Epistolario*, 775.
274 left on his scarf: Ernesto Giménez Caballero in *El Sol*, February 6, 1926, 1.

274 *"I cannot get"*: Olmet and de Torres Bernal, 332.

274 *"The whole world"*: Ibid.

275 *"When your dear letter"*: Letter from Anna Retzius, July 13, 1919, *Epistolario*, 505–506.

275 *"I could bury"*: Letter to Carlos María Cortezo, January 9, 1926, *Epistolario*, 775.

276 *"unimaginable torture"*: Ramón y Cajal, *Recollections*, 593f3.

276 *"[Achúcarro] would have been"*: Quoted in José Ramón Alonso, "Achúcarro."

276 *His 1908 paper*: Ehrlich, 20.

277 *"At heart"*: Ibid., 24.

278 *"collective lies"*: Ibid., 26.

278 *"surly and somewhat"*: Ibid.

278 *"I am delivering"*: Ibid., 90.

278 *In a dream*: Ibid., 91.

279 Sea expedition: Ibid., 115.

28. "Those Poisoned Wounds"

280 *wine from his hometown*: See Durán Muñoz and Alonso Burón, *Cajal: Vida y obra*, 317.

280 *He liked chatting*: Ibid., 320.

280 *enjoyed bantering*: Ibid.

281 *"Ranero"*: del Río Hortega, *El maestro y yo*, 80; Lewy Rodríguez, "Plenario: Cajal."

281 *Cajal's personal address book*: Ibid.

282 *"attitude of insolence"*: Letter from del Río to Cajal, February 4, 1920, *Epistolario*, 132.

282 *"My dear friend"*: Letter from Cajal to del Río, February 8, 1920, *Epistolario*, 134.

282 *"You can always"*: Ibid., 135.

282 *"I feel as though"*: Letter from del Río to Cajal, October 15, 1920, *Epistolario*, 141.

282 *"In any case"*: Ibid., 144.

282 *"[Cajal's] precious name"*: del Río Hortega, *El maestro y yo*, 22.

282 *"His face with"*: Ibid., 29.

283 *"a long, tortuous"*: Tremblay et al.

283 *"I believe that your"*: del Río Hortega, *El maestro y yo*, 182.

283 *"the fear of"*: Ibid., 186.

284 *"bitter and severe"*: Ibid., 184.

284 *Del Río left*: See Durán Muñoz and Alonso Burón, *Cajal: Vida y obra*, 319.

284 *pigeon shit*: Giménez Caballero, *El Sol*, February 6, 1926, 1.

284 *"very great"*: Ibid.
285 *"Mortified pride"*: Letter to de Nó, April 10, 1926, *Epistolario*, 87.
285 *"Like every old man"*: Rapport, *Nerve Endings*, 191.
286 *"a thousand congratulations"*: Letter to del Río Hortega, May 4, 1926, *Epistolario*, 151.

29. "No Solemn Gatherings"

287 *"giant, suspended tears"*: Machado, 3.
287 *850 entries*: Sánchez-Perez.
287 *"unfortunate" word café*: Giménez-Roldán, "Cajal's Unbearable Cephalalgias," 36.
287 *"Why unfortunate?"*: Ibid.
288 *"El doctor Ramón y Cajal"*: Juez Vicente.
288 *so deaf that*: Durán Muñoz and Alonso Burón, *Cajal: Escritos inéditos*, 379.
289 *"I believe he"*: Valenciano Gayá.
289 *"Don Santiago, it's time"*: Lapuente Mateos.
289 *When the janitor erased*: Marañon Posadillo, 30.
289 *"I do not regret that"*: Ramón y Cajal, *Recollections*, 597.
289 *"painful odyssey"*: Quoted in Durán Muñoz and Alonso Burón, *Cajal: Vida y obra*, 366.
289 *"no solemn gatherings"*: Letter to Carlos María Cortezo, March 4, 1922, *Epistolario*, 226.
289 *"A Great and Stimulating Example"*: *El Imparcial*, May 2, 1922, 1.
289 *"He was an exploring diver"*: M. R. Blanco-Belmonte, *ABC*, May 2, 1922.
289 *"modern Columbus"*: Ibid.
289 *"Columbus was a man"*: Letter to Juan B. Terán, June 22, 1927, *Epistolario*, 667.
290 *"All these sentimental outbursts"*: Ramón y Cajal, *El Sol*, June 19, 1921, *Epistolario*, 830.
290 *"not finding himself"*: "La jubilación de Cajal," *La Libertad*, May 3, 1922, 5.
290 *"he found himself"*: "La jubilación de Cajal," *La Correspondencia de España*, May 2, 1922, 4.
290 *"My mischievous pranks"*: Letter to Eduardo de Rute, June 13, 1922, *Epistolario*, 965.
291 *got into an accident*: Durán Muñoz and Alonso Burón, *Cajal: Vida y obra*, 351.
291 *"Let me do it"*: Ibid.
291 *"Santiago Ramón y Cajal Declared"*: Frances, *Nuevo Mundo*, June 24, 1921.

291 ABC *would compare*: *ABC*, January 13, 1931, quoted in Calvo Roy, 213.
291 *The famous bullfighter*: Ibid.
291 *Cajal's financial needs*: Durán Muñoz and Alonso Burón, *Cajal: Vida y obra*, 322.
292 *He gave her*: Cortezo.
292 *The Ramón y Cajal neurology clinic*: Tello Muñoz, "Recuerdos de Cajal."
292 *"It impressed him"*: Falcón.
292 *"picturesque"*: del Río Hortega, *El maestro y yo*, 64–65.
292 *perfect his grammar*: Durán Muñoz y Alonso Burón, *Cajal: Vida y obra*, 348.
292 *colder and lonelier*: See Ramón y Cajal, *El mundo visto*, 30.
293 *"All of him"*: Alfredo Guillen quoted in Jiménez, 35.
293 *"This little old man"*: El Caballero Audaz, *La Esfera*, November 26, 1921, 18.
293 *"you could study"*: Buñuel, 52.
293 *"We liked each other"*: Ibid., 62.
294 *"a redneck"*: Ibid.
294 *"Don Santiago"*: Aub, 75.
295 *"a multiform"*: Ramón y Cajal, *El mundo visto*, 120.
295 *"pompous names"*: Ibid.
295 *"deliberate nonsense"*: Ibid., 123.
295 *"In art, as in science"*: Durán Muñoz and Alonso Burón, *Cajal: Escritos inéditos*, 398.
295 *"microencephalic"*: Ibid., 399.

30. "Marvelous Old Man"

297 *"Long live"*: Glick, 299.
297 *"marvelous"*: Ibid., 326.
297 *"From existing memory"*: Montes Santiago, 113.
298 *"Visit with Cajal"*: Quoted in ibid., 116.
298 *"Chicken meat"*: Quoted in Woolman, 102.
298 *"a loaded rifle"*: Harrington, 242.
299 *"brief parenthesis"*: Quoted in Humlebaek, 31.
299 *"every single rebel"*: Quoted in Quiroga, 34.
299 *"jeopardized the throne"*: Luis Ramón y Cajal, 74.
299 *"my Mussolini"*: Ben-Ami, "The Dictatorship of Primo de Rivera."
299 *important symbol*: Durán Muñoz and Alonso Burón, *Cajal: Escritos inéditos*, 10.
299 *"He is also"*: Van-Baumberghen, 323.
300 *"The politics of all or nothing"*: Letter to José Castillejo Duarte, June 28, 1926, *Epistolario*, 701.
300 *"yeast"*: Ibid.

300 *"Better a hair"*: Ibid.
300 *"Are you an Athenaeum member?"*: Cortezo.
301 *"An invitation"*: De Juan.
301 *"I have good relations"*: Letter to de Nó, October 12, 1926, *Epistolario*, 95.

31. "Statues of the Living"

302 *"new consciousness"*: Quoted in Johnson, 135.
302 *"My illustrious"*: Letter from Margarita Nelken, November 18, 1925, *Epistolario*, 637.
302 *75 percent*: Letter to Nelken (undated), *Epistolario*, 642. (After Nelken released the book early, which Cajal attributed to a "lack of courtesy," he renounced the project.)
302 *"My child"*: Quoted in Calvo Roy, 87.
303 *"A model of jealous"*: See Ramón y Cajal, *Cajal: Escritos inéditos*, 138.
303 *Cajal also welcomed women*: See Giné, et al.
303 *"cordial thanks"*: Ibid.
303 *"Wait for society"*: Ramón y Cajal, "La capacidad intelectual de las mujeres," 16.
303 *"In our home"*: Lucientes.
304 *"as a modest scientist"*: Ramón y Cajal, *La mujer*, 12.
304 *"The female type"*: Ibid., 13.
304 *"There are women"*: Ibid.
304 *"militant"*: Letter to Nelken (undated), *Epistolario*, 636.
304 *"What feminist extremists"*: Ramón y Cajal, *La mujer*, 155.
304 *"ugliness and old age"*: Ibid., 158.
304 *"You know your respectful"*: Letter to Margarita Nelken, 1925, *Epistolario*, 638.
304 *"a little sexist"*: Falcón.
304 *"weakened heart"*: Cabot, "Silveria."
305 *"fatigue and discouragement"*: Letter to Lenhossek, July 24, 1925, *Epistolario*, 420.
305 *"lacking equilibrium"*: Letter to Margarita Nelken, 1925, *Epistolario*, 638.
305 *"We live like queens"*: Quoted in Naño.
306 *"Very conservative"*: Ibid.
306 *"infantile puerility"*: García Quiepo de Llano, 265.
307 *"odd and deplorable"*: Letter to del Río, May 4, 1926, *Epistolario*, 152.
307 *"annoying and hyperbolic"*: Ibid.
307 *"undeserved and exorbitant"*: Letter to Lorenzo Millares, May 15, 1926, *Epistolario*, 953.
307 *"insincere and undeserved"*: Letter to Gonzalo Lafora, June 1926, *Epistolario*, 161.

307 *sent his daughters*: Luis Ramón y Cajal, 74.

307 *"They rob me"*: Letter to del Río Hortega, May 4, 1926, *Epistolario*, 152.

307 *"replete with cheap rhetoric"*: Letter to Lorente de Nó, May 18, 1926, *Epistolario*, 91.

308 *Crazy people*: "Ramón y Cajal, visto por su secretaria," *Estampa* (1934).

308 *he took pride in*: Durán Muñoz and Alonso Burón, *Cajal: Vida y obra*, 322.

308 *special hat*: Ibid.

308 *"Do you believe"*: Letter from Granell Herrero, October 10, 1926, *Epistolario*, 1022.

308 *sick cat*: Ibid.

308 *man from Petilla*: Letter from Sixto Sánchez, October 14, 1925, *Epistolario*, 1014.

308 *A schoolteacher*: Letter from Ramón Piquer, May 30, 1929, *Epistolario*, 1026.

308 *A Cuban man*: Letter from José F. Triana y Triana, January 1, 1929, *Epistolario*, 1027.

308 *"As for me"*: Letter to José Díaz Triana, January 22, 1929, *Epistolario*, 1027.

308 *"the fever and pains"*: Letter to Jean Turchini, July 15, 1926, *Epistolario*, 533.

308 *"Even though"*: Letter to Cándido Ruiz Martínez, September 22, 1926, *Epistolario*, 857.

308 *"Every day"*: Letter to de Nó, February 5, 1927, *Epistolario*, 100.

309 *"I disapprove"*: *Epistolario*, 949.

32. "The Self Has No Mirror"

310 *"Sleeping Pills"*: Lewy Rodríguez, "Santiago Ramón y Cajal y la juventud española."

310 *"a vortex"*: Ramón y Cajal, *El mundo visto*, 33.

310 *inactive neurons*: See Triarhou and Vivas, 84.

310 *Early pharmaceutical trials*: López-Muñoz et al., 329.

311 *A special desk*: Durán Muñoz and Alonso Burón, *Cajal: Vida y obra*, 380.

311 *"We ask God"*: Ramón y Cajal, *Charlas*, 80.

311 *The next morning*: Pucho.

311 *fourteen hours a day*: Giménez-Roldán, "Cajal's Unbearable Cephalalgias," 6.

311 *I cannot sleep*: Ehrlich, 110.

312 *"Believing that"*: Ibid., 91.

312 *"an energy"*: Ibid.

312 *"In sum"*: Ramón y Cajal, *Charlas*, 104.

312 *"Farewell tour"*: Letter to de Nó, March 28, 1927, *Epistolario*, 106.

312 *"My grandfather"*: María Ángeles Ramón y Cajal Junquera quoted in Moreno.

312 *yellow Buick*: Ibid.

312 *"It's him!"*: Ibañez.

313 *"We see him alone"*: Ibid.

313 *delivering fresh trout*: Garces Romeo.

313 *Pedrín*: See Sender, 1.

313 *"When she is ill"*: Díaz Morales, 11.

313 *"My summer house"*: Letter to de Nó, June 21, 1929, *Epistolario*, 113.

314 *"It is nothing"*: Cabot, "Silveria."

314 *she professed*: See Durán Muñoz and Alonso Burón, *Cajal: Escritos inéditos*, 325.

314 *"Half of Cajal"*: Ramón y Cajal, *Recollections*, 272.

314 *"deeply and fully"*: Luis Ramón y Cajal, 75.

314 *"condemned herself"*: Ramón y Cajal, *Recollections*, 272.

314 *"Only the unsurpassable"*: Ibid.

314 *"You exaggerate"*: See Durán Muñoz and Alonso Burón, *Cajal: Escritos inéditos*, 203.

33. "Searching for Themselves in the Secret"

315 *"with an air"*: Díaz Morales, 11.

315 *"Don't you know him?"*: Ibid.

315 *"There is no one"*: Ignacio Bolívar quoted in Lewy Rodríguez, "Las ideas sociales de Cajal."

316 CAFÉ ELIPA: Giménez-Roldán, "Cajal's Unbearable Cephalalgias," 8–9.

316 *lump of sugar*: "La herencia de Cajal," *El Periódico de Aragon*, November 28, 1999.

316 *The waiters stood*: Lewy Rodríguez, *Santiago Ramón y Cajal*, 115.

316 *"with its miseries"*: See Ramón y Cajal, *Charlas*, 83.

316 *"My Religion"*: Unamuno.

317 TBO: Calvo Roy, *Cajal*, 207.

317 *"Absent, fine"*: See Ramón Jiménez.

317 *"What is the nature"*: Pucho.

317 *"melancholic and careless"*: Ibid.

318 *"all Spanish intellectuals"*: *El Sol*, February 10, 1931.

319 *"a bottle of champagne"*: Quoted in Ben-Ami, *The Origins of the Second Republic in Spain*, 23.

320 *"Soldiers, the homeland"*: Lyrics from "Himno de Riego."

320 *"I do not know"*: See Durán Muñoz and Alonso Burón, *Cajal: Escritos inéditos*, 188.

320 *life expectancy*: Gómez-Redondo and Boe, 525.

320 *bowel movements*: Durán Muñoz and Alonso Burón, *Cajal: Escritos inéditos*, 375.

321 *"When does old age"*: Ramón y Cajal, *El mundo visto*, 3.

321 *the Himalayas*: Letter to de Nó, May 11, 1933, *Epistolario*, 119.

321 *rest on a bench*: Jiménez.

321 *On hot days*: Lewy Rodríguez, *Santiago Ramón y Cajal*, 71.

322 *"will soon get"*: Cobb, 174.

322 *"granular cement"*: Quoted in Jacobson, *Foundations of Neuroscience*, 15.

322 *the word* synapse: Ibid.

323 *orthodox determinist*: See Ramón y Cajal, "Leyes de la morfología y dinamismo de las células nerviosas."

34. "My Strength Is Exhausted"

325 *"a republic"*: Romero Salvadó, *The Spanish Civil War*, 31.

325 *"the Russia of the West"*: Quoted in Vincent, 106.

325 *"enlace the parts"*: Letter to Joaquín Costa, March 1901, *Epistolario*, 711.

325 *"If I could go back"*: Ramón y Cajal, *El mundo visto*, 115.

326 *"one thousand times"*: Quoted in Lewy Rodríguez, "Las ideas sociales de Cajal."

326 *Cajal offered*: Ibid.

326 *Cajal was not pleased*: See *Epistolario*, 427.

327 *on letterhead*: Ibid., 428.

327 *"I can ask"*: Merchán et al., 14.

327 *"I serve Spain"*: González-Rigabert, 6.

328 *"like Don Quixote"*: Leoz Ortín, "Homenaje Cajal en el primer centenario de su Nacimiento."

328 *Springer Press*: Letter to Castillejo, August 1930, *Epistolario*, 117.

328 *"In general"*: Pérez de Tudela Bueso, "Algunas reminiscencias de Santiago Ramón y Cajal."

328 Country life is refuge: Ramón y Cajal, *El mundo visto*, 182.

328 *"humiliated"*: Ibid.

329 *"a hermetic box"*: Durán Muñoz and Alonso Burón, *Cajal: Vida y obra*, 378.

329 *"You can say"*: Criado y Romero, "El sabio, visto por su ama de llaves," *Heraldo de Madrid*, October 18, 1934, 5.

329 *"spiritual grandchildren"*: del Río Hortega, *El maestro y yo*, 31.

329 *"I want the garlic soup"*: "La herencia de Cajal," *El Periódico de Aragon*, November 28, 1999.

329 *"How are my chickens?"*: Lucientes.

330 *"No more food"*: Durán Muñoz and Alonso Burón, *Cajal: Vida y obra*, 391.

330 at ten-forty-five: Ibid.

330 *"I leave you"*: "Ha muerto D. Santiago Ramón y Cajal," *El Sol*, October 18, 1934.

Epilogue

331 *mahogany coffin*: "El cadaver del sabio Ramón y Cajal ha recibido sepulture esta tarde," *La Nación*, October 18, 1934, 18.

331 *no crucifix*: Criado y Romero, "El sabio, visto por su ama de llaves," *Heraldo de Madrid*, October 18, 1934, 16.

331 *a book was placed*: "El entierro de Santiago Ramón y Cajal," *El Sol*, October 19, 1934, 8.

331 *little children marching*: "El entierro de don Santiago Ramón y Cajal," *Ahora*, October 19, 1934, 7.

331 *"purely civil"*: See Durán Muñoz and Alonso Burón, *Cajal: Escritos inéditos*, 318.

331 *A vanguard*: Ibid.

331 *carrying the coffin*: "El entierro de don Santiago Ramón y Cajal," 7.

332 *"with no type"*: Lucientes.

332 *"in this beloved land"*: Ibid.

332 Bury me: See Durán Muñoz and Alonso Burón, *Cajal: Escritos inéditos*, 334.

332 *"The burial has been"*: "El entierro de Santiago Ramón y Cajal."

333 *"It is a spiritual pain"*: "La pregunta del día," *La Libertad*, October 19, 1934, 5.

333 *"when she was a little girl"*: Ibid.

333 *"I did not understand"*: Ibid.

333 *"It can be said that"*: Quoted in *Ramón y Cajal, 1852–1934: Expedientes administrativos*, 36.

333 *"Be careful"*: "La herencia de Cajal," *El Periódico de Aragon*, November 28, 1999.

333 *On the nightstand*: Lucientes.

333 *Cajal asked his chauffeur*: Ibid.

333 *The last will*: See Durán Muñoz and Alonso Burón, *Cajal: Escritos inéditos*, 328–33.

334 *"which should have seen"*: Ramón y Cajal, *El mundo visto a los ochenta años*, 174n114.

334 *"who would have thought it?"*: Quoted in Durán Muñoz and Alonso Burón, *Cajal: Vida y obra*, 162.

334 *"Every man"*: Ramón y Cajal, *Advice for a Young Investigator*, xv.

336 *"The greatest dream"*: García-Albea.

336 *"to restore"*: Ansede.

336 *"sane nationalism"*: Sánchez Ron.

337 *"As long as"*: Quoted in Swanson, 11.
338 *"The job of the anatomist"*: Ramón y Cajal, *Neuron Theory*, 143.
338 *"The absence of"*: Palay, 198.
339 *"a general property"*: Jacobson, *Developmental Neurobiology*, 168.
339 *"The findings of Cajal"*: Huerta de Soto, 41.
339 *"With each year"*: Daston and Galison, 189.
339 *"It is not enough"*: Laín Entralgo, *Cajal por sus cuatro costados*, 45.
339 *Cajal's metaphors*: See Garcia-Lopez.
340 *sixty-six*: "Nobel Prize—Neuroscience."
342 *"We know the landscape"*: Ramón y Cajal, *Histology*, xviii.
342 *"soul atoms"*: See Solmsen.
342 *"soul of the soul?"*: Ibid.
342 *"There is nothing"*: Ramón y Cajal, *Charlas*, 171.
343 *His tomb*: "La Real Academia de Ciencias denuncia vandalism en la tumba de Ramón y Cajal."
344 *According to de Nó*: Larriva Sahd, 8.
344 *"a drawing is never"*: Jacobson, *Foundations of Neuroscience*, 14.
344 *"By virtue of an"*: Ibid.

Bibliography

Abadías y Santolaria, León. "Importancia y necesidad del dibujo aplicado a las artes." *Revista de Bellas Artes*, June 23, 1867.

Adee, Alvey A. "Reminiscences of Castelar." *The Century*, vol. 31, 1886, pp. 792–94. www.google.com/books/edition/The_Century/osdZAAAAYAAJ ?hl=en&gbpv=1&bsq=castelar.

Akhtar Khan, Mohd. "Individualism: Origin and Evolution." *Indian Journal of Political Science*, vol. 48, no. 1, January–March 1987.

Aldecoa de González, Carmen. "En el centenario de Cajal." *Revista Hispánica Moderna*, vol. 19, 1953.

Alonso, José Ramón, and Juan Andrés de Carlos. *Cajal: Un grito por la ciencia.* Next Door, 2018. Kindle edition.

Alschuler, William. "Santiago Ramón y Cajal." In *Encyclopedia of Nineteenth-Century Photography*, edited by John Hannay. Taylor and Francis, 2008.

Álvarez-Junco, José. *Spanish Identity in the Age of Nations.* Manchester University Press, 2013.

Alvira Banzo, Fernando. *León Abadías: Pintor, escritor y didacta.* Instituto de Estudios Altoaragoneses, 2014.

Anctil, Michael. *Dawn of the Neuron: The Early Struggles to Trace the Origin of Nervous Systems.* McGill-Queen's University Press, 2015.

Ansede, Manuel. "Cuando Ramón y Cajal iba en taparrabos." *El País*, December 10, 2018. elpais.com/elpais/2018/12/07/ciencia/1544203336_618933 .html.

Antón del Olmet, Luis, and José de Torres Bernal. *Cajal: Historia intima y resumen cientifico del español más ilustre de su época.* Imprenta de Juan Pueyo, 1918. archive.org/details/cajalhistoriaint00antuoft/mode/2up.

Armillas Vicente, José Antonio. "Introducción (Zaragoza en los siglos XIX y XX)," pp. 10–55. *Zaragoza, 1808–2008: Dos siglos de progreso.* Lunwerg Editores, 2008.

Armstrong, William Jackson. "Castelar, the Orator." *Century Illustrated Monthly Magazine,* vol. 31, 1886. www.google.com/books/edition/The_Century/uWUT c6ta3gcC?hl=en&gbpv=0.

Artigas Sanza, Jose Antonio. "España y el Dechado de Cajal." *Amanecer,* December 20, 1951.

Atanasio Fuentes, Manuel. *Lima: Or, Sketches of the Capital of Peru, Historical, Statistical, Administrative, Commercial and Moral.* F. Didot, 1866. www.google.com/books/edition/Lima/aLIQQQAACAAJ?hl=en&gbpv=1.

Aub, Max. *Conversations with Buñuel: Interviews with the Filmmaker, Family Members, Friends and Collaborators.* McFarland, 2017.

"Ayer tarde se verificó el entierro de Don Santiago Ramón y Cajal." *Ahora,* October 19, 1934, p. 9. hemerotecadigital.bne.es/issue.vm?id=0029959810 &search=&lang=en.

Azmitia, Efrain. "Cajal and Brain Plasticity: Insights Relevant to Emerging Concepts of Mind." *Brain Research Reviews,* vol. 55, 2007, pp. 395–405. www .academia.edu/26959505/Cajal_and_brain_plasticity_Insights_relevant_to _emerging_concepts_of_mind.

Azorín. "Un libro de Ramón y Cajal." In *Los valores literarios,* pp. 75–81. Renacimiento, 1913.

Baedeker, Karl. *Spain and Portugal: Handbook for Travellers.* K. Baedeker, 1898. www.google.com/books/edition/Spain_and_Portugal/QV9TAAAAMAAJ?hl =en&gbpv=0.

Bailey, Diane. *Cholera.* Rosen Publishing Group, 2010. www.google.com /books/edition/Cholera/7rvLPx33GPgC?hl=en&gbpv=0.

Balfour, Sebastian. *The End of the Spanish Empire, 1898–1923.* Clarendon Press, 1997.

———. *El fin del Imperio Español, 1898–1923.* Crítica, 2004.

———. "Riot, Regeneration and Reaction: Spain in the Aftermath of the 1898 Disaster." *Historical Journal,* vol. 38, no. 2, June 1995, pp. 405–23. www .jstor.org/stable/i325012.

Balneario de Panticosa (Huesca). *Guia del bañista en Panticosa breve reseña acerca del origen de este establecimiento termal.* Imprenta de M. Minuesa, 1875. www.google.com/books/edition/Guia_del_ba%C3%B1ista_en_Panticosa /4Li5kXZBt5sC?hl=en.

Banerjee, J. C. *Encyclopaedic Dictionary of Psychological Terms.* M.D. Publications, 1994.

Barcelonauta, Miquel. "Café Pelayo. Pelai / Rambla. (1875–1895)." *Barcelofília,* January 31, 2015. barcelofilia.blogspot.com/2015/01/cafe-pelayo-pelai -rambla-1875-1895.html.

———. "La Pajarera (1888–1895)." *Barcelofília,* April 12, 2011. barcelofilia .blogspot.com/search?q=pajarera.

Barker, Lewellys F. "Travel Notes. III: Spain and Ramón y Cajal." *Journal of the American Medical Association,* vol. 43, no. 6, August 6, 1904. jamanetwork .com/journals/jama/article-abstract/461618.

Barnes, David S. *The Making of a Social Disease: Tuberculosis in Nineteenth-Century France.* University of California Press, 1995.

Bartual Pastor, Juan. *La amistad de Santiago Ramón y Cajal con Juan Bartual Moret a través de escritos y correspondencia inédita.* Real Academia de Medicina y Cirugía de Cádiz, 2011. www.researchgate.net/publication/269464452_La _amistad_de_Santiago_Ramon_y_Cajal_con_Juan_Bartual_Moret_a_traves _de_escritos_y_correspondencia_inedita.

Bartual Vicéns, Rafael. "Introducción a la obra de Cajal, anatómico." In *Actos inaugurales del curso de 1970 en la tarde del 20 de febrero, en la Real Academia de Medicina de Valencia,* pp. 11–40. Imprenta Cantos, 1970.

Bausells, Marta. "Story of Cities #13: Barcelona's Unloved Planner Invents Science of 'Urbanisation.'" *The Guardian,* April 1, 2016. www.theguardian .com/cities/2016/apr/01/story-cities-13-eixample-barcelona-ildefons-cerda -planner-urbanisation.

Bekkers, John M. "Pyramidal Neurons." *Current Biology,* vol. 21, no. 24, December 20, 2011. https://doi.org/10.1016/j.cub.2011.10.037.

Bell, Matthew. *The German Tradition of Psychology in Literature and Thought, 1700–1840.* Cambridge University Press, 2005.

Belloc, Hillaire. *The Pyrenees.* Methuen, 1928.

Beltrán de Lis, Fernando. *Reglas de urbanidad para uso de las señoritas.* Librerías "París-Valencia," 1995.

Ben-Ami, Shlomo. "The Dictatorship of Primo de Rivera: A Political Reassessment." *Journal of Contemporary History*, vol. 12, no. 1, January 1977, pp. 65–84. www.jstor.org/stable/260237?seq=1.

———. *The Origins of the Second Republic in Spain*. Oxford University Press, 1978.

Bennett, Max R. *History of the Synapse*. CRC Books, 2003.

Berlucchi, Giovanni. "Some Aspects from the History of the Law of Dynamic Polarization of the Neuron. From William James to Sherrington, from Cajal and Van Gehuchten to Golgi." *Journal of the History of Neuroscience*, vol. 8, no. 2, 1999. www.tandfonline.com/doi/abs/10.1076/jhin.8.2.191.1844 ?journalCode=njhn20.

Berman, Sheri. *Democracy and Dictatorship in Europe: From the Ancien Régime to the Present Day*. Oxford University Press, 2019.

Bernheim, Hippolyte. *Suggestive Therapeutics: A Treatise on the Nature and Uses of Hypnotism*. Translated by Christian A. Herter. G. P. Putnam's Sons, 1880. www.google.com/books/edition/Suggestive_Therapeutics/5jw7AAAAYAAJ ?hl=en&gbpv=0.

Bhattacharyya, Kalyan B. *Eminent Neuroscientists: Their Lives and Works*. Academic Publishers, 2011.

Blanco-Belmonte, M. R. "Homenaje a Ramón y Cajal." *ABC*, May 2, 1922. www.abc.es/archivo/periodicos/abc-madrid-19220502-9.html.

Blanco y Negro. August 18, 1900.

Bobo Martínez, Miguel. "Santiago Ramón y Cajal: Algo más que un fotógrafo." *Ambitos: Revista internacional de comunicación*, nos. 11–12, 2004. dialnet.unirioja.es/servlet/articulo?codigo=1004266.

Bohren, Craig F., and Eugene E. Clothiaux. *Fundamentals of Atmospheric Radiation: An Introduction with 400 Problems*. Wiley, 2006.

Bornside, George H. "Jaime Ferrán and Preventative Inoculation Against Cholera." Abstract. *Bulletin of the History of Medicine*, vol. 55, no. 4, 1981, pp. 516–32. www.jstor.org/stable/44441415?seq=1.

Bosch y Miralles, Antonio. "Un recuerdo." S.n. Cajal Institute library, alphabetized files.

Boyd, Carolyn P. *Historia Patria*. Princeton University Press, 1997.

Brioso, Julio. "Ramón y Cajal en Huesca." *4 Esquinas*, no. 147, June 2002.

Brugmann, Jeb. *Welcome to the Urban Revolution: How Cities Are Changing the World.* Bloomsbury, 2009.

Buñuel, Luis. *My Last Sigh.* Translated by Abigail Israel. Vintage Books, 2013.

Cabot, José Tomás. "Cajal frente a Ferrán." *Natura medicatrix: Revista médica para el estudio y difusión de las medicinas alternativas,* nos. 46–47, 1997, pp. 77–85. dialnet.unirioja.es/servlet/articulo?codigo=4989349.

———. "Silveria, la compañera abnegada." *Historia y vida,* no. 287, 1992, pp. 110–18.

"Café Suizo." *Flaneando por Madrid.* flaneandopormadrid.wordpress.com /2014/02/13/cafe-suizo/.

Calvo Roy, Antonio. *Cajal: Triunfar a toda costa.* Alianza, 1999.

Calzada, José M. *Visiones esteticas contrapuestas: Cajal o Baroja.* Cajal Institute library, alphabetized files.

Cameron, Charles. "Anti-Cholera Inoculation." *The Nineteenth Century,* vol. 18, pp. 338–52. www.google.com/books/edition/The_Nineteenth_Century /umE4AQAAMAAJ?hl=en&gbpv=0.

Campion, J. S. *On Foot in Spain.* Chapman and Hall, 1879. www.google .com/books/edition/On_Foot_in_Spain/wWMBAAAAQAAJ?hl =en&gbpv=1.

Campos-Bueno, José Javier. "Neuron Doctrine and Conditional Reflexes at the XIV International Medical Congress of Madrid of 1903." *Psychologia Latina,* vol. 3, no. 1, 2012, pp. 13–22. www.academia.edu/1536093/Title _Neuron_Doctrine_and_Conditional_Reflexes_at_the_XIV_International _Medical_Congress_of_Madrid_of_1903.

Cannon, Dorothy. *Explorer of the Human Brain.* Henry Schuman, 1949.

Capdevila, J. Goyanes. "Cajal, hombre." *Gaceta Médica Española,* no. 9, 1934, pp. 129–32.

"Carne Fresca." *Heraldo de Madrid,* February 19, 1915. hemerotecadigital.bne .es/issue.vm?id=0000667918&search=&lang=en.

Carr, Matthew. *The Savage Frontier: The Pyrenees in History and Imagination.* New Press, 2018.

Carr, Raymond. *Spain: A History.* Oxford University Press, 2001.

Castelar, Emilio. *Discursos parlamentarios de Don Emilio Castelar en la asamblea constituyente.* A. de San Martín, 1871. www.google.com/books/edition /Discursos_parlamentarios_de_Don_Emilio_C/geoCAAAAYAAJ?hl=en.

Castell. "Por la ciencia." *El Imparcial,* August 11, 1900. hemerotecadigital.bne .es/issue.vm?id=0000817779&search=&lang=en.

Catani, Marco, and Stefano Sandrone. *Brain Renaissance: From Vesalius to Modern Neuroscience.* Oxford University Press, 2015.

Cervantes Saavedra, Miguel de. *The History and Adventures of the Renowned Don Quixote.* Translated by T. Smollett. Effingham Wilson, 1833. www.google .com/books/edition/The_history_and_adventures_of_the_renown /E8QNAAAAQAAJ?hl=en&gbpv=0.

———. *The Ingenious Gentleman Don Quixote of La Mancha.* Vol. 3. Translated by Henry Edward Watts. Adam and Charles Black, 1895.

Cheung, Tobias. "Omnis Fibra ex Fibra: Fibre Economies in Bonnet's and Diderot's Models of Organic Order." In *Transitions and Borders Between Animals, Humans and Machines, 1600–1800,* edited by Tobias Cheung. Brill, 2010.

"Cholera's Seven Pandemics." CBC News, May 9, 2008. www.cbc.ca/news /technology/cholera-s-seven-pandemics-1.758504.

"Cholera. The Cholera in Spain." *British Medical Journal,* vol. 1, no. 1275, June 6, 1885, pp. 1170–71. www.jstor.org/stable/i25272663.

"Cholera. The Cholera in Spain." *British Medical Journal,* vol. 2, no. 1279, July 4, 1885, p. 33. www.jstor.org/stable/i25272969.

"Cholera. The Cholera in Spain." *British Medical Journal,* vol. 2, no. 1280, July 11, 1885, pp. 77–78. www.jstor.org/stable/i25273041.

Chryst, Margaret Waters. *Santiago Ramón y Cajal: His Personality and the Scope of His Work.* University of California, 1923. www.google.com/books /edition/Santiago_Ram%C3%B3n_Y_Cajal/Xd8yAAAAIAAJ?hl =en&gbpv=0.

Chudler, Eric H. "Nobel Prize—Neuroscience." University of Washington faculty page. faculty.washington.edu/chudler/nobel.html.

Clare, Israel Smith. *Nineteenth Century.* Werner Company, 1893. www.google .com/books/edition/Nineteenth_century/lTYQAAAAYAAJ?hl=en&gbpv=0.

Clodfelter, Michael. *Warfare and Armed Conflicts: A Statistical Encyclopedia of Casualty and Other Figures, 1492–2015.* McFarland, 2017.

Cobb, Matthew. *The Idea of the Brain: The Past and Future of Neuroscience*. Basic Books, 2020.

Colucci D'Amato, Luca, et al. "The End of the Central Dogma of Neurobiology: Stem Cells and Neurogenesis in Adult CNS." *Neurological Sciences*, vol. 27, no. 4, October 2006, pp. 266–70. https://doi.org/10.1007/s10072-006 -0682-z.

Companys Monclús, Julián. *España en 1898: Entre la diplomacia y la guerra*. Ministerio de Asuntos Exteriores, 1991.

"Consejo de Ministros." *La Época*, August 11, 1900, p. 3. hemerotecadigital .bne.es/issue.vm?id=0000660319&search=&lang=en.

Cortezo, Carlos María. *Cajal: Su personalidad, su obra, su escuela*. Imprenta del sucesor de E. Teodoro, 1922.

Costero, Isaac. "Centenario de Santiago Ramón y Cajal." *Academia Nacional de Medicina*, May 7, 1952.

Craigie, Edward Horne, and William Carleton Gibson. *The World of Ramón y Cajal with Selections from His Nonscientific Writings*. Thomas, 1968.

Criado y Romero. "El sabio, visto por su ama de llaves." *Heraldo de Madrid*, October 18, 1934. hemerotecadigital.bne.es/issue.vm?id=0001066223&search =&lang=en.

Crow, John A. *Spain: The Root and the Flower*. 3rd ed. University of California Press, 2005.

Cruz, Jesus. *The Rise of Middle-Class Culture in Nineteenth-Century Spain*. Louisiana State University Press, 2011.

Daniel, Thomas M. "The History of Tuberculosis." *Respiratory Medicine*, vol. 100, no. 11, pp. 1862–70. https://doi.org/10.1016/j.rmed.2006.08.006.

Darwin, Charles. *The Origin of Species*. P. F. Collier and Son, 1909. www .google.com/books/edition/The_Origin_of_Species/YY4EAAAAYAAJ?hl =en&gbpv=0.

Daston, Lorraine, and Peter Galison. *Objectivity*. Zone Books, 2010.

Davenport-Hines, Robert. *The Pursuit of Oblivion: A Global History of Narcotics*. W. W. Norton, 2001.

Davillier, Jean-Charles. *L'Espagne*. Librairie Hachette, 1874.

de Carlos Segovia, Juan Andrés. *Los Ramón y Cajal: Una familia aragonesa*. Gobierno de Aragón, Departamento de Cultura y Turismo, 2001.

————. *Los Ramón y Cajal: Una familia Aragonesa*. Gobierno de Aragón, Departamento de Cultura y Turismo, 2011.

de Carlos Segovia, Juan Andrés, and Alonso Peña. *Cajal: Un grito para la ciencia*. Next Door Publishers, 2018.

de Castro, Fernando. "The Cajal School in the Peripheral Nervous System: The Transcendent Contributions of Fernando de Castro on the Microscopic Structure of Sensory and Autonomic Motor Ganglia." *Frontiers in Neuroanatomy*, vol. 10, no. 43, April 20, 2016, pp. 1–11. https://doi.org/10.3389 /fnana.2016.00043.

de Castro, Fernando, et al. "Cajal: Lessons on Brain Development." *Brain Research Reviews*, vol. 55, no. 2, November 2007, pp. 481–89. https://doi.org /10.1016/j.brainresrev.2007.01.011.

De Espronceda, José. *Obras poéticas de d. José de Espronceda*. Edited by Juan Eugenio Hartzenbusch and Antonio Ferrer del Río. Baudry, 1870. www .google.com/books/edition/Obras_po%C3%A9ticas_de_d_Jos%C3%A9_de _Espronceda/a_ANAAAAQAAJ?hl=en&gbpv=0.

DeFelipe, Javier. "Brain Plasticity and Mental Processes: Cajal Again." *Nature Reviews Neuroscience*, vol. 7, no. 10, November 2006, pp. 811–17. https://doi .org/10.1038/nrn2005.

————. *Cajal's Butterflies of the Soul: Science and Art*. Oxford University Press, 2009.

————. *Cajal's Neuronal Forest: Science and Art*. Oxford University Press, 2018.

Defoe, Daniel. *Robinson Crusoe*. Macmillan, 1868. www.google.com/books /edition/Robinson_Crusoe/XoxYJCwQAoYC?hl=en&gbpv=0.

De Hoyos Sainz, Luis. "Crónica científica." *La españa moderna*, vol. 8, no. 65, May 1894.

De Juan, Pascual. "Recuerdo en la conmemoración del centenario de Cajal." *Anales de la Casa de Salud, Valdecilla*, vol. 13, 1952, pp. 27–33.

De La Rue, Warren. *On the Total Solar Eclipse of July 18th, 1860: Observed at Rivabellosa, Near Miranda de Ebro in Spain*. Taylor and Francis, 1862. www .google.com/books/edition/On_the_Total_Solar_Eclipse_of_July_18th/OIo _AAAAcAAJ?hl=en&gbpv=1.

Del Río, Angel. "Congreso Internacional de Medicina: Premio á Cajal." *El Liberal*, August 10, 1900. hemerotecadigital.bne.es/issue.vm?id=0001339250& search=&lang=en.

del Río Hortega, Pío. "Arte y artificio de la ciencia histológica." *Residencia: Revista de la Residencia de Estudiantes*, vol. 3, no. 6, December 1933, pp. 191–206. arhipa.org/documentos/Rev_Res_Est-1933-6.pdf.

—————. *El maestro y yo*. Ariel, 2015.

del Río Hortega, Pío, and Clemente Estable. *Ramón y Cajal. Homenaje en el décimo aniversario de su muerte*. Institución Cultural Española del Uruguay, 1944.

De Montalbán, A. "Los maestros: Ramón y Cajal." *Por Esos Mundos*, November 1904, pp. 5–17. hemerotecadigital.bne.es/issue.vm?id=0003123894& search=&lang=en.

De Quesada, Gonzalo. *Free Cuba: Her Oppression and Struggles for Liberty; History and Description of the Island*. Publishers' Union, 1898. www.google .com/books/edition/Free_Cuba/vZxEAQAAMAAJ?q=&gbpv=1#f=false.

De Quesada, Gonzalo, and Henry Davenport Northrop. *The War in Cuba: Being a Full Account of Her Great Struggle for Freedom, Containing a Complete Record of Spanish Tyranny and Oppression; Scenes of Violence and Bloodshed; Frequent Uprisings of a Galllant and Long Suffering People; Revolutions of 1868, '95–'96*. W. H. Ferguson, 1896. www.google.com/books/edition/The_War_in _Cuba/3nFHAQAAMAAJ?hl=en.

De Unamuno, Miguel. "My Religion." Translated by Armand F. Baker. www .armandfbaker.com/translations/unamuno/my_religion.pdf.

Díaz Morales, José. "El veraneo de la gente conocida." *Estampa*, August 19, 1930, p. 11. hemerotecadigital.bne.es/issue.vm?id=0003421250&search=& lang=en.

Donovan, William H. "Spinal Cord Injury—Past, Present, and Future." *Journal of Spinal Cord Medicine*, vol. 30, no. 2, 2007, pp. 85–100. www.ncbi.nlm .nih.gov/pmc/articles/PMC2031949/.

Doré, Gustave. *Doré's Spain: All 236 Illustrations from Spain*. Dover, 2013. www.google.com/books/edition/Dor%C3%A9_s_Spain/gLvCAgAAQBAJ ?hl=en&gbpv=0.

Dumas, Alexandre. *The Count of Monte Christo: A Romance*. Simms and M'Intyre, 1848. www.google.com/books/edition/The_Count_of_Monte_Christo _A_Romance/YV0VAAAAQAAJ?hl=en&gbpv=0.

Dunfort i Coll, Mercè. "Introducció." *Una mostra de la biologia i la patologia cellulars del sistema nerviós a Catalunya cent cinquantè aniversari del naixement de Santiago Ramón y Cajal*, pp. 7–8. Institut d'Estudis Catalans, 2003. www

.google.com/books/edition/Una_mostra_de_la_biologia_i_la_patologia /izWBf8em6mEC?hl=en&gbpv=0.

Durán Espuny, Salvador. "Don Santiago Ramón y Cajal y Victoriano Cajal, su primo hermano más allegado." *Serrablo*, no. 131, 2004, pp. 18–22.

Durán Muñoz, García, and Francisco Alonso Burón. *Cajal: Vida y obra.* Vol. 1. Institución Fernando el Católico, 1960.

———. *Cajal: Escritos inéditos.* Vol 2. Institución Fernando el Católico, 1960.

Ealham, Chris. *Class, Culture and Conflict in Barcelona, 1898–1937.* Taylor and Francis, 2004.

Eccles, J. C., and W. C. Gibson. *Sherrington: His Life and Thought.* Springer International, 1979.

"1888–1895 La Pajarera." Història de Barcelona, May 6, 2015. www.historia debarcelona.org/la-pajarera/.

Ehrlich, Benjamin. *The Dreams of Santiago Ramón y Cajal.* Oxford University Press, 2016.

El Caballero Audaz. "Don Santiago Ramón y Cajal." *La Esfera*, November 26, 1921, pp. 17–18. hemerotecadigital.bne.es/issue.vm?id=0003227554&search =&lang=en.

"El cadáver del sabio Ramón y Cajal ha recibido sepultura esta tarde." *La Nación*, October 18, 1934, p. 12. hemerotecadigital.bne.es/issue.vm?id=002633 3082&search=&lang=en.

"El Café del Siglo XIX—La Pajarera (1888–1895)." *La Barcelona de antes.* www.labarcelonadeantes.com/la-pajarera.html.

"El doctor Ramón y Cajal." *Blanco y Negro*, February 24, 1906.

"El entierro de don Santiago Ramón y Cajal." *El Sol*, October 19, 1934.

"El estudiante en acción." *El Sol*, February 28, 1927, p. 2. hemerotecadigital .bne.es/issue.vm?id=0000343748&search=&lang=en.

"El Gran Pecado." *El Imparcial*, July 3, 1898. hemerotecadigital.bne.es/issue .vm?id=0000795934&search=&lang=en.

El Imparcial. July 8, 1876. hemerotecadigital.bne.es/issue.vm?id=0000531572 &search=&lang=en.

"El Premio Nobel." *El País*, October 28, 1906, p. 1. hemerotecadigital.bne.es /issue.vm?id=0002057092&search=&lang=en.

The Encyclopaedia Britannica; Or, Dictionary of Arts, Sciences, and General Literature. Vol. 18. Adam and Charles Black, 1842. www.google.com/books /edition/The_Encyclopaedia_Britannica_Or_Dictiona/pko0AQAAMAAJ ?hl=en&gbpv=0.

"En Honor de Ramón y Cajal." *La Correspondencia Militar*, October 4, 1900, p. 2. hemerotecadigital.bne.es/issue.vm?id=0001848795&search=& lang=en.

"En honor de Ramón y Cajal." *La Época*, October 29, 1906. hemerotecadigital .bne.es/issue.vm?id=0000738725&search=&lang=en.

Espoz y Mina, Francisco. *Memorias del general don Francisco Espoz y Mina: Escritas por el mismo*. Vols. 1–2. M. Rivadeneyra, 1851. babel.hathitrust.org /cgi/pt?id=uc1.ax0000383372&view=1up&seq=7.

Esteban, José. *La generación de 98 en sus anecdotas*. Renacimiento, 2012.

Everdell, William. *The First Moderns: Profiles in the Origins of Twentieth-Century Thought*. University of Chicago Press, 2009.

Fahie, John Joseph. *Galileo: His Life and Work*. James Pott, 1903. www.google .com/books/edition/Galileo/xSJWAAAAMAAJ?hl=en&gbpv=1.

Falcón, Irene. *Irene Falcón, asalta a los cielos: Mi vida juna a pasionario*. Temas de Hoy, 1996.

Fernández Aldama, M. "La dinastia Cajal: Al cumplir don Pedro Ramón y Cajal los ochenta años, nos cuenta la vida de su hermano don Santiago." *La Voz de Aragon*, October 25, 1934.

Fernández de Molina, Antonio R. "Cajal y su cigarral de Amaniel." *Lamosaenfosálica* (blog), February 6, 2011. lamosaenfosalica.blogspot.com/2011/02 /cajal-y-su-cigarral-de-amaniel.html.

Fernández-Medina, Nicolás. *McGill-Queen's Studies in the Hist of Id*. 3rd ed. McGill-Queen's University Press, 2018.

Ferreira, Francisco R. M., et al. "The Influence of James and Darwin on Cajal and His Research into the Neuron Theory and Evolution of the Nervous System." *Frontiers in Neuroanatomy*, January 29, 2014. https://doi.org/10 .3389/fnana.2014.00001.

Fetridge, William Pembroke. *The American Traveller's Guide: Harper's Handbook for Travellers in Europe and the East . . . Twelfth Year*. Harper and Brothers, 1873. www.google.com/books/edition/The_American_Traveller_s_Guide /ypwwAQAAMAAJ?hl=en&gbpv=1.

Figueres-Oñate, María, et al. "Unraveling Cajal's View of the Olfactory System." *Frontiers in Neuroanatomy*, July 2, 2014, pp. 1–12. https://doi.org 10.3389/fnana.2014.00055.

Finck, Henry Theophilus. *Spain and Morocco: Studies in Local Color.* Charles Scribner's Sons, 1891.

Finger, Stanley. *The Origins of Neuroscience: A History of Explorations into Brain Function.* Oxford University Press, 1994.

Fischer, Christian August. *A Picture of Valencia: Taken on the Spot.* H. Colburn, 1811. www.google.com/books/edition/A_Picture_of_Valencia/VAwMAA AAYAAJ?hl=en&gbpv=0.

Ford, Richard. *Gatherings from Spain.* J. Murray, 1851. www.google.com/books /edition/Gatherings_from_Spain/OY1GAQAAMAAJ?hl=en&gbpv=1.

———. *A Handbook for Travellers in Spain:* Part 1. *Andalucia, Ronda and Granada, Murcia, Valencia, and Catalonia; the portions best suited for the invalid—a winter tour.* John Murray, 1855. www.google.com/books/edition /A_handbook_for_travellers_in_Spain/6kiv1IDPIOwC?hl=en&gbpv=0.

———. *A Handbook for Travellers in Spain.* Part 2. *Estremadura, Leon, Gallicia, the Asturias, the Castiles (old and new), the Basque provinces, Arragon, and Navarre—a Summer Tour.* John Murray, 1855. www.google.com/books /edition/A_Handbook_for_Travellers_in_Spain/65I2AAAAMAAJ?hl =en&gbpv=1.

———. *A Handbook for Travellers in Spain.* John Murray, 1878. https://www .google.com/books/edition/A_Handbook_for_Travellers_in_Spain /DEtexEMO-rYC?hl=en&gbpv=0.

Forel, Auguste. *From My Life and Work.* W. W. Norton, 1937.

Frances, José. "El perfil de los días." *Nuevo Mundo,* June 24, 1921, p. 3. hem erotecadigital.bne.es/issue.vm?id=0001782189&search=&lang=en.

Freire, Miguel. "Cajal escritor: La divulgación histológica." *Pasaje a la Ciencia,* no. 11, June 2008. www.pasajealaciencia.es/2008/a08-n11.html.

Frerichs, Ralph R. "Competing Theories of Cholera." UCLA Department of Etiology: Fielding School of Public Health. www.ph.ucla.edu/epi/snow /choleratheories.html.

Fresquet Febrer, José L. "The Image of Santiago Ramón y Cajal Through the Daily Press: The Awarding of the Nobel Prize." *European Journal of Anatomy,* vol. 23, no. S1, June 2019. www.eurjanat.com/data/pdf/eja.23s10001.pdf.

Frith, John. "History of Tuberculosis: Part 1—Phthisis, Consumption and the White Plague." *Journal of Military and Veterans' Health*, vol. 22, no. 2. jmvh.org /article/history-of-tuberculosis-part-1-phthisis-consumption-and-the -white-plague/.

Frixione, Eugenio. "Cajal's Second Great Battle for the Neuron Doctrine: The Nature and Function of Neurofibrils." *Brain Research Reviews*, vol. 59, no. 2, March 2009, pp. 393–409. https://doi.org/10.1016/j.brainresrev.2008.11.002.

Gallego, Antonio. *Cajal, opositor a catedras universitarias, visto a través de sus apuntes inéditos.* Santander, 1964.

Garcés Romeo, José. "Descendencia de Ramón y Cajal en Larres." *Serrablo*, vol. 124, September 2002.

García-Albea, Esteban. *Su majestad el cerebro: Historia, enigmas y misterios de un órgano prodigioso.* La Esfera de los Libros, 2017.

Garcia Camarero, E. "Las polemicas de la ciencia Española." In *Cajal y la modernidad*, edited by Alejandro R. Díez Torre. Ateneo de Madrid, 2008.

García Carraffa, Alberto, and Arturo García Carraffa. *Españoles ilustres: Cajal.* A. Marzo, 1918.

García del Real, E. "Cajal, profesor." *Gaceta Médica Española*, no. 9, 1934, pp. 111–12.

Garcia-Lopez, Pablo. "Sculpting the Brain." *Frontiers in Human Neuroscience*, February 9, 2012. https://doi.org/10.3389/fnhum.2012.00005.

García-Marin, Virginia, et al. "Cajal's Contributions to Glia Research." *Trends in Neurosciences*, vol. 30, no. 9, September 2007, pp. 479–87.

García-Mercadal y García-Loygorri, Fernando. "Nuevos datos genealógicos sobre los Causada un linaje aragonés de médicos y naturalistas." *Emblemata*, no. 5, 1999, pp. 371–78. https://ifc.dpz.es/recursos/publicaciones/21/63 /15garciamercadal.pdf.

García Quiepo de Llano, Genoveva. *Los intelectuales y la dictadura de Primo de Rivera.* Allianza Editorial, 1988.

García Sanz, Carolina. "Making Sense of the War (Spain)." In "1914–1918 Online," January 5, 2018, *International Encyclopedia of the First World War*, edited by Ute Daniel et al. encyclopedia.1914–1918-online.net/article /making_sense_of_the_war_spain.

"Germans Burn Belgian Town of Louvain." History.com. www.history.com /this-day-in-history/germans-burn-belgian-town-of-louvain.

Geyer-Kordesch, Johanna, and Fiona MacDonald. *Physicians and Surgeons in Glasgow, 1599–1858*. Vol. 1. *The History of the Royal College of Physicians and Surgeons of Glasgow*. Bloomsbury, 1999.

Gilmour, David. *Cities of Spain*. Random House, 2012. www.google.com /books/edition/Cities_of_Spain/ht1XQQELuqwC?hl=en&gbpv=0.

Giménez, M. R. "El Café de Levante." *Antiguos Cafés de Madrid y Otras Cosas de la Villa* (blog). antiguoscafesdemadrid.blogspot.com/2012/08/el-cafe-de -levante.html.

Giménez-Roldán, Santiago. "Cajal's Unbearable Cephalalgias: The Consequences of a Misdiagnosis." *Revue Neurologique*, vol. 172, no. 11, October 2016. https://doi.org/ 10.1016/j.neurol.2016.09.013.

———. "Monuments to Cajal in Madrid, Spain: Rejection of Public Tributes." *Revue Neurologique*, vol. 175, nos. 1–2, October 2018. https://doi.org/10 .1016/j.neurol.2018.02.086.

Giné, Elena, et al. "The Women Neuroscientists in the Cajal School." *Frontiers in Neuroanatomy*, July 16, 2019. https://doi.org/10.3389/fnana.2019 .00072.

Gitlitz, David M., and Linda Kay Davidson. *The Pilgrimage Road to Santiago: The Complete Cultural Handbook*. St. Martin's, 2000.

Glick, Thomas F. *The Comparative Reception of Darwinism*. University of Chicago Press, 1988.

Golgi, Camillo. "The Neuron Doctrine—Theory and Facts." Nobel Lecture, December 11, 1906. www.nobelprize.org/uploads/2018/06/golgi-lecture.pdf.

Gómez-Redondo, Rosa, and Carl Boe. "Decomposition Analysis of Spanish Life Expectancy at Birth: Evolution and Changes in the Components by Sex and Age." *Demographic Research*, vol. 13, no. 20, November 2005, pp. 521–46. https://doi.org/10.4054/DemRes.2005.13.20.

Goñi, Fermin. "Petilla, una deuda reyes." *El País*, March 11, 1979. elpais.com /diario/1979/03/18/espana/290559620_850215.html.

González de Pablo, Ángel. "The New Lifeblood of the Spanish Nation: The Regenerationist Movement and the Genesis of the Cajal Institute in the Late 19th Century." *European Journal of Anatomy*, vol. 23, no. S1, June 2019, pp. 29–38. www.eurjanat.com/data/pdf/eja.23s10001.pdf.

———. "El Noventayocho y las nuevas instituciones científicas: La creación del Laboratorio de Investigaciones Biológicas de Ramón y Cajal." *Dynamis: Acta Hispanica ad medicinae scientiarumque historiam illustrandam*, no. 18, 1998.

González Hernández, María Jesús. *Raymond Carr: The Curiosity of the Fox.* Translated by Nigel Griffin. Sussex Academic, 2013.

González Recio, José Luis. "Ariadna's Thread in the Labyrinth of Nerve Action or Santiago Ramón y Cajal's Law of Dynamic Polarization." *Ludus Vitalis,* vol. 15, no. 27, 2007. www.ludus-vitalis.org/ojs/index.php/ludus/article /view/381/375.

González-Rigabert, F. "Mentalidades españolas: Cajal." *Nuevo Mundo,* March 17, 1922, p. 6. hemerotecadigital.bne.es/issue.vm?id=0001786715&search =&lang=en.

Gott, Richard. *Cuba: A New History.* Yale University Press, 2005.

Gran Enciclopedia Aragonesa. "Solano Torres, Bruno." April 15, 2009. www .enciclopedia-aragonesa.com/voz.asp?voz_id=11875.

Grant, Gunnar. "Gustaf Retzius and Camillo Golgi." *Journal of the History of the Neurosciences,* vol. 8, no. 2, September 1999, pp. 151–63. www.researchgate .net/publication/11727661_Gustaf_Retzius_and_Camillo_Golgi.

———. "How the 1906 Nobel Prize in Physiology or Medicine Was Shared Between Golgi and Cajal." *Brain Research Reviews,* vol. 55, no. 2, November 2007, pp. 490–98. www.researchgate.net/publication/6499913_How_the _1906_Nobel_Prize_in_Physiology_or_Medicine_was_shared_between _Golgi_and_Cajal.

Grant, Susan T. "Reflections: Fabre and Darwin; A Study in Contrasts." *Bioscience,* vol. 26, no. 6, June 1976, pp. 395–98. www.jstor.org/stable/1297413.

Greef, R. "Introduction." In *The Structure of the Retina,* by Santiago Ramón y Cajal. Translated by Sylvia A. Clarke. C. C. Thomas, 1972.

Gross, Charles C. "Three Before Their Time: Neuroscientists Whose Ideas Were Ignored by Their Contemporaries." *Experimental Brain Research,* vol. 192, no. 3, January 2009, pp. 321–24. https://doi.org/10.1007/s00221-008-1481-y.

Guerra y Sánchez, Ramiro. *Guerra de los diez años 1868–1878.* Editorial de Ciencias Sociales, 1972.

Guía de Zaragoza ó sea breve noticia de las antigüedades, establecimientos públicos, oficinas y edificios que contiene, precedida de una ligera reseña histórica de la misma. Imprenta de Vicente Andrés, 1860. www.google.com/books/edition /Gu%C3%ADa_de_Zaragoza_%C3%B3_sea_breve_noticia_d/c7s4XreN hhkC?hl=en&gbpv=0.

"Habla el país: Lo que dice el Dr. Cajal." *El Liberal,* October 26, 1898. hem erotecadigital.bne.es/issue.vm?id=0001301241&search=&lang=en.

"Hablando con Cajal." *Heraldo de Madrid*, August 9, 1900. hemerotecadigital .bne.es/issue.vm?id=0000446146&search=&lang=en.

Haines, Duane. "Santiago Ramón y Cajal at Clark University, 1899: His Only Visit to the United States." *Brain Research Reviews*, vol. 55, no. 2, November 2007. https://doi.org/10.1016/j.brainresrev.2007.02.002.

Hale, Charles. *The Transformation of Liberalism in Late Nineteenth-Century Mexico*. Princeton University Press, 2014.

Hall, Granville Stanley, et al., eds. *American Journal of Psychology*. Vol. 7. J. H. Orpha, 1895. www.google.com/books/edition/The_American_Journal_of _Psychology/SXo4AQAAMAAJ?q=&gbpv=1#f=false.

Hall, Trowbridge. *Spain in Silhouette*. Macmillan, 1923. www.google.com /books/edition/Spain_in_Silhouette/XXCCAAAAIAAJ?hl=en&gbpv=0.

"Ha muerto D. Santiago Ramón y Cajal." *El Sol*, October 18, 1934. hemero tecadigital.bne.es/issue.vm?id=0000536417&search=&lang=en.

Hannay, David. *Don Emilio Castelar*. archive.org/details/donemiliocastela00 hannrich.

Harrington, Thomas S. *Public Intellectuals and Nation Building in the Iberian Peninsula, 1900–1925: The Alchemy of Identity*. Bucknell University Press, 2014.

Hart, Ernest. "How Cholera Can Be Stamped Out." *North American Review*, vol. 157, no. 441, August 1893, pp. 186–96. www.jstor.org/stable/i25103177.

Hays, J. N. *Epidemics and Pandemics*. ABC-CLIO, 2005. www.google.com /books/edition/Epidemics_and_Pandemics/GyE8Qt-kS1kC?hl=en&gbpv=0.

Heinbockel, Thomas, and Vonnie D. C. Shields, eds. *Histology*. IntechOpen, 2019.

Henschen, Folke. "From the First Nobel Prize Award Ceremony, 1901." No- bel Prize. www.nobelprize.org/ceremonies/from-the-first-nobel-prize-award -ceremony-1901/.

Heraldo de Aragon. August 18, 1900.

Heraldo de Aragon. November 28, 1999.

Heraldo de Madrid. July 3, 1898. hemerotecadigital.bne.es/issue.vm?id=00310 50758&search=&lang=en.

Heraldo de Madrid. July 29, 1914. hemerotecadigital.bne.es/issue.vm?id=00006 59631&search=&lang=en.

Herophilus. *The Art of Medicine in Early Alexandria*. Edition, translation, and essays by Heinrich von Staden. Cambridge University Press, 1989. www .google.com/books/edition/Herophilus_The_Art_of_Medicine_in_Early /rGhIfJZkVoC?hl=en&gbpv=0.

Herrera y Ruiz, José. *Memoria acerca de las aguas y baños minerales de Panticosa, que comprende la descripción topográfica del valle de Tena, la historia de dichas aguas . . .* Imprenta de la Viuda de Jordan e Hijos, 1845. www.google.com /books/edition/Memoria_acerca_de_las_aguas_y_ba%C3%B1os_min /VpKxoN1SezEC?hl=en&gbpv=0.

Herrero Pérez, José Vicente. *The Spanish Military and Warfare from 1899 to the Civil War: The Uncertain Path to Victory*. Palgrave, 2017.

Herzog, Lawrence A. *Return to the Center: Culture, Public Space, and City Building in a Global Era*. University of Texas Press, 2010.

Hippocrates. *The Genuine Works of Hippocrates*. Edited by Charles Darwin Adams. Perseus Digital Library. www.perseus.tufts.edu/hopper/text?doc =Perseus%3Atext%3A1999.01.0248%3Apage%3D365.

"Homenaje a un sabio español." *La Época*, March 9, 1905, p. 3. hemeroteca digital.bne.es/issue.vm?id=0000710940&search=&lang=en.

Homer. *The Iliad*. Translated by Anthony Verity. Oxford University Press, 2011.

Hooke, Robert. *Micrographia: or Some Physiological Descriptions of Minute Bodies Made by Magnifying Glasses with Observations and Inquiries Thereupon*. Jo. Martyn and Ja. Allestry, 1665. www.gutenberg.org/files/15491/15491-h /15491-h.htm.

House, William. "The Therapeutics of Veronal." *Therapeutic Gazette*, vol. 37, 1913, pp. 327–32. Edited by H. A. Hare and Edward Martin.

Hubbard, Ben. *Bloody History of Paris: Riots, Revolution and Rat Pie*. Amber Books, 2017.

Hubbard, Jacqueline A., and Devin K. Binder. *Astrocytes and Epilepsy*. Academic Press, 2016.

Huerta de Soto, Jesús. *Socialism, Economic Calculation and Entrepeneurship*. Edward Elgar, 2010.

Humlebaek, Carsen. *Inventing the Nation*. Bloomsbury, 2014.

Hunter, Michael. *Robert Hooke: Tercentennial Studies*. Ashgate Publishing, 2006.

Huntington, Archer Milton. *A Note-book in Northern Spain*. Putnam, 1898. www.google.com/books/edition/A_Note_book_in_Northern_Spain/ -tDJAAAAMAAJ?hl=en&gbpv=1.

Ibañez, Augustín. "Obligado y justo hombre." *La Casa de Médico*, November 1934.

Ibarra, Eduardo, and Julián Ribera. "Al que leyere." *Revista de Aragon*, January 1900. hemerotecadigital.bne.es/issue.vm?id=0005172124&search=&lang=en.

Ille Coque Internacional. *Azorín 1904–1924*. Université de Pau et des Pays de l'Adour, 1996.

"Ingresa en la sección 6ª (primer sorteo). April 10 al núm. 2." *Senado de España*. www.senado.es/cgi-bin/verdocweb?tipo_bd=IDSH&Legislatura=1909 -1910&Pagina=12&Bis=NO&Apendice2=&Boletin2=&Apendice1 =&Boletin1=.

"Instantaneas: Cajal." *El Globo*, December 5, 1897, p. 2. hemerotecadigital.bne .es/issue.vm?id=0001200702&search=&lang=en.

"Íntimos y curiosidades." *La Libertad*, June 6, 1932.

Isaac, Alistair M. C. "Realism Without Tears II: The Structuralist Legacy of Sensory Physiology." *Studies in History and Philosophy of Science Part A*, vol. 79, February 2020, pp. 15–29. https://doi.org/10.1016/j.shpsa.2019.01.003.

Ishizuka, Hisao. "Visualizing the Fibre-Woven Body: Nehemiah Grew's Plant Anatomy and the Emergence of the Fibre Body." In *Anatomy and the Organization of Knowledge, 1500–1850*, edited by Brian Muñoz and Matthew Landers. Pickering and Chatto, 2014.

Jabr, Ferris. "The Neuron's Secret Partner." *Nautilus*, August 13, 2015. www .scribd.com/article/338357622/The-Neuron-S-Secret-Partner-Glial -Cells-Are-The-Brain-S-Architects-Doctors-Police-Janitors-And -Gardeners.

Jacobson, Marcus. *Developmental Neurobiology*. 3rd ed. Springer, 1991.

———. *Foundations of Neuroscience*. Springer, 1993.

Javier, Krauel. *Imperial Emotions: Cultural Responses to Myths of Empire in Fin-de-Siècle Spain*. Liverpool University Press, 2013. www.google.com/books /edition/Imperial_Emotions/u4rnCwAAQBAJ?hl=en&gbpv=0.

Jensen, Anthony K. "Johann Wolfgang von Goethe (1749–1832)." *Internet Encyclopedia of Philosophy*, iep.utm.edu/goethe/.

Jiménez, Eduardo M. "Cajal en los cafés de Madrid." *Comarca*, vol. 33, 2002.

———. "La Honda: Don Santiago, don Miguel y el tío Mariano." *Comarca*, vol. 33, 2002.

"José de Espronceda y Delgado: Spanish Poet." *Encyclopaedia Britannica.* www.britannica.com/biography/Jose-de-Espronceda-y-Delgado.

Johnson, Roberta. *Major Concepts in Spanish Feminist Theory.* State University of New York Press, 2019.

Jones, Edward G. "Cajal's Debt to Golgi." *Brain Research Reviews,* vol. 66, nos. 1–2, January 7, 2011, pp. 83–91. https://doi.org/10.1016/j.brainresrev.2010 .04.005.

———. "The Neuron Doctrine 1891." *Journal of the History of the Neurosciences: Basic and Clinical Perspectives,* vol. 3, no. 1, 1994, pp. 3–20.

Juaristi, Vince J. *Basque Firsts.* University of Nevada Press, 2016.

Juarros, César. *Ramón y Cajal: Vida y milagros de un sabio.* Editorial Maxtor, 2019. www.google.com/books/edition/Ram%C3%B3n_y_Cajal/1q6xDwAAQBAJ ?hl=en&gbpv=0.

Juderías, Julián. "The Black Legend (1914)." In *Modern Spain: A Documentary History,* edited by Jon Cowans. University of Pennsylvania Press, 2003.

Juez Vicente, E. "Gran talento y gran corazón." *La Casa del Médico,* November 1934.

Kamen, Henry. *The Spanish Inquisition: A Historical Revision.* Yale University Press, 1998.

Keats, John. *The Complete Works of John Keats.* Vol. 5. T. Y. Crowell, 1820. www .google.com/books/edition/The_Complete_Works_of_John_Keats /Nr48AAAAYAAJ?hl=en&gbpv=0.

Kiddle, Lawrence B. "Las Oposiciones—an Old Spanish Custom." *Modern Language Journal,* vol. 46, no. 6, October 1962, pp. 255–58. https://doi.org/10 .2307/320000.

Kock, Wolfram. "Santiago Ramón y Cajal och Nobelpriset." *Nordisk Medicinhistorik Aarsbok,* 1981.

Konstantine, P. P., and K. P. Peter. "The Ancient Greek Discovery of the Nervous System: Alcmaeon, Praxagoras and Herophilus." *Journal of Neurology and Neuroscience,* nos. 2171–6625, 2015, pp. 1–5. https://doi.org/10.21767/2171 -6625.S10003.

Koob, Andrew. *The Root of Thought: Unlocking Glia—the Brain Cell That Will Help Us Sharpen Our Wits, Heal Injury, and Treat Brain Disease.* FT Press, 2009.

Køppe, Simo. "The Psychology of the Neuron: Freud, Cajal and Golgi." *Scandinavian Journal of Psychology,* vol. 24, 1983.

Kousoulis, Antonis A. "Etymology of Cholera." *Emerging Infectious Diseases*, vol. 18, no. 3, March 2012, p. 540. wwwnc.cdc.gov/eid/article/18/3/11-1636 _article.

Kurzban, Robert. "Cartesian Hydrolicism." Edge. www.edge.org/response -detail/25411.

Labanyi, Jo. "Relocating Difference: Cultural History and Modernity in Late Nineteenth-Century Spain." In *Spain Beyond Spain: Modernity, Literary History, and National Identity*, edited by Brad Epps and Luis Fernández Cifuentes, pp. 168–89. Bucknell University Press, 2005.

"La Catedra de Cajal." *La Voz Medica*, vol. 2, no. 1.25, 1922.

"La explosión del imperialismo europeo." *El País*, August 14, 1914, p. 1. hem erotecadigital.bne.es/issue.vm?id=0002473556&search=&lang=en.

"La guerra europea." *El Imparcial*, August 14, 1914, p. 1. hemerotecadigital .bne.es/issue.vm?id=0000327181&search=&lang=en.

"La guerra europea parece inevitable." *El Imparcial*, August 1, 1914. hemero tecadigital.bne.es/issue.vm?id=0000326606&search=&lang=en.

"La herencia de Cajal." *El Periódico de Aragon*, November 28, 1999.

Laín Entralgo, Pedro. *Cajal por sus cuatro costados*. Biblioteca Virtual Miguel de Cervantes, 2012. www.cervantesvirtual.com/obra/cajal-por-sus-cuatro -costados/.

———. "Estudios y apuntos sobre Ramón y Cajal." *España como problema*. Aguilar, 1957.

"La escuela de Cajal: La creación del primer Servicio de Anatomía Patológica en España por D. Francisco Tello." *Revista de Patología*, vol. 35, no. 4, 2002. www.patologia.es/volumen35/vol35-num4/35-4n04.htm.

"La jubilación de Cajal." *La Correspondencia de España*, May 2, 1922, p. 4. hemerotecadigital.bne.es/issue.vm?id=0000830839&search=&lang=en.

"La jubilación de Cajal." *La Libertad*, May 3, 1922, p. 5. hemerotecadigital .bne.es/issue.vm?id=0002663220&search=&lang=en.

"La jubilación del maestro: Ramón y Cajal." *El Imparcial*, May 2, 1922. hem erotecadigital.bne.es/issue.vm?id=0000460631&search=&lang=en.

"La letra con sangre entra." *Centro Virtual Cervantes*. cvc.cervantes.es/lengua /refranero/ficha.aspx?Par=58868&Lng=0.

"La pregunta del día." *La Libertad*, October 19, 1934, p. 5. hemerotecadigital .bne.es/issue.vm?id=0003121910&search=&lang=en.

Lapuente Mateos, A. "Cajal, maestro." *Medicamenta*, no. 18, 1952, p. 402.

"La Real Academia de Ciencias denuncia vandalismo en la tumba de Ramón y Cajal." *El Diario*, March 22, 2018. www.eldiario.es/tecnologia/real-academia -ciencias-ramon-cajal_1_2207775.html.

Larriva Sahd, Jorge. "Some Contributions of Rafael Lorente de Nó to Neuroscience: A Reminiscence." *Brain Research Bulletins*, vol. 59, no. 1, 2002.

Las provincias: Diario de Valencia—almanaque para el año 1885. Domenech, 1885. www.google.com/books/edition/Las_provincias_diario_de_Valencia /TZtPAAAAMAAJ?hl=en&gbpv=0.

Lathrop, George Parsons. *Spanish Vistas*. Harper and Brothers, 1883. www .google.com/books/edition/Spanish_Vistas/AI1JAAAAMAAJ?hl =en&gbpv=0.

Latimer, Elizabeth Wormeley. *Spain in the Nineteenth Century*. A. C. McClurg, 1903. www.google.com/books/edition/Spain_in_the_Nineteenth_Century /zJwLAAAAYAAJ?hl=en&gbpv=1.

Lazar, J. Wayne. "Acceptance of the Neuron Theory by Clinical Neurologists of the Late-Nineteenth Century." *Journal of the History of the Neurosciences*, vol. 19, no. 4, October 2010, pp. 349–64. https://doi.org/10.1080/09647041003661638.

Lee, Arthur Bolles. *The Microtomist's Vade-mecum: A Handbook of the Methods of Microscopic Anatomy*. J. and A. Churchill, 1885. www.google.com/books/edition /The_Microtomist_s_Vade_mecum/DctOp6aX2jwC?hl=en&gbpv=0.

Lenhoff, Howard M., and Sylvia G. Lenhoff. "Abraham Tremblay and the Origin of Research on Regeneration in Animals." In *A History of Regeneration Research: Milestones in the Evolution of a Science*, edited by Charles Dinmore. Cambridge University Press, 2007.

Leonardo da Vinci. *Da Vinci Notebooks*. Profile, 2001.

Leoz, Galo. "Mis recuerdos de Cajal." *Jano*, no. 657-H, 1985, pp. 39–47.

Leoz Ortín, Galo. "A Cajal en el primer centenario de su nacimiento." *Archivos de la Sociedad Oftalmológica Hispano-Americana*, vol. 12, no. 10, October 1952.

Lerma, Juan, and Juan A. De Carlos. "Epilog: Cajal's Unique and Legitimated School." *Frontiers in Neuroanatomy*, July 2, 2014. https://doi.org/10.3389 /fnana.2014.00058.

Levine, Catherine, and Alexander Marcillo. "Origin and Endpoint of the Olfactory Nerve Fibers: As Described by Santiago Ramón y Cajal." *Anatomical Record: Advances in Integrative Anatomy and Evolutionary Biology*, vol. 291, no. 7, July 2008. https://doi.org/10.1002/ar.20739.

Lewis, Orly. "Dissection as a Method of Discovery." In *The Soul Is an Octopus*, edited by Uta Kornmeier, pp. 24–29. Berliner Medizinhistorisches Museum der Charité Excellence Cluster Topoi, 2016. www.topoi.org/wp-content /uploads/2016/05/sb_01_Kornmeier.pdf.

Lewy, Kety. "Ramón y Cajal, visto por su secretaria." *Estampa*, November 17, 1934, pp. 23–25. hemerotecadigital.bne.es/issue.vm?id=0003466857&page =23&search=cajal+secretaria&lang=en.

Lewy Rodríguez, Enriqueta. *Asi era Cajal*. Espasa-Calpe, 1977.

———. "Las ideas sociales de Cajal." *Nuestra Bandera*, vol. 124, 1984, pp. 73–76.

———. "Plenario: Cajal; Antecedientes históricos." Cajal Institute library, alphabetized files.

———. *Santiago Ramón y Cajal: El hombre, el sabio, y el pensador*. Extensión Científica y Acción Cultural del CSIC, 1987.

Library of Congress. "The World of 1898: The Spanish-American War: Introduction." www.loc.gov/rr/hispanic/1898/intro.html.

Liddell, E.G.T. "Cajal and Sherrington." *Lectures on the Scientific Basis of Medicine*, vol. 6. www.google.com/books/edition/Lectures_on_the_Scientific _Basis_of_Medi/wTsgAQAAIAAJ?hl=en&gbpv=0.

Loewy, Arthur D. "Ramón y Cajal and Methods of Neuroanatomical Research." *Perspectives in Biology and Medicine*, vol. 15, no. 1, Fall 1971, pp. 7–36. muse.jhu.edu/article/406390.

López-Muñoz, Francisco, C. Álamo, and G. Rubio. "The Neurobiological Interpretation of the Mental Functions in the Work of Santiago Ramón y Cajal." *History of Psychiatry*, vol. 19, no. 1, 2008. www.semanticscholar.org /paper/The-neurobiological-interpretation-of-the-mental-in-L%C3%B3pez -Mu%C3%B1oz-%C3%81lamo/7afc55687881bac1e2008147362322f50ed4 7b50?p2df.

López-Muñoz, Francisco, Ronaldo Uha-Udabe, and Cecilio Alamo. "The History of Barbiturates a Century After Their Clinical Introduction." *Neuropsychiatric Disease and Treatment*, vol. 1, no. 4, December 2005, pp. 329–43. www.ncbi.nlm.nih.gov/pmc/articles/PMC2424120/.

López Piñero, José María. *La Facultad de Medicina de la Universidad de Valencia: Aproximación a su historia*. Universidad de Valencia, Facultad de Medicina, 1980.

———. *Santiago Ramón y Cajal*. 2nd ed. Publicacions Universidad de Valencia, 2006.

Lougheed, Kathryn. *Catching Breath*. Bloomsbury, 2017.

Lucientes, Francisco. "Una visita a la casa de Ramón y Cajal." *Diario de Madrid*, November 18, 1934, p. 9.

Machado, Manuel. "Filosofías de verano: El cine; El Café Suizo." *El Liberal*, July 18, 1919, p. 3. hemerotecadigital.bne.es/issue.vm?id=0001876480&search =&lang=en.

Macpherson, John. *The Baths and Wells of Europe: Their Action and Uses, with Notices of Climatic Resorts and Diet Cures*. Macmillan, 1873. www.google.com /books/edition/The_Baths_and_Wells_of_Europe/ENV0IpCm_XYC?hl =en&gbpv=1.

Madoz, Pascual. *Diccionario geográfico-estadístico-historico de España y sus posesiones de ultramar*. S.n., 1846–1850. Biblioteca Virtual Andalucía. www .bibliotecavirtualdeandalucia.es/catalogo/es/catalogo_imagenes/grupo.cmd ?path=1004959.

"Madrid: From Our Regular Correspondent." *American Journal of Medicine*, vol. 75, 1920, p. 425.

Mainer Baqué, Juan. "El Instituto Ramón y Cajal de Huesca entre 1845 y 1970: De la construcción de elites a la escolarización de masas." www.nebraskaria .es/wp-content/uploads/2016/09/INSTITUTO-RAM%C3%93N-Y -CAJAL-2009.pdf.

Majno, Guido, and Isabella Joris. *Cells, Tissues, and Disease: Principles of General Pathology*. 2nd ed. Oxford University Press, 2004.

Makari, George. *Revolution in Mind*. HarperCollins, 2009.

Marañon Posadillo, Gregorio. "Conmemoración del centenario del nacimiento de Cajal." *Efemérides y comentarios*. Espasa Calpe, 1955, pp. 29–33.

Marijuán, Pedro C. "Cajal and Consciousness: Introduction." In *Cajal and Consciousness: Scientific Approaches to Consciousness on the Centenary of Ramón y Cajal's* Textura, edited by Marijuán. Johns Hopkins University Press, 2002.

Márquez, Manuel. "Ramón y Cajal." *Revista Ibero-Americana de ciencias médicas*, vol. 4, 1900.

Martínez del Campo, Luis G. "How to Be an Intellectual." In *Spain in the Nineteenth Century*, edited by Andrew Ginger and Geraldine Lawless. Manchester University Press, 2018.

Martínez Tejero, Vicente. "Santiago Ramón y Cajal y la tesis doctoral de su padre (II)." *Andalán*, March 11, 2019. www.andalan.es/.

Mas, Poco. *Scenes and Adventures in Spain from 1835 to 1840.* Vol. 2. Richard Bentley, 1845. www.google.com/books/edition/Scenes_and_Adventures_in _Spain/Z7bWjwEACAAJ?hl=en&gbpv=1.

Matto de Turner, Clorinda. *Viaje de recreo: España, Francia, Inglaterra, Italia, Suiza y Alemania.* F. Sempere y Compañía, 1909.

Mayence, Fernand. "Blame for the Sack of Louvain: The Belgian Rejoinder." *Current History,* July 1928, pp. 556–71. net.lib.byu.edu/estu/wwi/PDFs /Sack%20of%20Louvain.pdf.

Mazzarello, Paolo. "Camillo Golgi's Scientific Relationships with Scandinavian Scientists." In *A Galvanized Network: Italian-Swedish Scientific Relations from Galvani to Nobel,* edited by Marco Beretta and Karl Grandin. Royal Swedish Academy of Sciences, 2011.

———. *Golgi: A Biography of the Founder of Modern Neuroscience.* Oxford University Press, 2010.

———. *The Hidden Structure: A Scientific Biography of Camillo Golgi.* Oxford University Press, 1999.

McComas, Alan J. *Galvani's Spark: The Story of the Nerve Impulse.* Oxford University Press, 2011.

McDonogh, Gary, and Sergi Martinez-Rigol. *Barcelona.* Wiley, 2019. www .google.com/books/edition/Barcelona/kMaKDwAAQBAJ?hl=en&gbpv=0.

McKinley, William. "April 11, 1898: Message Regarding Cuban Civil War." Speech. University of Virginia: Miller Center. millercenter.org/the-presidency /presidential-speeches/april-11–1898-message-regarding-cuban-civil-war.

McNair, John A. *Education for a Changing Spain.* Manchester University Press, 1984.

Menéndez Pidál, Gonzalo. *La España del siglo XIX vista por sus contemporáneos.* Centro de Estudios Constitucionales, 1988.

Merchán, Miguel A., et al. *Cajal and de Castro's Neurohistological Methods.* Oxford University Press, 2016.

Mew, James. *Manners and Customs in Spain.* Worthington, 1891. www.google .com/books/edition/Manners_and_Customs_of_Spain/onYWAAAAYAAJ ?hl=en&gbpv=1.

Millán Bermejo, Marina. "Spain: Restoration and Civil War." Timetoast, cache.timetoast.com/timelines/spain-5a1f49a9-e9e2–436b-93e8-ce892aade 4cb.

Minot, Charles. "The Relation of Embryology to Medical Progress." *Transactions of the Maine Medical Association*, vol. 15, 1906, pp. 478–98.

Moigno, F. "The Abbé Moigno." *Study and Stimulants; Or, The Use of Intoxicants and Narcotics in Relation to Intellectual Life, as Illustrated by Personal Communications on the Subject, from Men of Letters and of Science*. Edited by A. Arthur Reade. Abel Heywood and Son, 1883. www.google.com/books /edition/Study_and_stimulants_or_The_use_of_intox/mxkDAAAAQAAJ ?hl=en&gbpv=1.

Monlau, Pedro Felipe. *Elementos de psicologia*. Imprenta Esterrotipia y Galvanoplastia de Aribau, 1881. www.google.com/books/edition/Elementos_de _psicologia/U1BeAAAAcAAJ?hl=en&gbpv=0.

Montes Santiago, Julio. "El encuentro de Einstein y Cajal (Madrid, 1923): Un olvidado momento estelar de la humanidad." *Revista de neurología*, vol. 43, no. 2, 2006, pp. 113–17. dialnet.unirioja.es/servlet/articulo?codigo=2040086.

Mora, Carla, et al. "The Neural Pathway Midline Crossing Theory: A Historical Analysis of Santiago Rámon y Cajal's Contribution on Cerebral Localization and on Contralateral Forebrain Organization." *Journal of Neurosurgery*, vol. 47, no. 3, September 2019. https://doi.org/https://doi.org/10.3171/2019 .6.FOCUS19341.

Morales, F. "Cajal's Educational Environment in His Adolescence and Student Years." *Neurosciences and History*, vol. 1, no. 3, 2013, pp. 104–13. nah.sen .es/vmfiles/abstract/NAHV1N32013104_113EN.pdf.

Morales Pérez, A. "Mis impresiones referentes a Cajal." *Clínica y Laboratorio*, vol. 2, 1906, pp. 307–309.

Moreno, Antonio. "Cajal y el regeneracionismo científico, político y social." *Historia, medicina y ciencia en tiempos de . . . Cajal*, pp. 89–103. Ergon, 2006. www.fcs.es/images/publicaciones/CAJAL.pdf.

Moret, Segismundo. *Discurso leído por el Excmo. Señor D. Segismundo Moret y Prendergast, el día 4 de noviembre, en el Ateneo científico y literario de Madrid con motive de la apertura de sus cátedras*. Imprenta Central a Cargo de Victor Saiz, 1884.

Mumford, James Gregory. "Narrative of Surgery; A Historical Sketch." In *Surgery, Its Principles and Practice*, vol. 1, edited by William Williams Keen, pp. 17–78. www.google.com/books/edition/Surgery_Its_Principles_and _Practice/Hb4hAQAAMAAJ?hl=en&gbpv=1.

Nicholson, Daniel J. "Biological Atomism and Cell Theory." *Studies in History and Philosophy of Science, Part C: Studies in History and Philosophy of Biological and Biomedical Sciences*, vol. 41, no. 3, 2010, pp. 202–11. www.academia.edu /929504/Biological_Atomism_and_Cell_Theory.

Nicolelis, Miguel. *Beyond Boundaries: The New Neuroscience of Connecting Brains with Machines and How It Will Change Our Lives.* Times Books, 2011.

Nieto, M. Angela. "Molecular Biology of Axon Guidance." *Neuron*, vol. 17, December 1996, pp. 1039–48. www.cell.com/neuron/pdf/S0896-6273(00) 80237-8.pdf.

Nieto-Galan, Augustí. "Scientific 'Marvels' in the Public Sphere: Barcelona and Its 1888 International Exhibition." *Journal of History of Science and Technology*, vol. 6, 2012. www.johost.eu/vol6_fall_2012/agusti_galan.htm.

Nieto Amada, José Luis. "El bachillerato en Artes de don Santiago Ramón y Cajal en el instituto de Huesca." Congreso Nacional de Historia de la Medicina 1989, Zaragoza, no. 2, pp. 703–708.

Niño, Alex. "Muy cerca del Nobel." *El País*, April 9, 1996. elpais.com/diario /1996/04/09/madrid/829049073_850215.html.

"Nobel Banquet Menu 1906." Nobel Prize. www.nobelprize.org/ceremonies /nobel-banquet-menu-1906/.

"Noticias Generales." *Heraldo de Madrid*, October 5, 1900, p. 3. hemeroteca digital.bne.es/issue.vm?id=0000448151&search=&lang=en.

Novales, Alberto Gil. *La Revolución de 1868 en el Alto Aragón.* Guara, 1980.

Nubiola, Jaime. "The Reception of William James in Continental Europe." *European Journal of Pragmatism and American Philosophy*, vol. 3, no. 1, 2011. https://doi.org/10.4000/ejpap.869.

Nuñez Florencio, Rafael. "Culture." *The History of Modern Spain: Chronologies, Themes, Individuals*, edited by Adrian Shubert and José Álvarez Junco, pp. 229–46. Bloomsbury, 2017.

O'Connor, James P. B. "Thomas Willis and the Background to Cerebri Anatome." *Journal of the Royal Society of Medicine*, vol. 96, no. 3, April 2003, pp. 139–43. https://doi.org/10.1258/jrsm.96.3.139.

Odelberg, Shannon J. "Unraveling the Molecular Basis for Regenerative Cellular Plasticity." *PLoS Biology*, vol. 2, no. 8, pp. 1068–71. journals.plos.org /plosbiology/article/file?type=printable&id=10.1371/journal.pbio.0020232.

Offner, John L. *An Unwanted War: The Diplomacy of the United States and Spain over Cuba, 1895–1898.* University of North Carolina Press, 1992.

Olmedo de Cerdà, María Francisca. *Anecdotario histórico español*. Carena, 2004.

Orme, Edward B. "A History of *The Illustrated London News*." John Weedy's *Illustrated London News* collection. https://www.iln.org.uk/iln_years/historyofiln .htm.

O'Shea, Henry George. *O'Shea's Guide to Spain and Portugal*. Edited by John Lomas. Adam and Charles Black, 1889. www.google.com/books/edition/O _Shea_s_Guide_to_Spain_and_Portugal/EV1EAAAAIAAJ?hl=en&gbpv=1.

Otis, Laura. *Membranes: Metaphors of Invasion in Nineteenth-Century Literature, Science, and Politics.* Johns Hopkins University Press, 2000.

———. *Networking*. University of Michigan Press, 2001.

Ovid. *Metamorphoses*. Vol. 9. ovid.lib.virginia.edu/trans/Metamorph9.htm.

Palanca, J. A. "Cajal, ciudadano." *Gaceta Médica Español*, no. 9, 1934, pp. 126–28.

Palay, Sanford L. "Synapses in the Central Nervous System." *Journal of Biophysical and Biochemical Cytology*, vol. 2, no. 4, July 25, 1956, pp. 193–202. www .jstor.org/stable/i273928.

Palay, Sanford L., and Victoria Chan-Palay. *Cerebellar Cortex: Cytology and Organization*. Springer-Verlag, 1974.

Pallares-Barbera, Montserrat, et al. "Cerdà and Barcelona: The Need for a New City and Service Provision." *Urbani izziv*, vol. 22, no. 2, 2011, pp. 122–36. https://doi.org/10.5379/urbani-izziv-en-2011-22-02-005.

Parker, G. H. "The Neurofibril Hypothesis." *Quarterly Review of Biology*, vol. 4, no. 2, June 1929, pp. 144–78.

Parkin, J. "On the State of Medicine and Surgery in Spain." *The Lancet*, vol. 30, no. 767, May 12, 1838, pp. 223–24. www.thelancet.com/journals/lancet /issue/vol30no767/PIIS0140–6736(00)X1012–4.

Pascal, Blaise. *Pensées*. Dover, 2003.

PBS. "Yellow Journalism." *Crucible of Empire: The Spanish-American War*. www .pbs.org/crucible/frames/_journalism.html.

Pearce, J.M.S "The Development of Spinal Cord Anatomy." *European Neurology*, vol. 59, 2008, pp. 286–91. www.karger.com/Article/Fulltext/121417#:~: text=Galen%20described%3A%20'the%20cord%20like, consequences%20of %20transection%20and%20tuberculosis.

Perez, Andrew W. "*Life in the Year 6000* by Santiago Ramón y Cajal: A Translation of an Unpublished 'Vacation Story.'" *Literature and Medicine*, vol. 35, no. 1, 2017, pp. 203–28. muse.jhu.edu/article/659113/pdf.

Pérez, Fernán. "El Instituto Cajal: El glorioso sabio y sus colaboradores." *ABC*, 1929.

Pérez, Joseph. *The Spanish Inquisition: A History*. Yale University Press, 2005.

Pérez, Louis A., Jr. *Cuba in the American Imagination: Metaphor and the Imperial Ethos*. University of North Carolina Press, 2008.

Pérez de Tudela Bueso, M. A. "Algunas reminiscencias de Santiago Ramón y Cajal." Cajal Institute library, alphabetized files.

———. "Publicaciones del Prof. Dr. Santiago Ramón y Cajal existentes en los fondos de la Biblioteca del Instituto de Neurobiología." *Trabajos del Instituto Cajal de investigaciones biológicas*, vol. 74, nos. 1–4, 1983.

Peters, Alan, and Kathleen S. Rockland, eds. *Primary Visual Cortex in Primates*. Springer, 1994.

Piccolino, Marco. "Cajal and the Retina: A 100-Year Retrospective." *Trends in Neurosciences*, vol. 11, no. 12, 1988, pp. 521–25. https://doi.org/10.1016/0166-2236(88)90175-0.

Pieribone, Vincent, et al. *Aglow in the Dark: The Revolutionary Science of Biofluorescence*. Belknap Press of Harvard University Press, 2005.

Pittaluga, Gustavo. "Cajal." *Anales de Medicina Interna*, November 3, 1934.

Pitzer, Andrea. "Concentration Camps Existed Long Before Auschwitz." *Smithsonian*, November 2, 2017. www.smithsonianmag.com/history/concentration-camps-existed-long-before-Auschwitz-180967049/.

Ponce de León, Néstor. *The Book of Blood: An Authentic Record of the Policy Adopted by Modern Spain to Put an End to the War of Independence of Cuba*. M. M. Zarzamendi, 1871.

"Presupuesto de gastos de la Sección 7ª, Ministerio de Instrucción Pública y Bellas Artes, para 1911. Pág. 1219; 1221." Senado de España. www.senado.es/cgi-bin/verdocweb?tipo_bd=IDSH&Legislatura=1910&Pagina=1219&Bis=NO&Apendice2=&Boletin2=&Apendice1=&Boletin1=.

Primo de Rivera, José Antonio. "Discurso de José Antonio Primo de Rivera exponiendo los puntos fundamentales de Falange española, pronunciado en el Teatro de la Comedia de Madrid, el día 29 de octubre de 1933." *Segunda Republica*. www.segundarepublica.com/index.php?opcion=6&id=78.

Prockop, Leon D. "Cerebrospinal Fluid Alterations in Spinal-Cord Injury." In *Neurobiology of Cerebrospinal Fluid*, edited by James H. Wood, vol. 2, pp. 481–96. Springer, 2013. www.google.com/books/edition/Neurobiology_of _Cerebrospinal_Fluid_2/1PrpBwAAQBAJ?hl=en&gbpv=0.

Pucho, José. "La biblioteca personal de Cajal." Cajal Institute library, alphabetized files.

Quadrado, José María. *Recuerdos y bellezas de España: Aragon.* Jose Repullés, 1844. www.google.com/books/edition/Recuerdos_y_bellezas_de_Espa%C3% B1a/cRpA_KedPm0C?hl=en&gbpv=0.

Quiroga, Alejandro. *Making Spaniards: Primo de Rivera and the Nationalization of the Masses, 1923–30.* Palgrave Macmillan, 2007.

"A Ramón y Cajal." *El Imparcial,* October 4, 1900. hemerotecadigital.bne.es /issue.vm?id=0000819314&search=&lang=en.

"Ramón y Cajal." *El Liberal,* October 28, 1906, p. 1. hemerotecadigital.bne.es /issue.vm?id=0001489920&search=&lang=en.

"Ramón y Cajal." *La Correspondencia Médica,* August 18, 1900.

Ramón y Cajal, 1852–1934: Expedientes Administrativos de Grandes Españoles. Ministerio de Educación y Ciencia, 1978.

Ramón Alonso, José. "Achúcarro." *Neurociencia: El blog de José Ramón Alonso,* August 11, 2014. jralonso.es/2014/08/11/achucarro/.

————. "Cajal, explorador." *Neurociencia: El blog de José Ramón Alonso.* jralonso .es/2015/08/27/cajal-explorador/.

————. "Cajal, político." *Neurociencia: El blog de José Ramón Alonso.* jralonso.es /2014/07/21/cajal-politico/.

————. "Cajal, profesor." *Neurociencia: El blog de José Ramón Alonso.* jralonso .es/2014/08/12/cajal-profesor/.

————. "Cajal y el dinero." *Neurociencia: El blog de José Ramón Alonso.* jralonso .es/2014/09/18/cajal-y-el-dinero/.

Ramón Casasús, Justo. *Consideraciones acerca de la doctrina organicista memoria de los ejercicios del doctorado de D. Justo Ramon Casasus.* 1878. PhD dissertation. Universidad Complutense de Madrid. dioscorides.ucm.es/proyecto_digitaliza cion/index.php?531540912X.

Ramón Jiménez, Juan. "Santiago Ramón y Cajal (viene)." Fundación para el conocimiento Madrid. www.madrimasd.org/cienciaysociedad/poemas/poesia .asp?id=353.

Ramón y Cajal, Luis. "Cajal, as Seen by His Son." *Proceedings of the Cajal Club*, vol. 4, 1996, pp. 73–79. cajalclub.org/proceedings-of-the-cajal-club/.

Ramón y Cajal, Pedro. "La juventud de Cajal contada por su hermano Pedro." *Psicología de los artistas*. Espasa-Calpe, 1972.

Ramón y Cajal, Santiago. *Advice for a Young Investigator*. MIT Press, 1999.

———. "Caciquismo universitario." *El Siglo Médico*, 1919. www.iact.ugr-csic
.es/personal/julyan_cartwright/caciquismo.html.

———. *Cajal on the Cerebral Cortex: An Annotated Translation*. Edited by Javier DeFelipe and Edward G. Jones. Oxford University Press, 1988.

———. *Cajal's Degeneration and Regeneration of the Nervous System*. Edited by Javier DeFelipe and Edward G. Jones. Translated by Raoul M. May. Oxford University Press, 1991.

———. *Charlas de café: Pensamientos, anécdotas, y confidencies*. Espasa-Calpe, 1966.

———. "Coloración por el método de Golgi de los centros nerviosos de los embriones de pollo." *Gaceta Médica Catalana*, January 1, 1889.

———. *Concepto, metodo y programa de anatomia descriptiva y general*. Valencia Cultura, 1978.

———. "Contribución al estudio de la estructura de la médula espinal." *Revista Trimestral Histología Normal y Patológica*, vol. 1, 1889, pp. 79–106.

———. *Discursos leídos ante la Real Academia de Ciencias exactas, físicas y naturales en la recepción pública de Sr. D. Santiago Ramón y Cajal el día 5 de Diciembre de 1897*. Imprenta de L. Aguado, 1897. bdh-rd.bne.es/viewer.vm?id
=0000078227&page=1.

———. *Discursos leídos en la solemne sesión celebrada bajo la presidencia de S. M. el rey Alfonso XIII para hacer entrega de la Medalla Echegaray al Excmo. Sr. D. Santiago Ramón y Cajal el día 7 de mayo de 1922*. Real Academia de Ciencias Exactas Físicas y Naturales, Madrid, 1922.

———. *Elementos de histología normal y de técnica micrográfica: Para uso de estudiantes*. Nicolás Moya, 1897. archive.org/details/b28066303.

———. "El más seguro y sencillo de los métodos de coloración de los microbios." *La Crónica Médica*, vol. 9, 1885.

———. *El mundo visto a los ochenta años*. Espasa Calpe, 2000.

———. *Fotografía de los colores: Bases científicas y reglas practices*. Prames, 2007.

————. *Histology of the Nervous System of Man and Vertebrates.* Translated by Larry W. Swanson and Neely Swanson. Oxford University Press, 1995.

————. *In the Aftermath of Peace: Essay on the Future of Europe.* Translated by Lazaros C. Triarhou. Corpus Callosum, 2019.

————. "Investigaciones experimentales sobre la genesis inflamatoria y especialmente sobre la emigración de los leucocitos." *Trabajos escogidos.* Antoni Bosch, 2006, pp. 1–52.

————. "La capacidad intellectual de las mujeres." *Voluntad,* October 12, 1919.

————. *La mujer.* Editorial Glem, 1944.

————. "La teoria cellular," 1883.

————. "La verdad contra el error: La vacunación anticolérica." *La veterinaria Española,* vol. 33, 1890.

————. "Lecture III: The Sensori-Motor Cortex." *Clark University, 1889–1899: Decennial Celebration.* Norwood Press, 1899.

————. "Letter from a Slave-Making Ant." Translated by Emily Tobey. *Sci Phi Journal: A Universe of Wonder,* June 19, 2019. www.sciphijournal.org/index .php/2019/06/19/letter-from-a-slave-making-ant/.

————. "Leyes de la morfología y dinamismo de las células nerviosas." *Revista Trimestral Micrográfia,* vol. 2, 1897, pp. 1–12.

————. *Life in the Year 6000: A Fantasy Dream.* Translated by Lazaros C. Triarhou. Corpus Callosum, 2017.

————. *Manual de anatomía patológica general: Seguida de un resumen de microscopia aplicada a la histología y bacteriología patológicas.* Moya, 1896. archive .org/details/manualdeanatomap00ramn.

————. *Manual de histología normal y técnica micrográfica.* Libreria de Pascual Aguilar, 1893. archive.org/details/b20392941.

————. "Morfología y conexiones de los elementos de la retina de las aves." *Revista Trimestral de Histología Normal y Patológica,* vol. 1, 1888, pp. 11–16.

————. *Neuron Theory or Reticular Theory? Objective Evidence of the Anatomical Unity of Nerve Cells.* Translated by M. Ubeda Purkiss and Clement A. Fox. Consejo Superior de Investigaciones Científicas, 1954.

————. "Observaciones microscópicas sobre las terminaciones nerviosas en los músculos estriados de la rana." *Trabajos escogidos,* pp. 53–96. Antoni Bosch, 2006.

————. *Patogenia de la inflamación: Discurso para los ejercicios del grado de Doctor de Santiago Ramón y Cajal.* 1877. PhD thesis, Universidad Complutense Madrid, eprints.ucm.es/16729/.

————. "Post Scriptum." *Reglas y consejos sobre investigación científica.* Instituto Cervantes. cvc.cervantes.es/ciencia/cajal/cajal_reglas/post_scriptum.htm.

————. "Prólogo" to *Introducción al estudio de la psicología positive,* by Tomás Maestre Pérez. Bailly-Baileri, 1904.

————. "Prólogo" to *Superorganic Evolution,* by Enrique Lluria Despau. Translated by Lazaros C. Triarhou. Imprenta Ricardo Fe, 2005.

————. "Prólogo al segundo edición." Instituto Cervantes, cvc.cervantes.es /ciencia/cajal/cajal_recuerdos/recuerdos/prologos.htm.

————. *Recollections of My Life.* Translated by E. Horne Craigie and Juan Cano. MIT Press, 1989.

————. *Recollections of My Life: Part 1.* Translated by E. Horne Craigie. American Philosophical Society, 1937.

————. *Recuerdos de mi vida.* 2nd ed., vol. 1. Nicolás Moya, 1917. www .gutenberg.org/files/58331/58331-h/58331-h.htm.

————. "Recuerdos de mi vida: Advertencia al lector." *Revista de Aragon,* November 1901, pp. 1–9. hemerotecadigital.bne.es/issue.vm?id=0005176875& search=&lang=en.

————. "The Structure and Connections of Neurons." Nobel Lecture, December 12, 1906. www.nobelprize.org/uploads/2018/06/cajal-lecture .pdf.

————. *The Structure of the Retina.* Translated by Sylvia A. Clarke. Charles C. Thomas, 1972.

————. *Studies on Vertebrate Neurogenesis.* Charles C. Thomas, 1960.

————. *Texture of the Nervous System of Man and the Vertebrates.* Translated by Pedro Pasik and Tauba Pasik. Springer, 2002.

————. *Vacation Stories: Five Science Fiction Tales.* Translated by Laura Otis. University of Illinois Press, 2001.

Ramón y Cajal, Santiago, and Juan Fernández Santarén. *Santiago Ramón y Cajal: Epistolario.* La Esfera de los Libros, 2014.

Ramón y Cajal Junquera, Santiago. "D. Santiago, mi abuelo." *Revista Española de Patología,* vol. 35, no. 4, 2002. www.patologia.es/volumen35/vol35-num4 /35-4n14.htm.

Rapport, Richard. *Nerve Endings: The Discovery of the Synapse*. W. W. Norton, 2005.

"Recepción del Sr. Cajal." *El Día*, December 5, 1897, p. 2. hemerotecadigital .bne.es/issue.vm?id=0002384727&search=&lang=en.

Reinoso-Suarez, F. "Cajal's Concepts on Plasticity in the Central Nervous System Revisited: A Perspective." In *Neuroplasticity: A New Therapeutic in the CNS Pathology*, edited by R. L. Masland et al. Liviana Press, 1987.

Resina, Joan Ramon. *Barcelona's Vocation of Modernity: Rise and Decline of an Urban Image*. Stanford University Press, 2008. www.google.com/books /edition/Barcelona_s_Vocation_of_Modernity/TKfo_pR02U0C?hl =en&gbpv=0.

Rezaie, Payam, and Uwe-Karsten Hanisch. "Historical Context," ch. 2. In *Microglia in Health and Disease*, edited by Marie-Éve Tremblay and Amanda Sierra. Springer, 2014.

Ridaura Cumplido, Concha. *Vida cotidiana y confort en la Valencia burguesa, 1850–1900*. Biblioteca Valenciana, 2006.

Río-Hortega, J. "The Discoveries of Microglia and Oligodendroglia: Pío del Río-Hortega and His Relationships with Achúcarro and Cajal (1914–1934)." *Neurosciences and History*, vol. 1, no. 4, 2013, pp. 176–90. nah.sen.es/vmfiles /abstract/NAHV1N42013176_190EN.pdf.

Rivas Santiago, Natalio. *Gran triunfo de Cajal*. Editorial Nacional, 1953.

Roca, Josep Maria. *Tribut al mestre*. Institut d'Estudis Catalans, 2007.

Rocha Barral, Elvira, and Miquel Àngel Estévez i Torrent. "Documents dels Arxius Municipals: Nota sobre els domicils Barcelonins de Santiago Ramón y Cajal." *Gimbernat*, vol. 47, 2007, pp. 127–29. https://raco.cat/index.php/ Gimbernat/article/view/123201.

Rodríguez Lafora, Gonzalo. "Cajal en su total humanidad." *Foco*, March 17, 1952.

———. "Cajal: Pensamiento social, espiritu democrático." Cajal Institute library, alphabetized files.

———. "Ramón y Cajal: Madrid, 'tierra de amigos.'" *Alfoz*, no. 10, 1984, pp. 63–88.

———. "Santiago Ramón y Cajal y la juventud española." Cajal Institute library, alphabetized files.

Rodríguez Mourelo, José. "El Doctor Cajal." *La ilustración española y americana*, vol. 38, no. 9, March 8, 1894, pp. 146–47.

Rodríguez Quiroga, Alfredo. "El pensamiento regeneracionista de Santiago Ramón y Cajal." *Boletín de la Institución Libre de Enseñza*, nos. 32–33, 1998, pp. 77–94.

Romanes, George J. "The Beginning of Nerves in the Animal Kingdom." *Fortnightly Review*, vol. 30, 1878, pp. 509–26. www.google.com /books/edition/The_Fortnightly_Review/dMdCAQAAMAAJ?hl=en& gbpv=0.

Romero Salvadó, Francisco J. "Spain and the First World War: The Logic of Neutrality." *War in History*, vol. 26, no. 1, 2019, pp. 44–64. journals.sagepub .com/doi/pdf/10.1177/0968344516688931.

———. *The Spanish Civil War: Origins, Course and Outcomes.* Macmillan International Higher Education, 2005.

Rosenblatt, Helena. *The Lost History of Liberalism: From Ancient Rome to the Twenty-First Century.* Princeton University Press, 2020.

Rowland, Martin. *Biology.* Nelson Thomas, 1992.

Ruestow, Edward G. *The Microscope in the Dutch Republic: The Shaping of Discovery.* Cambridge University Press, 2004.

"Ruta a la Peña Oroel." Jaca.com. https://www.jaca.com/oroel.php.

"'Sack of Louvain—Awful Holocaust' (*Daily Mail* headline, Monday 31 August 1914)." European Studies Blog, British Library, September 19, 2014. blogs.bl.uk/european/2014/09/sack-of-louvain.html.

Sagasta, Praxedes Mateo. *Diario de las sesiones de Cortes: Congreso de los disputados, legislature de 1898, esta legislature dió principio el 20 de Abril de 1898.* Vol. 1, Hijos de J.A., 1898.

Sala Catalá, J. "Cambio de paradigma y polemica cientifica entre los biologos expañoles (1860–1922)." *Asclepio*, vol. 34, 1982, pp. 239–63.

Sánchez Granjel, Luis. "Cajal y la generación del noventa y ocho." *Imprensa Médica*, no. 22, 1958.

Sánchez Martin, M. L. "Recuerdo de Don Santiago Ramón y Cajal a los veinte años de su muerte." *Revista Sociedad Venezolana de la Historia de Medicina*, vol. 3, 1995, pp. 53–70.

Sánchez-Perez, J. M. "Cajal the Philosopher, as Revealed by His Literary Works." *Bulletin of the Los Angeles Neurological Society*, vol. 17, nos. 1–2, 1952.

Sánchez Ron, José Manuel. "Cajal y la comunidad neurocientífica internacional de su tiempo." In *Historia, medicina y ciencia en tiempos de . . . Cajal*, pp. 117–39. Ergon, 2006. www.fcs.es/images/publicaciones/CAJAL.pdf.

Sancho, J. "Santiago Ramón y Cajal and His Father Justo Ramón Casasús: The Valencia Years." *Neurosciences and History*, vol. 4, no. 4, 2016, pp. 130–39. nah.sen.es/vmfiles/abstract/NAHV4N42016130_139EN.pdf.

Sancho Menjón Ruiz, María. *Los Pirineos*. Caja de Ahorros de la Inmaculada, 2001.

San Rafael. "Cintarazos." *El Correo Militar*, October 28, 1898. hemeroteca digital.bne.es/issue.vm?id=0003227649&search=&lang=en.

"Santiago Ramón y Cajal." *Alma Historia*, vol. 16, 1999.

Sartorius, David. *Ever Faithful: Race, Loyalty, and the Ends of Empire in Spanish Cuba*. Duke University Press, 2013.

Satué Oliván, Enrique. *"As Crabetas": Libro-museo sobre la infancia tradicional del Pirineo*. Instituto de Estudios Altoaragoneses, 2011.

Scarborough, John. *Medical and Biological Terminologies: Classical Origins*. University of Oklahoma Press, 1992.

Schulz, Julia. *The Influence of the Spanish Language on the English Language Since 1801: A Lexical Investigation*. Cambridge Scholars Publishing, 2018.

Science History Institute. "Louis Pasteur." www.sciencehistory.org/historical -profile/louis-pasteur.

Sender, Ramón J. "Ramón y Cajal, montañés de Alto Aragon." *La Libertad*, October 19, 1934, p. 1. hemerotecadigital.bne.es/issue.vm?id=0003121910 &search=&lang=en.

Shape of Barcelona. "The Takeoff of Industrialization and Overpopulation 1820–1840 c.e." shapeofbarcelona.weebly.com/the-19th-century.html.

Sharpey-Schäfer, Edward Albert. *A Course of Practical Histology*. Smith, Elder, 1897. www.google.com/books/edition/A_Course_of_Practical_Histology/f4 kLAQAAIAAJ?hl=en&gbpv=0.

Shepherd, Gordon. *Foundations of the Neuron Doctrine*. Oxford University Press, 2016.

Sheppard, Samuel Edward. *Gelatine in Photography*. Vol. 1. D. Van Nostrand, 1923. www.google.com/books/edition/Gelatine_in_Photography/ZINCA AAAIAAJ?hl=en.

Sherrington, Charles. "A Memoir of Dr. Cajal." In *Explorer of the Human Brain*, by Dorothy Cannon. H. Schuman, 1949.

———. "Santiago Ramon y Cajal 1852–1934." *Obituary Notices of Fellows of the Royal Society*, vol. 1, no. 4, December 1935, pp. 424–41. www.jstor.org /stable/i231358.

"Silveria Fañanás García." *Heraldo de Aragon*. Cajal Institute library, alphabetized files.

Sime, William. *To and Fro; Or, Views from Sea and Land*. Stock, 1884. www .google.com/books/edition/To_and_Fro_Or_Views_from_Sea_and_Land /fFcbAQAAMAAJ?hl=en&gbpv=0.

Smith, Angel. "The Rise and Fall of 'Respectable' Spanish Liberalism, 1808– 1923: An Explanatory Framework." *Journal of Iberian and Latin American Studies*, vol. 22, no. 1, 2016, pp. 55–73. https://doi.org/10.1080/14701847 .2016.1212977.

Smith, Joseph. *The Spanish-American War 1895–1902: Conflict in the Caribbean and the Pacific*. Routledge, 1994.

Snyder, Laura J. *Eye of the Beholder: Johannes Vermeer, Antonio Van Leeuwenhoek, and the Reinvention of Seeing*. W. W. Norton, 2015.

Sobrer, Josep Miquel, ed. *Catalonia: A Self-Portrait*. Indiana University Press, 1992.

Soláns Manero, Ángel. *Los Cajal, su historia y su leyenda*. Imprenta Octavio y Félez, 1990.

Solmsen, Friedrich. "Greek Philosophy and the Discovery of the Nerves." *Museum Helveticum*, vol. 18, no. 3, 1961, pp. 150–67. www.jstor.org/stable /24812475?seq=1.

Solsona, F. "Pedro Ramón y Cajal (1854–1950)." *Serrablo Revista Trimestral (Sabiñánigo)*, vol. 30, no. 117, 2000. sanp.ch/journalfile/view/article/ezm _sanp/en/sanp.2008.01993/ca1360f05ee731b5fbed9501033d8ba27cacce98 /sanp_2008_01993.pdf/rsrc/jf.

"Spanish Concordat of March 16, 1851." Berkley Center for Religion, Peace and World Affairs. berkleycenter.georgetown.edu/quotes/spanish-concordat -of-march-16-1851.

Stanley, Matthew. *Einstein's War*. Dutton, 2019.

Stefanidou, Maria, et al. "Cajal's Brief Experimentation with Hypnotic Suggestion." *Journal of the History of the Neurosciences*, vol. 16, no. 4, October 2007, pp. 351–61. https://doi.org/10.1080/09647040600653915.

Sternlicht, Stanley. *McKinley's Bulldog: The Battleship Oregon*. Nelson-Hall, 1977.

Stoddard, Charles Augustus. *Spanish Cities: With Glimpses of Gibraltar and Tangier*. C. Scribner's Sons, 1892. www.google.com/books/edition/Spanish _Cities/tdsLAAAAYAAJ?hl=en&gbpv=0.

Strabo. *Geography*. Translated by H. L. Jones. Harvard University Press, 1917–1932. penelope.uchicago.edu/Thayer/E/Roman/Texts/Strabo/3A*.html.

Strobel, Edward Henry. *The Spanish Revolution, 1868–1875*. Small, Maynard, 1898. www.google.com/books/edition/The_Spanish_Revolution_1868_1875 /9JwLAAAAYAAJ?hl=en&gbpv=0.

Swanson, Larry. "Santiago Ramón y Cajal." In *The Beautiful Brain: The Drawings of Santiago Ramón y Cajal*, edited with commentaries by Eric A. Newman, Alfonso Araque, and Janet M. Dubinsky. Harry N. Abrams, 2017.

Tardieu, Ambroise. *Diccionario de higiene pública y salubridad, ó Repertorio de todas las cuestiones pertenecientes a la salu pública*. E. Rubiños, 1886. biblioteca digital.jcyl.es/es/consulta/registro.cmd?id=31606.

Tedeschi. "El premio Nobel: Ramón y Cajal." *El Imparcial*, October 27, 1906. hemerotecadigital.bne.es/issue.vm?id=0000899888&search=&lang=en.

Tello Muñoz, Jorge Francisco. "Recuerdos de Cajal." *Ibys Laboratories*, no. 19, 1952, pp. 87–98.

———. "Universal reconocimiento a la obra de Cajal." *Residencia: Revista cuatrimestral de la Residencia de Estudiantes*, vols. 1–5. Program de Extension Científica, Consejo Superior de Investigaciones Científicas, 1926, pp. 114–22.

Thomas, Perrín G. "Cajal, el hombre de genio." *Revista Mexicana de Biología*, vol. 2, 1923, pp. 248–54.

Three Wayfarers. *Roadside Sketches in the South of France and Spanish Pyrenees by Three Wayfarers*. Bell and Daldy, 1859. www.google.com/books/edition /Roadside_Sketches_in_the_South_of_France/jtULAAAAYAAJ?hl =en&gbpv=0.

Timoner Sampol, Gabriel. *Revista Trimestral Micrografica y la doctrina de la neurona*. Instituto Fernando el Católico, 2006.

Todes, Daniel P. *Ivan Pavlov: A Russian Life in Science*. Oxford University Press, 2014.

Tone, Andrew. *The Age of Anxiety: A History of America's Turbulent Affair with Tranquilizers*. Basic Books, 2009.

Tone, John Lawrence. *War and Genocide in Cuba, 1895–1898.* University of North Carolina Press, 2006.

Trallero Anoro, Salvador. *Barcelona Antigua.* Sariñena Editorial, 2017.

Tremblay, Marie-Ève, et al. "From the Cajal Alumni Achúcarro and Río-Hortega to the Rediscovery of Never-Resting Microglia." *Frontiers in Neuroanatomy,* vol. 9, no. 45, 2015. https://doi.org/10.3389/fnana.2015.00045.

Triarhou, Lazaros C. "Ramón y Cajal as an Analytical Chemist of Bottled Water? Use (and Misuse) of the Great Savant's Repute by the Industry." *SAGE Open,* vol. 3, no. 1, January 2013. https://doi.org/10.1177/21582440 13481357.

———. "Two Readings of Quixote: Cajal and Turgenev." *Neurosciences and History,* vol. 3, no. 4, 2015, pp. 154–65. www.researchgate.net/publication /301787876_Two_readings_of_Quixote_Cajal_and_Turgenev.

Triarhou, Lazaros C., and Ana B. Vivas. "Poetry and the Brain: Cajal's Conjectures on the Psychology of Writers." *Perspectives in Biology and Medicine,* February 2009.

Turner, Gerard L'Estrange. *Essays on the History of the Microscope.* Senecio, 1980.

Turner, Gerard L'Estrange, and Margaret Weston Turner. *Nineteenth-Century Scientific Instruments.* Sotheby Publications, 1983.

Ubieto Ausere, Emilio. *Santiago Felipe Ramón y Cajal—Altoaragones universal.* Ayuntamiento de Ayerbe, 2005.

United States Bureau of Foreign and Domestic Commerce. *Cholera in Europe in 1884: Reports from Consuls of the United States.* Government Printing Office, 1885. www.google.com/books/edition/Cholera_in_Europe_in_1884/fwg1A QAAMAAJ?hl=en&gbpv=0.

United States Department of State. *Foreign Relations of the United States.* Government Printing Office, 1873. www.google.com/books/edition/Foreign _Relations_of_the_United_States/jXxUA8bSDPEC?hl=en&gbpv=0.

———. *Papers Relating to the Foreign Relations of the United States [and Spain] Transmitted to Congress with the Annual Message of the President, 1871[–1877].* Government Printing Office, 1871. www.google.com/books/edition/Papers _Relating_to_the_Foreign_Relations/_OoCAAAAYAAJ?hl=en&gbpv=0.

"Un Premio Nobel para Ramón y Cajal." *La Época,* October 27, 1906, 3. hemerotecadigital.bne.es/issue.vm?id=0000738672&search=&lang=en.

University of California Museum of Paleontology. "Antony van Leeuwenhoek (1632–1723)." ucmp.berkeley.edu/history/leeuwenhoek.html.

———. "Robert Hooke (1635–1703)." ucmp.berkeley.edu/history/hooke .html.

Vaill, Amanda. *Somewhere: The Life of Jerome Robbins.* Crown, 2008. www .google.com/books/edition/Somewhere/hHmOOft9kOUC?hl=en&gbpv=0.

Valenciano Gayá, Luis. "Cajal, recuerdos y reflexiones de uno de sus últimos alumnos." *Archivos de Neurobiología,* vol. 46, no. 4, 1983.

Valenstein, Elliot S. *The War of the Soups and Sparks: The Discovery of Neurotransmitters and the Dispute over How Nerves Communicate.* Columbia University Press, 2005.

Van-Baumberghen, Augustín. "Cajal, médico militar." *Revista de Sanidad Militar,* vol. 24, 1934, pp. 323–31.

Varela Ortega, Jose. "Aftermath of Splendid Disaster: Spanish Politics Before and After the Spanish American War of 1898." *Journal of Contemporary History,* vol. 15, no. 2, April 1980, pp. 317–44. www.jstor.org/stable/260516?seq=1.

Velasco Pajares, José. "Mi impression de estudiante ante Cajal." *Gaceta Médica Española,* vol. 26, 1952.

Vera Sempere, Francisco. "Cajal, catedrático de anatomía en Valencia (1884–1887)." *Revista Española de Patología,* vol. 35, no. 4, 2002, pp. 395–408. www .patologia.es/volumen35/vol35-num4/pdf%20patologia%2035–4/35–4–05 .pdf.

Verkhratsky, Alexei, and Arthur Butt, editors. *Glial Physiology and Pathophysiology.* Wiley, 2012.

Verkhratsky, Alexei, and Vladimir Parpura. *Introduction to Neuroglia.* Morgan and Claypool Life Sciences, 2014.

Verne, Jules. *Twenty Thousand Leagues Under the Sea.* Butler Brothers, 1887. www.google.com/books/edition/Twenty_Thousand_Leagues_Under_the _Sea/470XAAAAYAAJ?hl=en&gbpv=0.

Verplaetse, Jan. *Localizing the Moral Sense: Neuroscience and the Search for the Cerebral Seat of Morality, 1800–1930.* Springer, 2009.

"Victimario Histórico Militar: Capítulo VI." *De Re Militari.* remilitari.com /guias/victimario6.htm.

Villacorta Baños, Francisco. "El Ateneo de Madrid (1896–1907): La escuela de estudios superiores y la extensión universitaria." *Hispania: Revista española de historia*, vol. 38, no. 141, 1979. digital.csic.es/bitstream/10261/16870/1 /20090909085902882.pdf.

Viñao Frago, Antonio. "The History of Literacy in Spain: Evolution, Traits, and Questions." *History of Education Quarterly*, vol. 30, no. 4, Winter 1990, pp. 573–99. www.jstor.org/stable/368947?seq=1.

Vincent, Mary. *Spain 1833–2002: People and State*. Oxford University Press, 2007.

Walker, Alan. *Fryderyk Chopin*. Farrar, Straus and Giroux, 2018.

Walker, Alex, and Bernardo Porraz. "The Case of Barcelona, Spain." In *Understanding Slums: Case Studies for the Global Report on Human Settlements*, 2003. www.ucl.ac.uk/dpu-projects/Global_Report/pdfs/Barcelona.pdf.

Wardropper, Bruce W. "Cadalso's 'Noches Lúgubres' and Literary Tradition." *Studies in Philology*, vol. 49, no. 4, October 1852, pp. 619–30. www.jstor.org /stable/4173035?seq=1.

Weld, Charles Richard. *The Pyrenees, West and East*. Longman, Brown, Green, Longmans and Roberts, 1859. www.google.com/books/edition/The_Pyrenees _West_and_East/-RMMAAAAYAAJ?hl=en.

Wick, Georg, and Cecilia Grundtman, eds. *Inflammation and Atherosclerosis*. Springer, 2012.

Wilson, Andrew. "Science Jottings." *Illustrated London News*, April 7, 1894, pp. 21+. www.britishnewspaperarchive.co.uk/titles/illustrated-london-news.

Woolman, David S. *Rebels in the Rif*. Stanford University Press, 1968.

Wuerth, Julian. *Kant on Mind, Action, and Ethics*. Oxford University Press, 2014.

Yuste, Rafael. "The Discovery of Dendritic Spines by Cajal." *Frontiers in Neuroanatomy*, April 21, 2015. https://doi.org/10.3389/fnana.2015.00018.

Zapata, Marcos. *Poesías, con un prólogo del Doctor S. Ramón y Cajal*. Translated by Lazaros C. Triarhou and Ana B. Vivas. Fernando Fé, 1902.

Zeigler, Vanessa Michelle. *The Revolt of the "Ever-Faithful Isle": The Ten Years' War in Cuba, 1868–1878*. PhD dissertation, 2007, University of California, Santa Barbara. www.latinamericanstudies.org/1868/Ten_Years_War.pdf.

Acknowledgments

At a certain point you realize that a book provides a good excuse to thank people. I would like to thank Molly Atlas for seeing the potential in this project when few people did. Thank you to Laird Gallagher for bringing the book to Farrar, Straus and Giroux. Thank you to Alex Star for his intellectual precision and to Ian Van Wye for his keen attention to detail as they shepherded me through the editing process. M. P. Klier did a superb job of copyediting the text.

Doron Weber had faith in me, and the Alfred P. Sloan Foundation's program in Public Understanding of Science, Technology, and Economics supported me with a generous grant. Thank you to my expert reviewers, Stuart Firestein and Larry Swanson, for ensuring that the manuscript was scientifically accurate. Special thanks to Larry for years of guidance beginning the moment he taught me the secret Cajal Club handshake. Thank you to Juan de Carlos, Fernando de Castro, Juan Del Río Hortega Bereciartu, Javier DeFelipe, Carol Mason, Jesus Cruz, Mary Vincent, Richard Rapport, Sam McDougle, John Krakauer, and John Marciari for answering my various technical and sometimes arcane questions. Thank you to Laura Otis for a timely conversation during a pivotal moment in which she encouraged me to pursue this book.

Heartfelt thanks to Jorge Larriva Sahd for so generously welcoming me to his laboratory, demonstrating histological processes to me, and reviewing passages from my book. Thank you to the Cajal Institute in Madrid and to its director, Ricardo Martínez Murillo, for facilitating my archival research, with special thanks to María Ángeles Langa Langa for enabling my access to all necessary materials. Thanks to Lisa Fenk, Meg Younger, and Leah Kelly for letting me tour their neuroscience laboratories and patiently teaching me about the visual system of insects. Thanks to NeuWrite for nurturing this project from its fledgling beginnings. Thank you to the Virginia Center for the Creative Arts, the Brush Creek Foundation for the Arts, and the Jentel Foundation for granting me writing residencies.

I wish to acknowledge the following Spanish institutions where I did my research, ranging from national libraries and universities to a small, remote museum dedicated to Cajal: Biblioteca Nacional de España, the University of Barcelona, the University of Valencia, the University of Zaragoza, Ayuntamiento de Huesca, Ayuntamiento de Jaca, Ayuntamiento de Zaragoza, Instituto de Estudios Altoaragoneses, and Centro de Interpretación Santiago Ramón y Cajal in Ayerbe.

My deepest thanks to Pablo López-García and Virginia García Marín and their wonderful family, who became my family, and especially to Marisa, who took such good care of me during my time in Madrid. To Jessica Ludwig, Paul Puciata, Michael Woodbury, and Diego Antoni for their friendship and help. To Arliss and Debra Howard for a lengthy private residency and for constantly urging me on. To Tim Hutton for his advice and hospitality during the early development of the project.

Thank you to my uncle, David Ehrlich, for reading a draft of the book and giving me helpful notes. Thanks to Noah Hutton and Babe Howard for their careful, serious reading and insightful suggestions. Thank you to Daniella Gitlin and Elianna Kan for reading sections of my work, helping me write in Spanish, and encouraging me along the way. Thank you to the MRO sangha for supporting my practice and especially to my teacher, Shugen Roshi. To my sister, Rachel Ehrlich, for her loyalty and love. To my mother, Marilyn Ehrlich, who carefully read and commented on the manuscript. I cannot imagine a better research assistant than my father, Alex Ehrlich, whose uncanny under-standing of the material and deep reading helped me through many drafts. There is no way I can thank my parents enough. This book is a tribute to the life they have given me, not just once but again and again.

Estimado D. Santiago:
Aunque nunca nos hemos concocido, soy estudiante suyo. Ud. me ha regalado part de me vida y mi propio ser, y por todo eso jamás podría agradecérselo lo suficiente. Espero que le otorguen todo el respeto y la admiración que tanto merece.
Q.b.s.m.,
Benjamin Ehrlich

Index

Page numbers in *italics* refer to photographs.

A Note About the Author

Benjamin Ehrlich is the author of *The Dreams of Santiago Ramón y Cajal*, the first translation of Cajal's dream journals into English. His work has appeared in *The Gettysburg Review*, *The Paris Review Daily*, *Nautilus*, and *New England Review*, where he serves on the editorial panel.